The Origin of the Concept of Nuclear Forces

The Origin of the Concept of Nuclear Forces

Laurie M. Brown

Northwestern University, Evanston, Illinois

and

Helmut Rechenberg

Max-Planck-Institut für Physik, Munich

Institute of Physics Publishing
Bristol and Philadelphia

British Library Cataloguing-in-Publication Data

A catalogue record for this book is available from the British Library.

ISBN 0 7503 0373 5

Library of Congress Cataloging-in-Publication Data

Brown, Laurie M.
    The origin of the concept of nuclear forces / Laurie M. Brown and Helmut Rechenberg.
        p.   cm.
    Includes bibliographical references and index.
    ISBN 0-7503-0373-5 (alk. paper)
    1. Nuclear forces (Physics)   I. Rechenberg, Helmut.   II. Title.
QC793.3.B5B76   1996   96-31758
539.7'54—dc20   CIP

Published by Institute of Physics Publishing, wholly owned by The Institute of Physics, London

Institute of Physics Publishing, Techno House, Redcliffe Way, Bristol BS1 6NX, UK

US Editorial Office: Institute of Physics Publishing, The Public Ledger Building, Suite 1035, 150 South Independence Mall West, Philadelphia, PA 19106, USA

Typeset by P&R Typesetters Ltd, Salisbury
Printed in the UK by J W Arrowsmith Ltd, Bristol

Contents

Preface   ix

Prologue   1

1 Nuclear Forces Before the Neutron   1
   1.1 The electron, radioactivity and the penetrating radiation   1
   1.2 The nuclear atom   6
   1.3 Nuclear structure, nuclear reactions and quantum mechanics   11
   1.4 Problems with the e–p model — and proposed solutions   17
   Notes to text   22

Part A: Towards a Unified Theory of Nuclear Forces   27

2 Nuclear Structure and Beta Decay   31
   2.1 Introduction   31
   2.2 Heisenberg's model of the nucleus (1932–33)   33
   2.3 The 1933 Solvay Conference   36
   2.4 Fermi's theory of β-decay (1933–34)   41
   Notes to text   44

3 The Fermi-field Theory   47
   3.1 Introduction   47
   3.2 Nuclear electrons, the neutron and the neutrino (1933–34)   48
   3.3 The origins of the Fermi-field theory of nuclear forces (1934)   53
   3.4 The Fermi-field theory and the charge independence of
        nuclear forces (1935–37)   57
   3.5 Conclusion: the Fermi-field and related theories (1938–41)   63
   Notes to text   65

4 Cosmic Rays, Quantum Field Theories and Nuclear Forces   69
   4.1 Introduction   69
   4.2 The cosmic rays and nuclear interaction (1934–36)   72
   4.3 The neutrino theory of light (1934–37)   76
   4.4 Electromagnetic shower theory and the interpretation
        of the hard component (1937)   80
   4.5 A fundamental length and cosmic-ray bursts (1937–39)   86
   Notes to text   89

Part B: Yukawa's Heavy Quantum and the Mesotron   95

5 The Origin of Yukawa's Meson Theory   97
   5.1 Introduction   97
   5.2 Yukawa takes up the problem of nuclear forces (up to 1933)   99

5.3 A new fundamental theory of nuclear forces (1933–34)   103
5.4 The *U*-quantum and the cosmic-ray "mesotron" (1934)   106
5.5 The meaning of the meson   111
Notes to text   112

6 The Discovery of the Mesotron (1935–37)   **115**
6.1 Introduction   115
6.2 Yukawa's researches in 1935 and 1936   116
6.3 The mesotron discovered (1936)   121
6.4 The discovery is confirmed (1937)   123
6.5 The mesotron and the Yukawa theory—hopes and doubts
     (1937)   129
6.6 Conclusions   135
Notes to text   137

**7 The Development of the Vector Meson Theory in Britain and
Japan (1937–38)**   **141**
7.1 Introduction   141
7.2 The formation of the Yukawa school and the scalar field
     theory   142
7.3 The vector meson theory in Japan   146
7.4 Three refugees from Hitler take up Yukawa's theory   151
7.5 The first vector meson theories of Kemmer and Bhabha   157
7.6 British papers on the vector field theory of nuclear forces 162
7.7 Bhabha's paper and the application of meson theory to
     cosmic-ray phenomena   166
Notes to text   169

Part C: The Meson Takes its Place Among the Elementary Particles **175**

8 Decay of the Meson—Experiment Versus Theory (1937–41)   **177**
8.1 Introduction   177
8.2 Estimating the lifetime of meson decay (1937–38)   178
8.3 Mesotron decay and the resolution of some cosmic-ray
     puzzles (1938)   183
8.4 The mesotron lifetime measurements of Rossi and
     Rasetti (1938–41)   187
8.5 Meson decay and β-decay (1938–41)   189
8.6 Meson decay versus meson capture (1939–42)   195
8.7 Preliminary conclusions and post-1942 development   197
Notes to text   198

9 The Meson Theory and Yukawa Circumnavigate the Globe   **202**
9.1 Introduction   202
9.2 European conferences take note of the mesotron and the
     meson theory (1937–38)   204

9.3 The vector meson theory in Switzerland and Germany
(1938)   207
9.4 New experiments and calculations on mesotron production
(1938–39)   213
9.5 Progress in theory until summer 1939: the Americans
enter the scene (1938–39)   215
9.6 Yukawa's trip to Europe and America (1939)   221
Notes to text   225

10 General Properties of Elementary Particles   **229**
10.1 Introduction: what is an elementary particle?   229
10.2 The description of particles of any spin and the
spin-statistics theorem (1938–39)   230
10.3 The classical approach to meson interaction: Bhabha and
Heisenberg (1939)   233
10.4 International physics conferences in the fateful year 1939   237
10.5 The 1939 Solvay report of Heisenberg and Pauli, and
Pauli's publications based on it   240
10.6 The spin of the meson from its electromagnetic effects
(1939–41)   243
Notes to text   248

Part D: Meson Physics from 1939 to 1950: the Meson Puzzle Resolved 251

11 Meson Theory During the War (West)   **253**
11.1 Introduction. Research in nuclear physics   253
11.2 Field theories with higher spins (1939–41)   255
11.3 Mixture and pair theories (1939–41)   258
11.4 Strong- and intermediate-coupling theories (1939–44)   260
11.5 Meson physics in Britain and Ireland (1941–45)   266
11.6 Pauli's evaluation of the work on meson theory in the
West during the Second World War (1945)   268
Notes to text   270

12 Meson Physics During the War (East)   **272**
12.1 The Soviet Union and Germany (1940–43)   272
12.2 Cosmic-ray physics in Germany and the theory of the
S-matrix (1941–44)   274
12.3 Japanese meson physics after Pearl Harbor (1941–45)   277
12.4 Nuclear and cosmic-ray physics in wartime Italy (1942–44) 282
12.5 From war to peace: the years 1945–47   284
Notes to text   288

**13 The Meson Paradox is Resolved — and a Clear View of the Nuclear
Forces Emerges**   **292**
13.1 Introduction   292

13.2   Cloud chambers and counter arrays: penetrating showers
           and strange particles (1940–47)   294
13.3   Experiments in Rome on the capture of mesons by nuclei
           and their interpretation (1943–47)   297
13.4   The nuclear emulsion technique for observing particle
           tracks (1939–48)   301
13.5   The discovery of the pion (1947)   304
13.6   Artificial production of mesons and the discovery of the
           neutral pion (1948–50)   307
Notes to text   312

Epilogue   **316**

14   The Strong Nuclear Forces after the Pion   **316**
14.1   Introduction   316
14.2   Experiments with mesons produced at accelerators:
           isospin amplitudes   318
14.3   Renormalized QED and meson theories   320
14.4   The new particles and their symmetry properties   323
14.5   Strong interactions without the pion field   325
14.6   Mesons and nuclear forces in the Standard Model   329
Notes to text   332

Bibliography   **335**

Index   **371**

Preface

This book is a study of the history of fundamental theories of nuclear forces, from 1932 to 1950 (with prehistory and epilogue), beginning with the discovery of the neutron and ending with the discovery of the neutral pion. By "fundamental theory of force" we mean (in a microscopic sense) that the force arises from the exchange of quanta. This exchange principle is embodied in the relativistic quantum theory of fields, the archetype of which is quantum electrodynamics, where the exchanged massless quanta are photons of spin 1. In the theory of nuclear forces that finally emerged during the period in question, the *symmetric pseudoscalar meson theory*, the quanta are charged and neutral massive particles of spin 0 (the pions). It is the discovery of the neutral pion, the experimental completion of the initial phase of the meson theory, that we take as the historical marker for the end of the period. Like all such markers it also, inevitably, marks the significant beginning of a new period.

The prehistory of the subject, dealing with nuclear physics before the discovery of the neutron, falls naturally into two sub-periods: from Becquerel's discovery of radioactivity in 1896 to Rutherford's discovery of the nucleus (and with it the nuclear atom) in 1911, and thence from 1911 to the emergence of the neutron in 1932. We deal with this introductory prehistory in our first chapter, in which we discuss the first indications that forces other than electromagnetism were operative in the nucleus, and the realization of a possible distinction between the strong forces responsible for nuclear structure and nuclear scattering and the much weaker force of β-decay. The main body of the book is divided into four parts, dealing with specific periods in the theory of nuclear forces, and concludes with an epilogue on subsequent related developments, especially with regard to the elementary particles.

In part A we begin with the first fundamental theory of nuclear forces, Heisenberg's so-called neutron–proton model of 1932, which we consider in detail. The Heisenberg theory also marks the beginning of *phenomenological* theories of the strong nuclear interaction and of nuclear structure. Such models introduce various ordinary and exchange potentials, essentially non-relativistic in character, to represent the forces between nuclear particles. The parameters characterizing them, such as the ranges and depths of the potentials, are adjusted to conform to the existing empirical knowledge of the time. (The phenomenological method is perforce still in common use in treating complex nuclei.) In the present work, nuclear-force models will be dealt with only in so far as they have had a bearing on possible fundamental theories. Following the successful treatment of β-decay by Enrico Fermi at the end of 1933, Heisenberg modified his model, replacing electron exchange by the exchange of an electron–neutrino pair. Known as the Fermi-field theory, it became the dominant fundamental theory of nuclear forces until the late 1930s, when it was supplanted by the meson theory of Hideki Yukawa. The Fermi-field theory was also applied to describe cosmic-ray phenomena that could not be accounted for by electromagnetic cascade showers, especially to explain the "penetrating showers" and the "Hoffmann bursts".

Part B deals with the origin of the meson theory and the cosmic-ray and nuclear physics developments that led physicists to take it seriously after two or more years of neglect. This came about in particular because of the discovery in the cosmic rays of particles which closely resembled the charged mesons postulated by Yukawa as the "heavy quanta" of the nuclear force field. Finding these particles, called "mesotrons" or "mesons", led to a tough competition between physicists in Japan and England to develop what emerged for a time as a "standard" quantum field theory of nuclear forces, the vector meson theory.

Part C is concerned with the period 1937–41, when theorists and experimentalists in the nuclear laboratories and the cosmic-ray observatories tried urgently to reconcile the requirements of the theory with the properties of the cosmic-ray mesons, such as their mean life and their high-energy interactions, while at the same time explaining the subtleties of nuclear binding and other low-energy nuclear behaviour. Although these efforts were mostly unsuccessful, they served as useful preliminary studies in the newly emerging field of elementary particle physics, introducing theoretical concepts (such as internal symmetry) and the formal description of particles of any spin.

In part D we treat the advances in the fundamental theories of nuclear forces which took place during and after the war, including alternative types of meson theories, strong- and intermediate-coupling meson theories, alternatives to meson theory (such as Heisenberg's S-matrix theory), and other advances in quantum field theory. Most importantly, we discuss the new experimental discoveries soon after the war, which changed the picture drastically, especially the identification of the true Yukawa mesons, the pions, charged and neutral, and the disclosure of their relation to the mesotrons referred to earlier.

Finally, in the epilogue, we discuss the picture of nuclear forces and elementary particles that emerged during the next four decades, with the mesons being embedded in the growing empire of elementary particles. We also mention the important advances in quantum field theory in the post-war years, leading to renormalized quantum electrodynamics and renormalized pseudoscalar meson theory. The difficulties of applying field theory to strong interactions led to an extension of S-matrix theory as the Regge formalism and to the temporary abandonment of field theory. However, the latter made a triumphant return with the non-abelian gauge theory formalisms of the present "Standard Model", in which super-strong interactions (*quantum chromodynamics* or *QCD*) occur between fractionally charged quarks interacting by the exchange of massless vector bosons, the gluons. In QCD, mesons and baryons (including the nucleons) are composites of quarks accompanied by their gluon fields. Ordinary nuclear interactions are secondary, and analogous to the chemical interactions between atoms. In the Standard Model, the weak interactions are treated by electroweak theory, a unification of weak and electromagnetic interactions. Like the QCD forces, these are also carried out by the exchange of gauge vector bosons: the massless photon for electromagnetism and the massive $W^{\pm}$ and Z bosons for the weak.

The developments that we deal with here were paralleled (between 1932 and 1950) by major progress in low-energy nuclear phenomenology, in cosmic-ray studies, and in the theory of relativistic quantum fields — advances which combined with those we discuss to bring about the birth of elementary particle physics. Because our subject is thus intimately related to the growth of some of the most important fields of modern physics, bearing on the innermost structure of matter, we believe it deserves the detailed treatment that we give it in this book.

One general observation seems to be called for in these days when both learned and semi-popular writers agree that a possible unified, grand unified or even super-unified theory of elementary particles (not to speak of a "theory of everything") is in the offing. Our story can be read as a cautionary tale: all of the earlier proposed fundamental theories of fundamental nuclear forces aspired to be "unified field theories", which unified at least the strong and weak nuclear forces — but during the two decades in question, Nature's response was always "no!", or at least, "not yet!".

Some of the chapters are based upon previously published articles: chapter 2, Brown and Rechenberg 1988; chapter 3, Brown and Rechenberg 1994; chapter 4, Brown and Rechenberg 1991a; chapter 5, Brown 1981 and 1985; chapter 6, Rechenberg and Brown 1996; chapter 7, Brown and Rechenberg 1991b.

The authors would like to acknowledge the contribution to this work of both Northwestern University in Evanston, Illinois and the Werner Heisenberg Institute of the Max Planck Society in Munich. Also LMB expresses his thanks to the National Science Foundation (USA), to the Japan Society for the Promotion of Science, to Kyoto, Tokyo and Keio Universities for their cordiality and to all his extremely helpful collaborators, especially the late Satio Hayakawa and the late Yasutaka Tanikawa, and to Yoichiro Nambu, Rokuo Kawabe, Michiji Konuma, and Ziro Maki. MR is very grateful to Kyoto University (Kazuo Yamasaki) and Tokai University (Seitaro Nakamura and the Matsumae Foundation) for their hospitality. We are both greatly indebted to Rokuo Kawabe for introducing us to the treasures of the Yukawa Hall Archival Library, and for his translations from the Japanese. Helpful suggestions and assistance have also been provided by Karl von Meyenn, Silvan S. Schweber, Roger Stuewer and Spencer Weart.

We are grateful to the archivists of the following institutions for supplying illustrations and giving permission to use them: Niels Bohr Archive

(Copenhagen), Heisenberg Archive (Munich), Yukawa Hall Archival Library (Kyoto), and especially the Niels Bohr Library of the American Institute of Physics (College Park, Maryland).

We also owe particular thanks to the pioneers in meson physics who have given us access to their private materials and thought, of whom we name only two: Nicholas Kemmer (Edinburgh) and Mituo Taketani (Tokyo).

It has been a great pleasure for us to work together on this project over the past decade. This book is dedicated to our wives, Brigitte Brown and Martha Rechenberg, for their cheerful support and encouragement.

Laurie M. Brown and Helmut Rechenberg
Evanston, Illinois, USA and Munich, Germany, July 1996

Prologue

Chapter 1

Nuclear Forces Before the Neutron

It is a paradox of nuclear science that the first evidence for the existence of nuclear forces, the discovery of radioactivity by Henri Becquerel in 1896, preceded by about 15 years Ernest Rutherford's discovery of the nucleus itself, which occurred in the years 1909–11. These two revolutionary experimental findings were accompanied by other outstanding ones, and also by two great theoretical upheavals: relativity and quantum theory [1]. Other important experimental discoveries were: x-rays in 1895, radioactivity in 1896, the electron in 1897 and a proof of the existence of molecules, between 1906 and 1908 [2].

We begin our study of the origin of the concept of nuclear forces by asking what indications there were before, say, 1910 for the existence of any fundamental force or forces other than electromagnetism and gravitation. There were several known phenomena for which existing theories could not provide even explanatory hints, which therefore indicated the possible existence of novel forces. These puzzling facts included the stability of the electron, the very high energies of radioactive radiations and a universally present penetrating radiation of even higher energy, later identified as being of extraterrestrial origin, given the name *cosmic rays.*

In section 2 we turn to the concept of the nuclear atom, discussing the emission of atomic spectra, on the one hand, and of radioactive radiations from the nucleus, on the other. Although quantum mechanics gave a good account of the former, it seemed unable to deal with the prevailing models of nuclear structure, especially with beta decay (sections 3 and 4). Finally, in section 5 we mention the developments in the *phenomenology* of nuclear physics which play an incidental role in our drama — that of the *fundamental* theory of nuclear forces.

1.1 The electron, radioactivity and the penetrating radiation

1.1.1 *The electron and its stability*
To see why the stability of the electron suggested a new fundamental force, we turn to the electron theory of Hendrik Antoon Lorentz [3]. Following the

H.A. Lorentz (1853–1928) about 1900. Photograph reproduced by permission of AIP Meggers Gallery of Nobel Laureates.

earlier evidence for an "atom of electricity" (Helmholtz 1881) based on Faraday's experiments on electrolysis, performed in 1834, Joseph John Thomson measured in 1897 the ratio of charge to mass (e/m) for the negative charge carriers of the so-called cathode rays. These rays are particle streams produced in an evacuated glass tube, having a heated wire at one end (the cathode) and a collecting metal plate at the other end (the anode), when a high voltage is applied across the tube. Thomson found the ratio e/m, determined with the aid of crossed electric and magnetic fields, to be about 1000 times larger than the corresponding ratio for ionized hydrogen atoms and inferred correctly that he had discovered a subatomic particle, with a mass very small compared with that of the lightest atom (Thomson 1897). In 1899 Thomson drew the general conclusion: "Electrification essentially involves the splitting up of the atom, a part of the mass of the atom getting free and becoming detached from the original atom." [4].

Additionally, on purely theoretical grounds, without even referring to the experiments of Thomson and others, Lorentz claimed that "the theory of electrons is an offspring of the great theory of electricity", due to Faraday and Maxwell [5]. Lorentz argued that, in order to understand the electric and magnetic properties of materials, some sort of microscopic theory was needed, and hence [6]:

It is by this necessity, that one has been led to the conception of *electrons*, i.e. of extremely small particles, charged with electricity, which are present in immense numbers in all ponderable bodies, and by whose distribution and motions we endeavor to explain all electric and optical phenomena that are not confined to the free ether.

Lorentz made the assumption that the electron has a finite size and that it is pervaded by ether (a substance supposed to be the seat of the electric and magnetic fields and the medium for the propagation of electromagnetic waves). He added the hypothesis that "though the particles may move, *the ether always remains at rest.*" [7]. In discussing the problem of the mass of the electron, which β-ray studies, using fast electrons emitted from radioactive substances, showed to vary with energy (Kaufmann 1901), Lorentz advocated the view that "negative electrons have no material mass at all", but only a mass associated with their electric charges. That is, the electromagnetic inertia, or mass, was ascribed to the necessity of "accelerating" the fields produced by the charge [8]. Lorentz concluded:

> I for one should be quite willing to adopt an electromagnetic theory of matter and of the forces between material particles. As regards matter... its ultimate particles always carry electric charges... We should introduce what seems to be an unnecessary dualism, if we considered these charges and what else there may be in the particles as wholly distinct from each other.

And again, "...all forces may be regarded as connected more or less intimately with those which we study in electromagnetism." [9].

In the Lorentz electron theory and in Einstein's later special theory of relativity, the moving electron is subject not only to a change in mass, but also to a deformation, being contracted along the direction of motion and thus assuming an ellipsoidal form. Max Abraham objected in 1904 that Lorentz had not proven that the ellipsoidal electron would be in stable equilibrium (Abraham 1904). To this, Lorentz replied [10]:

> That is certainly true, but I think the hypothesis need not be discarded for *this* reason. The argument proves only that the electromagnetic actions and the stress of which we have spoken [i.e., the normal self-stress of the electron's field at its own surface] cannot be the only forces which determine the configuration of the electron.

Put more simply: to keep the electron, viewed as a classical ball of charge, from exploding, a strong attractive internal force must be present and this force cannot be electromagnetic in nature. The electromagnetic "radius" of the electron (obtained by assuming all of the mass to be of electromagnetic origin) is

$$r_0 = a(e^2/mc^2)$$

where a is a dimensionless parameter of order unity that depends upon the assumed structure of the electron. The length r_0 is about 10^{-15} m, and it was within such a distance that the additional force was expected to be significant [11]. (The Newtonian and the electrostatic potentials have the same spatial dependence and opposite sign, but the gravitational force is too weak to stabilize the electron.)

1.1.2 *Radioactivity*

In early 1896, Henri Becquerel discovered radioactivity, an "accidental" observation that occurred in the course of an attempt to determine whether phosphorescent salts of uranium might be a source of the mysterious x-rays that Röntgen had discovered the previous year (Becquerel 1896). Besides uranium, several other heavy elements were found to emit the new radioactive rays, which turned out to be of three types: positively charged α-rays, negatively charged β-rays and neutral γ-rays. By measuring the deflection of the charged particles in magnetic fields and by other decisive experiments performed upon them, it was shown that the α-particles are identical with doubly ionized helium atoms (and thus have the charge of two positive electrons); the β-rays proved to be identical with the particles of the cathode rays, i.e., they are negatively charged electrons; and the γ-rays were found to be electromagnetic radiation, similar to x-rays, but of shorter wavelength [12].

All three types of radioactive rays are capable of disrupting matter by ionization, i.e., by stripping electrons from atoms (though with differing effectiveness). This property allows them to be detected by the use of fluorescent screens, photographic emulsions, electrical counters and by other devices. One of the most useful and revealing of the latter turned out to be the Wilson cloud chamber, which relies on the condensation of droplets from a supersaturated vapour on the ions produced along the track of a fast charged particle. It had been invented by Charles Thomson Rees Wilson for meteorological studies and was later applied by Wilson and others in radioactivity research [13].

In his very instructive book, *Inward Bound*, Abraham Pais has a chapter called "Radioactivity's three early puzzles" [14]. These are the salient questions:

(1) What is the source of the energy that is released by radioactive materials?
(2) What is the significance of the radioactive half-life (the time for half of the substance to decay)?
(3) What is it that makes some, but not all, elements radioactive?

Pais pointed out that several decades were required after the discovery of radioactivity in 1896 to begin to answer these questions [15].

We address the first one and consider what the answer to it implies about the possible existence of a new force. Observations showed that:

(i) radioactivity acts as an atomic phenomenon (since the number of decays per second is proportional to the amount of decaying matter) [16] and
(ii) "The energy release by a given volume of radon gas is more than a million times greater than the heat evolved by the same volume of hydrogen and oxygen when they explode to form water." [17]

Hence, since equal volumes of gas at the same temperature and pressure contain equal numbers of molecules, it becomes evident that radioactivity involves an energy release per atom that is six to seven orders of magnitude greater than previously known, even in chemical explosions. This would indicate the presence of a *new kind of force*, or at least it seems so in hindsight.

According to the most widely considered atomic models up to about 1911, the electric charges in atoms were spread more or less uniformly throughout the atomic volume. Assuming that all the effective forces were electric and magnetic, the physicist could understand, at least qualitatively, the energy scale of atomic radiations up to kilo-electronvolt energies (such as optical, ultraviolet and even x-ray emissions), assuming some form of energy quanta. What kind of structures, held together by what kind of forces, could result in the emission of particles whose kinetic energy amounted to millions of electronvolts?

Rutherford's announced discovery of the nuclear atom (1911) did not immediately resolve this difficulty, but it did lead to a reformulation of the problem. It had been shown by William Henry Bragg (who later won a Nobel Prize, together with his son William Lawrence Bragg, for their joint work on x-ray crystallography) that α-particles from a given nuclear transition have a unique energy (Bragg and Kleeman 1905).On the other hand, the situation for β-rays was very confused for more than two decades after Becquerel's discovery. James Chadwick proved that primary β-rays possess a continuous spectrum (Chadwick 1914) and showed that the earlier observed β-ray line spectrum is of secondary origin [18]. Chadwick's observations led to a quandary, to which we shall return below: the apparent failure of the law of conservation of energy. In the case of γ-rays, the question of their energy source had to be phrased in the language of the quantum theory (Planck 1900, Einstein 1905), but their energies were still much larger than seemed reasonable on the basis of electromagnetic interaction.

1.1.3 *The penetrating radiation*

A sheet of heavy cardboard will stop the most energetic α-rays, whereas most β-rays require about a centimetre of aluminium. Radioactive γ-rays, in contrast, are much more penetrating—being able to traverse a centimetre of lead, or 10 centimetres of aluminium. As indicated above, the high energies of these radiations posed a severe problem for atomic models before Rutherford's nuclear atom.

How much more challenging, therefore, was the discovery of a radiation that could penetrate more than 10 centimetres of lead? [19]. This penetrating radiation was observed quite early in the century, in 1903, and it was generally assumed to consist of γ-rays of especially high energy, extrapolating from the knowledge that radioactive γ-rays were much more penetrating than x-rays [20]. At first the penetrating rays, which produced a small, just barely measurable, ionization in the atmosphere, were thought to come from radioactive material, either embedded in the earth's crust or suspended in the atmosphere. Victor Franz Hess disproved this assumption by showing that the ionization increased at great heights (Hess 1912). Later the rays were proven to have an extraterrestrial origin and were therefore christened *cosmic rays*. For a number of years after the establishment of the Bohr–Rutherford atomic model, no one speculated about an *atomic* origin for this highly penetrating radiation. However, clearly there was a suggestion here of new and powerful forces in operation somewhere within the atom.

1.2 The nuclear atom

1.2.1 *The structure of the atom and atomic spectra*
During the second half of the nineteenth century, the establishment of chemical periodicity of the elements (Meyer and Mendeleev in 1869) and the regularities that were noted in the simpler line spectra (Balmer in 1885) stimulated growing interest in the construction of atomic models, especially in Great Britain. It was their eventual aim to account for the properties of atoms — their sizes, weights, stability and dynamical behaviour, including chemical behaviour, as well as the emission and absorption of light [21]. The most influential of these was the vortex model of William Thomson, Lord Kelvin (Silliman 1963). However, this is not the place to review the fascinating history of these models in the nineteenth and early twentieth centuries [22], except as forerunners of the Bohr–Rutherford nuclear atom, or insofar as they give hints of the existence of forces of a new type, which would eventually become recognized as nuclear forces.

The discovery of the electron in 1897 provided new challenges and opportunities for model building, as well as new constraints. According to John Heilbron, at least six problems needed attention: "The nature of the positive charge deemed necessary to neutralize and to retain the atomic electrons; the number n of these electrons; the fixing of the size of the atom; the prevention of radiative collapse; the form of the spectral series; and the identity of the spectral radiators" [23].

Various models were proposed, some involving conjectured positive electrons, in addition to the negative ones that had been observed as cathode rays. To account for the masses of atoms and to explain complex atomic spectra (if one assigned one vibrational mode to each spectral line), many thousands of electrons would have to be present in each atom. In other models, especially the famous "raisin–pudding" model of J.J. Thomson, the positive charge was distributed as a uniform thin jelly extending to the atom's outer periphery. (Incidentally, the model of Thomson was misnamed, because he quickly abandoned the original static version in favour of one with electrons circulating in rings.) Assuming that the inverse square Coulomb electric force held between point charges, electrons within the positive sphere would be attracted to its centre by a force directly proportional to the radial distance, as in the analogous gravitational case. The mutual repulsion of the electrons then put them into a kind of shell structure, from which Thomson attempted to develop a theory of atomic structure, spectra and chemical valency [24]. (We note that Thomson always used the word *corpuscle*, in preference to electron.)

Still other models, including the "Saturnian atom" of Hantaro Nagaoka, had the positive charge concentrated in the centre, with the electrons orbiting it [25]. With a quasi-continuous ring current made up of many electrons (as it was assumed), the radiation intensity calculated from classical electro-dynamics was reduced, compared with that of a single orbiting or vibrating electron, by the factor (a/λ), where a denotes the radius of the ring and λ the wavelength of the emitted radiation. Thus, to some extent, radiative collapse of the atom was delayed; but the Nagaoka atom was also *mechanically* unstable and it was soon abandoned. However, Rutherford referred to it in his famous paper of 1911 on the nuclear atom [26].

The story of Rutherford's request to a Manchester University student, Ernest Marsden, under the guidance of Rutherford's assistant, Hans Geiger, to look for "some effect from α particles directly reflected from a metal surface" (Marsden 1962) and their subsequent discovery of large-angle

E. Rutherford (1871–1937) about 1912. Photograph reproduced by permission of AIP Emilio Segrè Visual Archives.

scattering of α-particles has been told so often that it has become the stuff of legend [27]. Interestingly, although Geiger and Marsden had reported already by mid-1909 (Geiger and Marsden 1909) that one in every 8000 α-particles striking a thin platinum foil was scattered at an angle greater than 90°, the significance of their results was not appreciated until the analysis in 1911 by Rutherford showed that such large deflections could not occur from multiple scattering, but only through the single scattering of the α-particles by a large strongly concentrated electric charge. In 1912 he began to call this central charge the *nucleus*.

Rutherford's quantitative analysis of the scattering data, assuming a repulsive inverse square law for the force of interaction, allowed him to estimate the nuclear charge, and hence to infer that $n \approx \frac{1}{2}A$, where n is the number of atomic electrons and A is the nuclear mass number [28]. (The nuclear charge is evidently ne.) Thus, of the six atomic properties that were listed at the beginning of this section, Rutherford's analysis gave evidence only on two: the nature of the positive charge (concentrated in a small volume and having most of the atom's mass) and the approximate value of n. To reach this limited goal, as Heilbron has emphasized, Rutherford made several uncontrolled assumptions; for example, that the nucleus and the α-particle were both point-like charges and that the inverse square law held even for microscopic separations. As a result, the justification for the model was, at best, its self-consistency.

It is, therefore, not surprising that there was neither immediate acceptance of Rutherford's nuclear atom nor rejection of Thomson's very different model. Indeed, Thomson's atom provided a possible mechanism for producing a line spectrum, had a chance at stability and offered some rationale for chemical periodicity — all characteristics that Rutherford's nuclear atom did not possess. Before the latter could become the standard and replace the Thomson atom, one needed the introduction of the quantum condition by Niels Bohr, which occurred in 1913 (Bohr 1913).

One problem that none of the pre-Bohr atomic models could address was that of fixing the atomic size. For some time it had been realized that the only natural length that could be constructed out of the fundamental constants e, m and c was the so-called classical electron radius $e^2/(mc^2)$. This distance, about 3.8×10^{-13} cm, is too small (by five orders of magnitude) to relate to atomic size. Additionally, the electrostatic law of Coulomb contains no intrinsic scale of length. Thus, in Thomson's or Rutherford's atomic models, the atomic size was a freely assignable parameter. However, Arthur Erich Haas realized that, if one associated Planck's constant h with the vibrational states of Thomson's atom and assumed that the atomic radius was the maximum amplitude of a Planck oscillator (energy equal to $h\nu$), then one could express h in terms of the radius [29].

It was possible to invert the argument of Haas and find instead an expression for the atomic radius, a procedure recommended by Arnold Sommerfeld in 1911 and adopted by Bohr in 1912. As is well known, Bohr's principal innovation was to stabilize atomic states by imposing a non-mechanical condition (i.e., one that did not follow from classical mechanics), the quantization of angular momentum. In this way atomic states became "stationary" and "discrete". In the ground (or lowest) state of a one-electron atom, such as hydrogen, the electron's angular momentum was required to be $h/(2\pi)$ [30]. In general, Bohr specified the angular momentum of stationary states as $nh/(2\pi)$, where n is a positive integer. (In contrast to Bohr, in modern quantum mechanics, the orbital angular momentum of the hydrogen ground state is zero.) Since this procedure led to an appreciable energy gap between the allowed successive stationary states, continuous atomic radiation of the classical type was forbidden, resulting in a line spectrum.

Although the later developments of the Bohr–Sommerfeld model, especially as treated by the post-1925 quantum mechanics, addressed successfully all six of the issues mentioned by Heilbron, a further application not on his list, *the location of the seat of radioactivity*, became the subject of informed speculation, beginning around 1910.

1.2.2 *The nucleus as the source of radioactivity*

In the earlier ring models (Thomson, Nagaoka), the β-rays were assumed to consist of electrons from the outermost rings, which received (in a manner not specified) the high energy needed for ejection at high speed. Bohr, on the contrary, according to his own and others' testimony, believed that all forms of radioactivity originated in the atomic nucleus. In his Rutherford Memorial Lecture of 1961, Bohr stated that he "followed, on Rutherford's advice, an introductory course on the experimental methods of radioactive research" for some weeks [31]. However, he "rapidly became absorbed in the general theoretical implications of the new atomic model, and especially in the possibility it offered of a sharp distinction as regards the physical and chemical properties of matter, between those originating in the atomic nucleus itself" and those depending on the electron distribution.

Not only was it "evident" to him that radioactivity came from the nucleus, whereas the ordinary physical and chemical properties were due to the electrons, but also: "It was even clear that, owing to the large mass of the nucleus and its small extension compared with that of the whole atom, the constitution of the electron system would depend almost exclusively on the total electric charge of the nucleus." This suggested, as Bohr said, that all the ordinary physical and chemical properties should depend on a single integer Z, the nuclear charge in units of e, which became known as the *atomic number*.

The notion of a nuclear charge number was strongly reinforced by the concept of *isotopy* (section 3.1 below), as Bohr related:

> Thus, when I learned that the number of stable and decaying elements already identified exceeded the available places in the famous table of Mendeleev, it struck me that... chemically inseparable substances, to the existence of which Soddy had earlier called attention and which later by him were termed

N. Bohr (1885–1961) about 1917. Photograph reproduced by permission of AIP Emilio Segrè Visual Archives (W.F. Meggers Collection).

"isotopes", possessed the same nuclear charge and differed only in the mass and intrinsic structure of the nucleus. The immediate conclusion was that by radioactive decay the element, quite independently of any change in its atomic weight, would shift its place in the periodic table by two steps down or one step up, corresponding to the decrease or increase in the nuclear charge accompanying the emission of α- or β-rays, respectively.

The ideas referred to by Bohr were certainly "in the air" at the time — reinforced by chemical and spectroscopic studies — and the so-called displacement rules were stated by a number of authors independently. However, according to Bohr, the connection of the rules to the nucleus was not at all clear until the concept of atomic number became well established [32]. This was accomplished by a young associate of Rutherford, Henry Moseley, who measured the wavelengths of the principal x-ray series of many elements and showed that they exhibited the simple dependence on atomic number that was predicted by the Bohr–Rutherford atomic model (Moseley 1913, 1914) [33]. After this, it became possible to ask questions about the *structure* of the nucleus — and its implications for the nature of nuclear forces. Unfortunately, by then the Great War was in progress! (Moseley was killed in action in 1915 [34].)

1.3 Nuclear structure, nuclear reactions and quantum mechanics

During the years following the acceptance of the Bohr–Rutherford nuclear atom, there was considerable progress in determining the properties of atomic nuclei, mainly by accurately measuring atomic weights and by experiments on nuclear transformation, by α-particles and later by fast protons.

1.3.1 *Nuclear masses, isotopes and early nuclear models*

In studying the positions of radioactive elements in the periodic table of the chemical elements, Kasimir Fajans and Frederick Soddy discovered the property of isotopy [35]. That is they found elements of identical chemical properties but of different atomic weights. Earlier it had been a puzzle that the lead obtained as the products of different radioactive series had different atomic weights: 206.0 from radium and 207.2 from thorium. At about the same time J.J. Thomson and his research assistant Francis William Aston collaborated on the deflection of positively charged ions (canal rays) by crossed electric and magnetic fields and found the two isotopes of neon (20 and 22) which contribute to neon's atomic weight of approximately 20.2. These were the first identified isotopes not associated with radioactivity (Thomson 1913). Aston considerably refined Thomson's method in developing his "mass spectrograph", which allowed the determination of atomic masses to better than 0.1% (Aston 1922). This permitted a test of the hypothesis that atomic weights of individual isotopes should be approximate multiples of the mass of hydrogen, that their nuclei masses should be integral multiples of the hydrogen nucleus (or "proton" as Rutherford named it in 1920) [36].

The results of Aston and others—e.g., Arthur Jeffrey Dempster in the USA and Willy Wien in Germany—that the nuclear masses are approximately multiples of the proton mass, supported the nineteenth-century hypothesis of William Prout that atoms ultimately are built of hydrogen. Furthermore, the table of stable isotopes of moderate Z exhibited preferred atomic weights A. Those with even $Z = 2n$ frequently have $A = 4n$, those with odd $Z = 2n + 1$ frequently have $A = 4n + 3$; thus series of elements with even and odd nuclear charge Z exist, whose members differ by an α-particle. Sommerfeld noted these facts as he speculated how the two stable series might be related, through an analogy to radioactive transformation (Sommerfeld 1924, p 167):

> We hardly need emphasize that with this speculation, we leave for the moment the safe ground of facts, and that one can establish in the case of the elements considered here, neither an H[ydrogen-] transition nor the trace of a spontaneous transition. But the existence of isotopes in the case of a non-radioactive substance frankly demands a search for genetic connections in the periodic system and to extend the [radioactive] displacement law to the whole system; it implies most likely that atomic nuclei are composite and can be constructed [of smaller parts].

At the end of his chapter 3, he summarized: "There is no doubt that radioactive nuclei contain helium nuclei and electrons which they emit as α- and β-rays" and added (Sommerfeld 1924, p 206):

> Prout's hypothesis and its substantiation in Aston's experiments demands further that all atomic nuclei ultimately are built out of H-nuclei and electrons ("positive and negative"). In the case of the helium nucleus itself, this necessarily leads to the assumption that they consist of four H-nuclei bound by two electrons. Generally we can state that a nucleus having atomic weight A and atomic number Z must contain $K = A - Z$ electrons... Of these nuclear electrons a large number are built into He-nuclei.

Nuclear electrons which are not contained in helium nuclei might be (according to Lise Meitner (1921)) attached to α-particles; thus nuclei would consist of α-particles, neutralized α-particles (called α'-particles), protons and extra electrons.

Sommerfeld then turned to the important problem of the "mass-defect". The mass of helium is smaller than the mass of four protons by $\Delta m \approx 0.03$ units, corresponding to a huge binding energy, exceeding the largest energies observed in radioactive decay. This accounts for the exceptional stability of the α-particle and led to the suggestion that the sun's energy arises primarily from the formation of helium from hydrogen (Eddington 1923).

As a theoretician, Sommerfeld never considered himself to be a nuclear physicist. However, he encouraged his student Wilhelm Lenz to construct a model of the helium nucleus, using the quantum-mechanical rules that Bohr had used for the atom. Lenz considered four protons rotating (with equal spacing) on a circle, the binding force being supplied by two electrons lying on opposite sides of the axis of rotation. The radius of this "helium nucleus" turned out to be 5×10^{-4} as large as Bohr's hydrogen atom, still much larger than the result obtained from Rutherford's scattering experiments (Lenz 1918). Sommerfeld felt that: "The important aspect of this consideration is certainly not the specific picture... but the method of applying the quantum rules to the nucleus." He ended this section on an optimistic note (Sommerfeld 1924, p 217):

> We are therefore convinced *that the build-up (Aufbau) of the nuclei from elementary constituents can be accomplished according to the same construction principles, namely the rules of quantum theory, as the build-up of atoms from nuclei and electrons.*

1.3.2 *Nuclear transformations by α-particles and "anomalous scattering"*

In a review article of 1914 on the structure of the atom as it was deduced from the scattering of α-particles, Rutherford said [37]:

> Special interest attaches to the effect of a collision of a swift α particle with a light atom like that of hydrogen... It can be simply calculated that in a close encounter between the α particle and the hydrogen atom, the latter should in rare cases be set in motion with a velocity about 1.6 times that of the α particle and should travel about four times as far through a gas as the α particle itself.

From the data, Rutherford estimated that to get the observed hydrogen recoil velocity, the distance of approach between α-particle and hydrogen nucleus must be less than 1.7×10^{-13} cm, thus the hydrogen nucleus must be even smaller than the "electron radius" of 3.8×10^{-13} cm.

Referring to the electron theory of Lorentz, in which a charge of small radius gives rise to a large electromagnetic mass, Rutherford suggested [38]:

> It thus appears possible that the hydrogen nucleus of unit charge may prove to be the positive electron, and that its large mass compared with the negative electron may be due to the minuteness of the volume over which the charge is distributed.... It would be natural on this view to suppose that the positive and the negative electrons are the two fundamental units of which all the elements are composed.

As we shall describe presently, in 1919 Rutherford found definite evidence that the hydrogen nucleus, his conjectured "positive electron", must be a constituent of the nitrogen nucleus (or at least, that it can be knocked out of this nucleus). This particle he later called the *proton*. In the paragraph just quoted, we have one of the earliest utterances of the electron–proton (e–p) model of the nucleus, which we have already mentioned, the model which dominated nuclear physics until after the neutron was discovered in 1932. We also see that, in view of the Lorentz-type model assumed for both "electrons", non-electromagnetic forces are required in order to stabilize them (i.e., to prevent their exploding).

Returning to Manchester from a trip to North America in the spring of 1917 (a trip made partly for discussions on anti-submarine warfare and partly for academic purposes), Rutherford again took up the study of α-particle scattering, using the scintillation method [39]. He repeated and extended Marsden's work on the scattering in hydrogen and other gases. By this time, the war had emptied the universities of their young men and so Rutherford had to perform the experiments on his own, with the help of a technician, William Kay. In 1919 he published a four-part paper containing the principal results.

Rutherford observed, as had Marsden, that an α-particle source (e.g., a metal strip coated with radium C) always gives "rise to a number of scintillations on a zinc sulphide screen far beyond the range of the α-particles" [40]. The particles producing these scintillations were similar in range (and thus also in energy) to the recoil hydrogen atoms produced when the tube contained hydrogen and they were thought to come from the source itself, either as direct radioactive decay products or from "occluded hydrogen" (Rutherford 1919a).

However, when the tube was filled with dry air, the effect *increased*—and further experimentation showed that it was due to nitrogen. Thus, a fast α-particle striking nitrogen produced (even faster) hydrogen plus a nuclear recoil. Thus was reported the first artificially induced nuclear reaction! Rutherford concluded his article as follows: "The results as a whole suggest that, if α particles—or similar projectiles—of still greater energy were available for experiment, we might expect to break down the structure of many of the lighter elements" [41].

In the last part of his paper of 1919, Rutherford already argued (13 years, be it noted, before the discovery of the deuteron) [42]:

> We should anticipate from radioactive data that the nitrogen nucleus consists of three helium nuclei each of atomic mass 4 and either two hydrogen nuclei or one of mass 2. If the H nuclei were outriders of the main system of mass 12, the number of close collisions with the bound H nuclei would be less than if the latter were free, for the α particle in a collision comes under the combined field of the H nucleus and of the central mass... *Without a knowledge of the laws of force at such small distances*, it is difficult to estimate the energy required to free the H nucleus.

In part I of the same paper, Rutherford studied the collisions of fast (≈ 5 MeV) α-particles with hydrogen, and found that the number and the distributions in angle and energy of "H particles" (i.e., ejected hydrogen nuclei) were very different from the theoretical expectations for the scattering of two point particles, indicating "that the forces involved in a close collision differ considerably from those to be expected on the simple theory" [43]. To his mind, however, this did not imply that a new type of force was operating. Instead, he reasoned [44]:

> This is not unexpected, for we have every reason to believe that the α particle has a complex structure consisting probably of four hydrogen nuclei and two negative electrons. If we assume, for simplicity, that the hydrogen nucleus act as a point charge for the distances under consideration, we still have a complicated system of forces near the nucleus [sic] of the α particle.

Later on in the year 1919, Rutherford succeeded J.J. Thomson as the Cavendish Professor at Cambridge, and brought with him a research student, James Chadwick. The latter (who had been interned in Germany during the war) continued the study of α-particle scattering with E.S. Bieler. They again found strong deviations from inverse-square law behaviour and they tried without success to account for their detailed results by trying various arrangements of the assumed nuclear constituents (four H nuclei and two negative electrons). They concluded from their failure that it was simplest to assume "that the law of force is not the inverse square in the immediate neighbourhood of an electric charge" [45]. Thus they attributed their anomalous results to a modification of *electric forces* at small distances [46].

Similar "anomalous" scattering, i.e., deviations from point-like behaviour, was found for α-particle scattering from other light elements, such as aluminium and magnesium. Again, it was not possible to distinguish the effects of a distributed charge (with both positive and negative charges assumed to be present in the nucleus) from those of a deviation of the short-range electrical force from pure Coulomb character. Sometimes, magnetic forces were invoked for the purpose [47]. Other authors considered the possible effects of electric polarization of the nucleus, caused by the approaching α-particle, which can produce deviations from the Coulomb law (Debye and Hardmeier 1926).

1.3.3 *Nuclear physics and the new quantum mechanics*

During the 1920s the main advances in nuclear physics were instrumental and experimental. Thus Rutherford, in opening a discussion on the structure of nuclei at the Royal Society in London in February 1929—the first meeting on that subject in 15 years—emphasized that there had been three new methods of attack (Rutherford 1929). The first was the accurate measurement of the masses of isotopes, especially by Aston (e.g., Aston 1924), which showed that the proton was the unit of mass, with the departure from integer values of the nuclear mass number A (called the "packing fraction") being a measure of the nuclear binding energy or "mass defect". The second line of attack was the bombardment of light elements with α-particles from radioactive decay, producing nuclear disruption (with the emission of protons) or demonstrating "anomalous scattering", the latter being jargon for non-Coulombic point-like scattering. In the 1920s, quantitative experiments of this type were still performed by the scintillation method, although electrical counting tubes were available and an important start was made on the use of cloud chambers for viewing nuclear reactions (Shimizu 1921a, b; Blackett 1922, 1925) [48]. The third line of attack was the observation of γ-ray line spectra, which showed the existence of discrete excited states in nuclei (e.g., by C.D. Ellis).

Experiments on β-decay were not discussed at the 1929 Royal Society Discussion Meeting. These had shown that each decay resulted in the emission of a single electron having a broad energy spread, which rather undermined the idea of quantum states. Neither were nuclear spins mentioned, of which many were known, which also raised serious doubts, as we shall see, about the application of quantum mechanics in nuclei.

In spite of Sommerfeld's optimism referred to above, atomic quantum theory in the first half of the 1920s offered little hope when applied to the nucleus. Edward M. Purcell has noted the difficulty in the early 1920s for physicists to go beyond the standard e–p model of the nucleus, even if they had known the forces. "In 1922 the Bohr–Sommerfeld quantum theory of atomic structure had still failed to make any headway with the two-body problem presented by the electrons in the helium atom", he said and continued: "One simple set of facts did command attention, the relative abundances of the isotopes—the now familiar isotope chart, which had already been filled in enough to reveal some tantalizing regularities." [49] This led to some descriptive schemes, which emphasized the stability of certain nuclear substructures, especially the α-particle (Harkins 1920; Stuewer 1983).

With the advent of quantum and wave mechanics in 1925 and 1926 the theoretical situation changed drastically, for now nearly all problems of atomic and molecular theory could be attacked successfully. The question arose of whether quantum mechanics could be applied to such nuclear problems as the anomalous scattering of α-particles from aluminium and magnesium nuclei, for which the explanation of electrostatic polarization had been proposed (Debye and Hardmeier 1926). Magnetic forces to account for

G. Gamow (1904–1968) (left) and P. Kapitza (1894–1984) in Cambridge, early 1930s. Photograph reproduced by permission of AIP Emilio Segrè Visual Archives (Frenkel Collection).

the non-Coulombic response, which had been suggested earlier, were now reconsidered in connection with the electron spin hypothesis. So, for example, it was noted that magnetic interaction between electron and proton spins provided attractive forces at distances less than 10^{-11} cm, and these were suggested to provide stability to nuclei made of these two kinds of particle (Frenkel 1926).

In 1928, the first application of quantum mechanics to the nucleus was announced: α-particle radioactivity was explained as quantum mechanical barrier penetration or "tunnelling" (Gamow 1928a, b; Gurney and Condon 1928). At the 1929 Royal Society Discussion, Gamow and Rutherford applied this idea to α-particle scattering in general (see also Gamow 1928c).

They also argued in favour of a new "liquid drop" model of moderate and heavier nuclei, which assumed that these consisted mainly of α-particles occupying the same state.

Quantum mechanics was also successful in confirming and extending Rutherford's formula of 1911 for Coulomb scattering from a point charge. Non-relativistic wave-mechanical calculations reproduced the classical result (Wentzel 1926; Gordon 1928; Mott 1928), although deviations were expected for large penetrations, occurring at higher energies, at which the effect of nuclear structure would be apparent. In addition, there would be important quantum interference effects when identical particles were scattered. For example, for the nuclear scattering of α-particles in helium, in the classical Rutherford scattering formula, as a result of the use of the symmetrical wavefunction required by Bose–Einstein statistics, the factor $\mathrm{cosec}^4\,\theta$, θ being the scattering angle in the centre-of-mass system, is replaced by

$$\mathrm{cosec}^4\,\theta + \sec^4\,\theta + 2\,\mathrm{cosec}^2\,\theta\,\sec^2\,\theta\,\cos u,$$

with $u = (8/137)(c/v)\log(\cot\theta)$, where v denotes the velocity of the α-particle. This result of Neville Mott's scattering theory was experimentally confirmed by his Cambridge colleagues [50].

Encouraged by these successes, Gamow gave α-particles a central role in the treatise on nuclear physics that he published a year later (Gamow 1931). He discussed the experiments which gave fair agreement with his picture. However, he also pointed out the difficulties facing a wave-mechanical description of the e–p model, and the grave failure to explain β-decay. Another status report on "the quantum theory of the nucleus", given by Fritz Houtermans a year before Gamow's book appeared, emphasized how preliminary the understanding was: "Up to now we do not yet know whether quantum mechanics can really explain the processes in atomic nuclei and their structure, or whether again a new physics, so to say a kind of 'super-quantum mechanics' is necessary" [51]. On the non-Coulombic attractive forces, Houtermans stressed the importance of magnetic forces for the make-up of nuclei, because "both fundamental constituents of matter, the electron as well as the proton, possess, as today can be taken as certain beyond doubt, a magnetic moment" [52]; thus one understands, "though for the moment only on a purely qualitative basis, the forces keeping the constituents of the nucleus together" [53].

1.4 Problems with the e–p model — and proposed solutions

The theory of α-particle radioactive decay confirmed the very small sizes of heavy nuclei that had been inferred from α-particle scattering, so that Rutherford said of uranium: "[In] this small nuclear volume 238 protons and 146 electrons have to be made room for". He concluded, "It sounds incredible but may not be impossible" [54]. This same confinement caused serious difficulties, however, in understanding the presence of electrons in nuclei.

1.4.1 *Spin-statistics, electron confinement and energy conservation problems*

The e–p model was the basis for all theories of nuclear structure until Chadwick discovered the neutron in 1932 — and even for a short time afterwards. Doubts had been raised by Ralph de Laer Kronig in 1928, who pointed out that any unpaired electron spin in the nucleus should contribute a magnetic moment of the order of the Bohr magneton [55]. The effect of such a magnetic field on atomic spectra (hyperfine structure) should be as large as the line splitting due to spin–orbit coupling of the electron (Kronig 1926). Instead, observation showed it to be 1000 times smaller [56]. Secondly, Kronig noted that measurements on the band spectrum of the ion N_2^+ showed that the nitrogen nucleus (14 protons plus 7 electrons) had spin 1, although it was presumed to consist of an odd number of spin-$\frac{1}{2}$ particles (Kronig 1928). That the spin of ^{14}N is indeed 1 was strongly confirmed by measurements in Rome of the Raman spectrum of diatomic nitrogen (Rasetti 1930). Furthermore, it appeared that the ^{14}N nucleus obeyed Bose–Einstein rather than Fermi–Dirac statistics, although supposedly composed of an odd number of fermions (Heitler and Herzberg 1929) [57].

Although the quantum-mechanical spin-statistics paradox was perhaps the most serious threat to the e–p model, the difficulty of keeping the light electron from escaping the small nuclear volume (in modern terms "confinement") ran a close second. As mentioned above, Rutherford wondered whether it was possible to push the required number of particles into so small a space. Quantum mechanics, specifically Heisenberg's uncertainty principle, demanded that a nuclear electron could have a kinetic energy in excess of 100 MeV, much larger than the characteristic nuclear energies exhibited in radioactive decay. Such an energetic electron must escape in much less than a second. Indeed, after Paul Dirac proposed his relativistic electron theory, Oscar Klein pointed out that even a very steep potential well could not confine an electron, for it would make use of negative energy states (later, of pair production) to escape. This became known as the "Klein paradox" (Klein 1929).

Finally, there was the problem of nuclear β-decay. On the one hand, this could be read as confirming the presence of electrons in heavy nuclei, since electrons were observed to be leaving them. On the other hand, if only the electron left the nucleus, then the process clearly failed to conserve energy, for the electron's energy could account for only a fraction of the energy given up by the nucleus. (This apparent violation of an important physical law disturbed neither Bohr nor some of his followers; for some time, during 1924–25, Bohr had held that energy conservation might not hold in "elementary processes", but only on the average. He had felt that this would restore a symmetry to the laws of thermodynamics, since the Second Law is valid only statistically [58].)

One of the Copenhagen school who did not agree with Bohr on energy conservation was Pauli; he was also troubled by the spin-statistics paradox. One suggestion had been offered to explain the lack of energy homogeneity of the electrons in β-decay, namely, that the electron radiated as it emerged through the strong Coulomb field of the decaying nucleus. However, when Cambridge experimenters meticulously tested this hypothesis, using a calorimeter surrounding a β-ray source (Ra E) to detect any possible radiation accompanying the electron through its heating effect (Ellis and Wooster 1927), they found that the detected energy was that of the electrons alone. After this result had been confirmed in Berlin (Meitner and Orthmann 1930), Wolfgang Pauli became convinced that the theory had a fatal flaw, and wrote to Lise Meitner and Hans Geiger, in a letter dated 4 December 1930: "...I have hit upon a desperate remedy to save the 'exchange theorem' of statistics and the energy theorem." [59] This remedy was the neutrino, which would save the energy (and momentum) conservation laws in β-decay and also indirectly resolve the statistics problem.

1.4.2 *The first steps towards an electron-free nucleus*

As we shall discuss in the next two chapters, physicists did not abandon the e–p model of the nucleus until 1934, and the electron was still being invoked as a carrier of nuclear force within the nucleus (forming with the neutrino the "quantum" of the nuclear force field) as late as 1937 (Bethe 1937). However, especially during the year 1932, several discoveries hastened the day when nuclear electrons and the problems to which they gave rise, became dispensable (Weiner 1972; Meyenn 1982).

Pauli at first called the neutrino the "neutron" and he visualized it as a light neutral particle, with mass less than ten electrons. He assumed it to be bound in the nucleus, possibly by magnetic forces acting on the "neutron's" non-zero magnetic moment [60]. It was to have spin $\frac{1}{2}$, and, if one "neutron" were present for each electron in the nucleus, that would solve the above-mentioned statistics paradox and could also resolve the nuclear spin paradox. In β-decay, each emitted electron would be accompanied by a "neutron" carrying off the missing energy and momentum. Being neutral, it could easily pass through the experimentalist's calorimeters and cloud chambers without leaving any trace. (However, the neutrino hypothesis did not in itself solve the nuclear confinement problem.)

Pauli did not publish his neutrino proposal until it appeared as a comment by him in the proceedings of the Solvay Conference, held in Brussels in October 1933 [61]. Still, it became known earlier to much of the physics community through talks that he gave, especially on an American visit in 1931. Generally considered to be highly speculative (Brown 1978), the neutrino was not intended to banish electrons from the nucleus, but rather to make their presence tolerable. Indeed, Pauli's friend and regular correspondent, Werner Heisenberg, was convinced, largely on the basis of cosmic ray evidence, that electrons were necessary, even in α-particles, in order to

W. Pauli (1900–1958) in 1931. Photograph reproduced by permission of AIP Emilio Segrè Visual Archives (Goudsmit Collection).

account for the observed large radiative interactions of fast charged particles with nuclei (Brown and Moyer 1984).

The cosmic ray phenomena that so impressed Heisenberg, the cascade showers, were soon accounted for by other discoveries of 1932: first the positron was observed, and then the production of e^+–e^- pairs as predicted by Dirac's hole theory [62]. However, the most significant discovery for nuclear structure was that of the neutron (Chadwick 1932). Rather than recount once again that story, often told [63], we can capture some of the flavour briefly from the summary paragraph of Chadwick's neutron paper [64]:

> The properties of the penetrating radiation emitted from beryllium (and boron) when bombarded by the α-particles of polonium have been examined. It is concluded that the radiation consists, not of quanta as hitherto supposed, but of neutrons, particles of mass 1, and charge 0. Evidence is given to show that the mass of the neutron is probably between 1.005 and 1.008. This suggests that the neutron consists of a proton and an electron in close combination, the binding energy being about 1 to 2×10^6 electron volts. From experiments on the passage of the neutrons through matter the frequency of their collisions with atomic nuclei and with electrons is discussed.

Chadwick evidently considered his newly discovered particle to be something like a nucleus of $Z = 0$, that is, the "neutron" whose existence had been proposed by Rutherford in his second Bakerian Lecture (Rutherford 1920). To emphasize this point, we quote an earlier passage from Chadwick's paper [65]:

> It has so far been assumed that the neutron is a complex particle consisting of a proton and an electron... Such a neutron would appear to be the first step in the combination of the elementary particles toward the formation of a nucleus. It is obvious that this neutron may help us to visualize the building up of more complex structures, but the discussion of these matters will not be pursued further for such speculations, though not idle, are not at the moment very fruitful. It is, of course, possible to suppose that the neutron is an elementary particle. This view has little to recommend it at present, except the possibility of explaining the statistics of such nuclei as N^{14}.

1.4.3 *Phenomenological theories of nuclear physics*

In this section we shall emphasize the empirical models of nuclei, using phenomenological potentials with adjustable parameters — that is, the part of nuclear physics with which this book is not directly concerned.

The discovery of the neutron immediately stimulated renewed interest in nuclear physics, as did the other major advances of 1932. Experimental and theoretical studies provided increasing insight into nuclear systematics, structure, nuclear reactions, and the nature of nuclear forces [66]. During the 1930s, many papers were published and several international conferences were devoted to this subject — including the Rome Conference in 1931, the Seventh Solvay Conference at Brussels in 1933, the London Conference in 1934, and smaller meetings held nearly annually in Copenhagen. Niels Bohr took a leading role in this field, working actively and publishing on many aspects, especially on neutron capture, on nuclear transmutation by collision and, in 1939, on nuclear fission [67]. At Bohr's institute, physicists from all over the world could always meet and exchange news: refugees from Nazi Germany, other guests from Europe, America and Japan, and physicists who remained in the Third Reich [68]. Other centres of importance were those in Rome, under the guidance of Enrico Fermi; in Paris, under the Joliot-Curies; in the famous Cavendish Laboratory in Cambridge as well as in other English Universities; in the United States, in the Soviet Union, and in Japan under Yoshio Nishina. Nuclear physics flourished especially in the United States, in the hands both of native Americans like Gregory Breit, Ernest O. Lawrence, John Wheeler and Robert Oppenheimer and of refugees Hans Bethe, Edward Teller, Eugene Wigner and Victor Weisskopf.

Most of the papers provided new data and analyses based on a few basic assumptions about the nuclear forces [69]. Important as they were for our detailed knowledge of the nuclear world, in this book we concentrate on the fundamental aspects of the nuclear forces, which have less to do with the peculiarities of large nuclei and much to do with the simpler nuclei [70]. Also, some fields that were regarded as quite separate from nuclear physics, such as cosmic rays and quantum field theory, are crucial to our story and will be discussed in some detail [71].

Our emphasis will be on theoretical ideas, but a few details of experimental methods and results will be mentioned at their appropriate places (e.g., in the introduction to part A, and in chapters 6, 8 and 13). Although our scope is thus limited, from the viewpoint of nuclear physics, our subject is a vital one from the standpoint of the theory of elementary particles. Indeed, we might have called this volume *The Development of High Energy Nuclear Physics*, since the fundamental aspects are concerned mainly with relativistic quantum theory. The "concept of nuclear forces" led to the birth of modern elementary particle physics.

Notes to text

[1] For a general history of the physics of this period, see Pais (1986), Mehra and Rechenberg (1982), Segrè (1980), Bunge and Shea (1979) and Trenn (1977). See also Brown *et al.* (1995), especially the chapters of Pais (1995), Stachel (1995), Rechenberg (1995) and Brown (1995).

[2] X-rays were discovered by Wilhelm Röntgen. Many books credit John Joseph Thomson as the discoverer of the electron; however, Pais (1986) gives equal credit to Walther Kaufmann and Emil Wiechert. Molecular reality was definitively shown by Jean Perrin in 1909, who verified Albert Einstein's theory of Brownian motion — but see also the earlier results of Marian von Smoluchowski in 1906.

[3] Lorentz (1916), being lectures delivered by Lorentz at Columbia University in 1906. A note in the second edition of 1916 says that it is nearly unchanged from the first edition of 1909.

[4] Thomson (1899, p 565).

[5] The idea of the electron goes back to the mid-nineteenth century. As Whittaker (1951, p 392) has noted:

> Some writers have inclined to use the term "electron theory" as if it were specially connected with Sir Joseph Thomson's justly celebrated discovery that all negative electrons have equal charges. But Thomson's discovery, though undoubtedly of the greatest importance as a guide to the structure of the universe, has hitherto exercised but little influence on general electromagnetic theory. The reason for this is that in theoretical investigations it is customary to denote the charges of electrons by symbols e_1, e_2, ...; and the equality or non-equality of these makes no difference to the equations. To take an illustration from celestial mechanics, it would clearly make no difference in the general equations of the planetary theory if the masses of the planets happened to be all equal.

[6] Lorentz (1916, p 8).

[7] Lorentz (1916, p 11) (original emphasis).

[8] The variation of mass with speed is, of course, a feature of Einstein's special theory of relativity, proposed in 1905.

[9] Lorentz (1916, p 45 and 46). The name *electron* for the fundamental unit of electric charge was coined by G.J. Stoney in 1891. We should point out that Thomson did not *prove* that all negative electrons have equal charge, but he

assumed it. The experimental proof was by Robert A. Millikan (beginning in 1909).

[10] Lorentz (1916, pp 214–5).

[11] See Dresden (1993).

[12] See, e.g., Pais (1986), Segrè (1980), Trenn (1977), and Bunge and Shea (1979).

[13] Galison and Assmus (1989).

[14] Pais (1986, chapter 6). See also Pais (1977).

[15] Pais identifies three "energy crises" during the early twentieth century, all of them related to the law of conservation of energy: radioactive decay of nuclei, the Bohr–Kramers–Slater suggestion concerning quantum phenomena in atomic decay and the continuous β-decay spectrum. We are concerned here, however, not with the conservation problem, but with that of the puzzlingly large energies of radioactive particles.

[16] Here we simply quote some facts noted in Pais (1986), where the original sources are referenced.

[17] Pais (1986, p 114).

[18] Pais (1986, chapter 8).

[19] For the early history of cosmic rays see Xu and Brown (1987) and Mehra and Rechenberg (1987, section I.4).

[20] It was not known until the 1930s that γ-ray absorption increases when the energy rises above the threshold for the production of electron–positron pairs (i.e., above about 1 MeV). By about 1930 it was established that the penetrating radiation at sea level consists mainly of charged particles, belonging to the secondary radiation produced by the primaries in traversing the atmosphere. After the Second World War, it was found that the primary cosmic rays consist mostly of protons with a small (but significant) admixture of other nuclides.

[21] J.C. Maxwell, in his article "Atom" in the ninth edition of the *Encyclopedia Britannica* (1878), gave three essential conditions that an atomic model must satisfy: "permanence in magnitude, capability of internal motion or vibration, and a sufficient amount of possible characteristics to account for the difference between atoms of various kinds". (These statements were quoted by Silliman (1963), who discussed nineteenth century atomism.)

[22] See Pais (1986), Purcell (1964), Heilbron (1977a, b) and Bunge and Shea (1979).

[23] Heilbron (1977a, p 45).

[24] Thomson (1904).

[25] Nagaoka (1903) and, for a historical account, Yagi (1964).

[26] According to Heilbron, "Rutherford almost certainly owed nothing to Nagaoka; his attention was apparently drawn to the work of the Japanese physicist by Bragg, who only learned about it himself from N.R. Campbell early in March 1911." (Heilbron 1968, p 300)

[27] However, see Heilbron (1968) for a more critical account. See also Marsden (1962) and Feather (1963).

[28] On the basis of his analysis of the scattering of x-rays and β-rays, and also on optical dispersion, J.J. Thomson had concluded that n was of the same order of magnitude as A (Thomson 1906).

[29] See Mehra and Rechenberg (1982, chapter II) for a discussion of Haas and of the Bohr–Sommerfeld theory. Also, see Heilbron (1977a) and Bohr (1963, 1981).

[30] In quantizing the angular momentum, Bohr followed a prescription of John William Nicholson, who between 1911 and 1914 elaborated a quantum version of an atomic model starting from a model close to Thomson's, but soon approaching Nagaoka's, in having a small positive massive nucleus.

[31] Bohr (1961), especially pp 1084–6. The succeeding Bohr quotations are from this source.

[32] According to Pais (1986): "This simple regularity was initially either

incompletely or incorrectly treated by all its originators."

[33] Of course, the matter did not appear as straightforward at the time. See Heilbron (1974).

[34] Concerning physicists in the First World War see Pais (1986, pp 234–7) and Mehra and Rechenberg (1987, section I.4).

[35] Fajans called the groups of radio-elements occupying the same place in the Periodic System "Plejade" whereas Soddy called them "isotopes" (Fajans 1913; Soddy 1913; von Hevesy and Paneth 1914). One should also mention Antonius Johannes van den Broeck, who earlier in 1913 inferred the concept of the nuclear charge from an organization of chemical elements based on their weights, chemistry and radioactivity.

[36] For the history of isotope research and nuclear masses, see Siegel (1978).

[37] Rutherford (1914); p 450 in Rutherford (1963).

[38] Rutherford (1914); p 451 in Rutherford (1963).

[39] Feather (1963, pp 31–3).

[40] Rutherford (1919b); p 585 in Rutherford (1963).

[41] Rutherford (1919b); p 590 in Rutherford (1963). As we shall discuss below, it was not possible to identify the products of the collision, other than the projected hydrogen nucleus, in this experiment.

[42] Rutherford 1919b; pp. 589–90 in Rutherford 1963 (emphasis added).

[43] Rutherford (1919a); p 562 in Rutherford (1963).

[44] Rutherford (1919a); p 566 in Rutherford (1963).

[45] Chadwick and Bieler (1921, p 939).

[46] A rather different interpretation of these words has also been presented (Pais 1986, p 240): "Chadwick and Bieler's final conclusion avoids all reference to a possible electromagnetic cause for the deviations from the simple theory: 'The present experiments do not seem to throw any light on the nature of the variation of the forces at the seat of an electric charge, but merely show that the forces are of very great intensity...It is our task to find some field of force which will reproduce these effects'."

Pais says about this: "I consider this statement, made in 1921, as marking the birth of the strong interactions". However, the beginning of the passage quoted above by Pais is on p 939 of Chadwick and Bieler, whereas the sentence following the ellipsis occurs earlier, on p 937! Between the two quoted sentences, there are two pages of attempted analysis, assuming different nuclear structures, and finally the phrase we have quoted in our text [45].

[47] See, e.g., Rutherford (1920, p 378): "It is to be anticipated that under the intense forces in the [nucleus], the electrons are much deformed and the forces may be of a very different character from those to be expected from an undeformed electron, as in the outer atom...At the same time if the electrons and parts composing the nucleus are in motion, magnetic fields must arise which will have to be taken into account in any complete theory of the atom."

[48] Besides Rutherford's Cavendish Laboratory, a major centre studying nuclear transformation by α-particle bombardment was the Institut für Radiumforschung in Vienna. However, the difficulty of observation of scintillation of phosphors led to more claims of artificial transformation than could be later substantiated.

[49] Purcell (1964). Recall that the "hydrogen atom" of nuclear physics, the deuteron, was discovered only in 1932.

[50] For α-particle scattering see Blackett and Champion (1931). For the analogous theory of the scattering of identical fermions, see Mott (1929, 1930).

[51] Houtermans (1930, p 124).

[52] According to Purcell (1964, p 127), "By 1931 the spins of some 30 nuclear species had been determined."

[53] Houtermans (1930, p 129).

[54] Rutherford (1929, p 379) (see also Rutherford *et al* 1930).

[55] Some of this discussion is based on Stuewer (1983), especially pp 34–42.

[56] The hyperfine structure of spectral lines had been discussed for several years. E.g., Hantaro Nagaoka and his associates in Japan had observed certain "satellites" of mercury lines (Nagaoka *et al.* 1924), which Pauli subsequently discussed as the effect of the compositeness and the magnetic moment of the nucleus (Pauli 1924).

[57] For a review of the situation with nuclear moments and the statistics of nuclei responsible for the satellites or the hyperfine structure of spectral lines, see Kronig and Frisch 1931. See also E. Amaldi 1987.

[58] For Bohr and energy conservation, see Pais (1986, pp 310–13).

[59] For the history of Pauli's neutrino proposal, see Brown (1978).

[60] The name neutrino was allegedly invented by Fermi, after the discovery by Chadwick of the neutron.

[61] Institut International de Physique (1934, p 324).

[62] The positron: Anderson (1932, 1933). Pair production: Blackett and Occhialini (1933) and for a historical account, Hendry (1984, pp 56–62).

[63] E.g., Pais (1986, pp 397–402), Stuewer (1983, pp 42–6) and Hendry (1984, pp 7–48).

[64] Chadwick (1932, p 708).

[65] Chadwick (1932, p 706).

[66] For the first use of accelerated deuterons as nuclear projectiles, see Davis (1968, pp 53–6).

[67] The papers and manuscripts on nuclear physics are in Bohr (1986), which has an extended introduction by Rudolf Peierls.

[68] Although many nuclear physicists were forced to leave Germany and modern fields of physics were little encouraged or financially supported, Heisenberg succeeded in maintaining a school of nuclear theory in Leipzig (Rechenberg 1993b).

[69] For phenomenological treatments see, e.g. Bethe (1937), Bethe and Morrison (1956) and Blatt and Weisskopf (1952). For light nuclei, especially, see Sachs (1953). For the history, see Stuewer (1979, 1984). Also see Stuewer (1983) and other articles in the same volume.

[70] Most of Bohr's work in nuclear physics does not bear on the fundamental concept of nuclear forces.

[71] In this book we do not tell anything like the full story of quantum field theory, only touching on the infinities and renormalization. Rather, our story of meson theory and its forerunners and competitors complements the history of quantum electrodynamics, as given, e.g., in Schweber (1994).

Part A

Towards a Unified Theory of Nuclear Forces

At the end of the first chapter we hinted at some important innovations in experiment and theory that came about in the early 1930s and initiated a new period in the investigation of the structure of matter (Weiner 1972). To introduce the three chapters that follow, a few remarks, especially on the new experimental tools, may be useful. Before 1930, some apparatus had been developed that could be used for high-energy studies, such as cloud chambers with strong magnetic fields and Geiger–Müller counters. These were applied after 1930, in combination with each other and with electronic coincidence and anti-coincidence circuits, to make more sophisticated detection systems. Also, methods were devised to accelerate electrons and charged nuclear particles, so that thus controlled beams were produced with particle energies of 1 MeV and higher. This led to the following results.

The neutrino (December 1930). In order to solve the difficulty of apparent lack of energy and angular momentum conservation in β-decay and also to remove a contradiction in the statistical description of some nuclei (e.g., ^{14}N and ^{6}Li), Wolfgang Pauli suggested the existence of a neutral particle of spin $\frac{1}{2}$ and small mass, later called the *neutrino*.

The prediction of the anti-electron or positron (May 1931). Because of problems connected with the previous interpretation of the "hole" solutions of his relativistic electron equation as protons, Paul Dirac predicted the existence of new particles having the same mass and spin as the electron, but of opposite charge.

The Van de Graaff generator (September 1931). At a meeting of the American Physical Society in Schenectady, New York, Robert J. Van de Graaff of Princeton University described a 1.5 MeV electrostatic generator for investigating the atomic nucleus and other fundamental problems as follows:

The machine is simple, inexpensive and portable. An ordinary lamp socket furnishes the only power needed. The apparatus is composed of two identical units, generating opposite potentials. The high potential electrode of each unit consists of a 24 inch hollow copper sphere mounted upon a 7 foot upright Pyrex rod. Each sphere is charged by a silk belt running between a pulley in its interior and a grounded motor driven pulley at the base of the rod. The ascending surface of the belt is charged near the lower pulley by a brush discharge maintained by a 10 000 volt transformer kenotron set, and is subsequently discharged by points inside the sphere.

Van de Graaff generators, the first high-energy particle accelerators, were used throughout the 1930s. One was constructed in Washington, D.C. by Merle Tuve (1933); Van de Graaff continued his programme at MIT. (See Van de Graaf (1931), especially p 1920; see also McMillan (1979).)

The discovery of a hydrogen isotope of mass 2 (December 1931). Harold C. Urey and his collaborators at Columbia University and the National Bureau of Standards announced the discovery of a heavy hydrogen isotope, following from the analysis of the atomic spectra in a discharge tube (Urey *et al.* 1932). Heavy hydrogen, or deuterium, has a relative abundance in ordinary hydrogen of about 1 : 4000. The stable mass-2 isotope would play an important role in the theoretical discussion of nuclear forces and in the experimental study of nucleon–nucleon scattering.

The discovery of the neutron (February 1932). Lord Rutherford had proposed as early as 1920 (in the same Bakerian Lecture in which he had named the hydrogen nucleus the "proton") the existence of a neutral nuclear particle, which he called the "neutron". He thought of this particle as a tightly bound proton–electron system (Rutherford 1920). After preliminary investigations using α-particle bombardment of beryllium, by Walther Bothe and Herbert Becker (1930) and Irène Curie and Frédéric Joliot (1932) — both teams having failed to identify the resulting massive particle — James Chadwick worked out the kinematics of the reaction from the range of the protons recoiling after being struck by the "beryllium radiation", noting (Chadwick 1932):

> The results, and others I have obtained in the course of a week, are very difficult to explain on the assumption that the radiation from beryllium is a [light] quantum radiation, if energy and momentum are to be conserved in the collision. The difficulties disappear, however, if it is assumed that the radiation consists of particles of mass 1 and charge 0, or neutrons.

Bothe and Becker had thought the new beryllium radiation to be γ-rays, as had the Joliot-Curies, imagining that the observed proton recoils were due to 50 MeV γ-rays.

The cyclotron (February 1932). In February 1932, Ernest O. Lawrence and M. Stanley Livingston (1932, p 19) of the University of California, Berkeley, described a new method of producing high speed ions:

> Semi-circular hollow plates, not unlike duants of an electrometer, are mounted with their diametral edges adjacent, in a vacuum and in a uniform magnetic field that is normal to the planes of the plates. High frequency oscillations are applied to the plate electrodes producing an oscillating electric field over the diametral region between the electrodes. The magnetic field is adjusted so that the time required for transversal of a semi-circular path within the electrodes equals a half period of the oscillations. In consequence, when the ions return to the region between the electrodes, the electric field will have reversed direction, and the ions receive second increments by passing into the other electrode... the time required for a transversal of a semi-circular path is

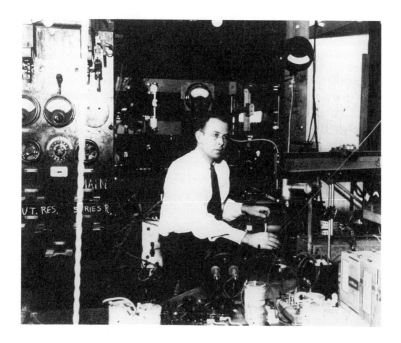

E.O. Lawrence (1901–1958) at his 27" cyclotron in 1934. Photograph from Lawrence Berkeley Laboratory and reproduced by permission of AIP Emilio Segrè Visual Archives.

> independent of their velocities. Hence they... spiral around in resonance until they reach the periphery of the apparatus. Their final kinetic energies are as many times greater than that corresponding to the voltage applied to the electrodes as the number of times they have crossed from one electrode to the other.

The Berkeley physicists achieved 1.2 MeV protons with their first circular accelerator, and shortly afterwards announced the observation of their first nuclear reaction (Lawrence *et al.* 1932).

The Cockcroft–Walton machine (February and June 1932). Three days after the cyclotron paper had been submitted, Ernest Rutherford communicated a report on the completion of another apparatus for producing high-energy ions to the Royal Society of London. His Cambridge students John Cockcroft and Ernest T.S. Walton had constructed it over several years; they now obtained protons of energy up to 0.7 MeV using a voltage multiplying system (Cockcroft and Walton 1932).

The discovery of the positron (September 1932). In September 1932, Carl Anderson of Caltech wrote a short note to *Science* (Anderson 1932) to announce a new particle discovery, which he described in a later publication as follows (Anderson 1933, p 491):

On August 2, 1932, during the course of photographing cosmic-ray tracks produced in a vertical Wilson chamber (magnetic field of 300 000 Gauss) designed in the summer of 1930 by Professor R.A. Millikan and the writer, the tracks were obtained, which seemed to be interpretable only on the basis of the existence in this case of a particle carrying a positive charge but having a mass of the same order of magnitude as that normally possessed by an electron. Later study of the photograph by a whole group of men of the Norman Bridge Laboratory only tended to strengthen that view.

In the course of the next weeks, Anderson found other photographs showing the same positively charged electron-like object. (It should be mentioned that the result was quite qualitative: only upper limits of twice the electron charge and twenty times the electron mass were given.)

The year 1932 has been repeatedly called an *annus mirabilis* of modern physics. Today's physicists are convinced that it opened the door to the nuclear age, or more specifically, to the fields of nuclear and elementary particle physics. In a historical study of the development of nuclear physics in the early 1930s, Erwin Hiebert (1988, p 76) reached an interesting conclusion:

In trying to establish a roster of fundamental steps in the genesis of nuclear physics, the most conspicuous observation to make, it seems, is the almost total omission of reference to theoretical papers on the part of investigators preoccupied with experimental nuclear phenomena. The theoreticians, by contrast, were swimming rather aimlessly in a sea of novel experimental phenomena about the nucleus that did not fit readily with any of their theoretical moves. There was an ever growing mass of experimental data, but very little that seemed in any way fundamental for the theoretician. Perhaps even the experimentalists—there actually were not very many investigators engaged in nuclear physics in the 1930s—were not having too much fun either. This state of affairs might well be chalked up to inherent complexities with nuclear phenomena, but also to a situation in which the theoreticians either were too confused to contribute something relevant to the experimenter's craft or too remote from the world of the laboratory, or both.

However, we shall show in our more detailed study that, in high-energy nuclear physics, the relationship between theorists and experimentalists was not that bad. Certainly, empirical data often revealed paradoxical features that could not be matched by (often erroneous) theoretical speculation, but overall there existed a good, often close, cooperation between experiment and theory.

Chapter 2 will discuss the consequences of the discovery of the neutron for theories of nuclear forces, as well as Fermi's theory of β-decay using Pauli's hypothesized neutrino. On the basis of the latter's success, a kind of "Standard Model" was developed to explain all nuclear forces, strong and weak (chapter 3). This Fermi-field model also dominated attempts by theoreticians to understand puzzling features of cosmic radiation, as we recount in chapter 4. This model was eventually replaced by Yukawa's meson theory, with some of its results being carried over to the latter.

Chapter 2

Nuclear Structure and Beta Decay

2.1 Introduction

As described above, the year 1932 saw the discovery of the neutron, the deuteron and the positron; in the same year, artificially accelerated protons and deuterons came into use for the production and study of nuclear reactions. To the existing trinity of fundamental particles — electron, proton and photon — a new trio was added: positron, neutron and neutrino, the last being a neutral particle conjectured by Pauli to save the conservation laws in β-decay. The nucleus was still thought to consist of protons and electrons, held together (also in the case of the neutron itself) by electric and magnetic forces. That view prevailed for some time, even though the presence of electrons in the nucleus was incompatible with quantum mechanics. Although almost all physicists, Chadwick included, believed at first that the neutron was a tightly bound composite of a proton and an electron, a "collapsed" hydrogen atom, whose existence had been anticipated (Rutherford 1920), Dmitri Iwanenko in Moscow urged that the neutron be considered an elementary particle and suggested that the electron in β-decay is produced at the moment of transition, as a photon is in the decay of an atomic state (Iwanenko 1932a, b).

Soon after Chadwick's announcement, Heisenberg proposed that, in spite of the neutron's composite structure, it could still behave largely as an "elementary particle" within the nucleus. As Heisenberg expressed himself in a letter to Bohr: "The basic idea is to shove all the difficulties of principle onto the neutron and to apply quantum mechanics within the nucleus" [1]. Heisenberg's theory introduced nuclear forces of the exchange type, analogous to those said to be responsible for homopolar chemical bonding [2]. That is, the force between neutron n and proton p involved their "sharing" an electron, the force between two neutrons, the sharing of two electrons. An "incomplete" exchange between n and p, in which the electron escaped from the nucleus, then provided a model for β-decay. At the cost of blurring the distinction between "composite" and "elementary", Heisenberg's theory was able to deal with nuclear states and their transitions.

Since Heisenberg's model for the *neutron* structure was incompatible with the usual quantum mechanics, this picture of the nucleus was not to everyone's taste. In addition, his description of β-decay still violated accepted

J. Chadwick (1891–1974) about 1935. Photograph reproduced by permission of AIP Meggers Gallery of Nobel Laureates.

conservation laws, including those of energy and angular momentum. Furthermore, pure charge-exchange forces did not give correctly the properties of the light nuclei, such as the saturation of forces observed there and in heavier nuclei as well [3]. Thus, other neutron-proton models of the nucleus were constructed by Ettore Majorana and Eugene Wigner, who used spin-dependent potentials, designed to fit the observed nuclear properties. These new potentials were introduced purely phenomenologi-cally — that is, they did not invoke any fundamental mechanism, as Heisenberg's exchange forces did. On the other hand they had the advantage that they could be readily incorporated into standard non-relativistic quantum mechanics. The theories of Majorana and Wigner treated the neutron as an unproblematic elementary particle; they *did not describe β-decay at all.*

The 1933 Solvay Conference in Brussels dealt with the structure and properties of nuclei. Heisenberg gave one of the principal talks (see section 3 below). In the discussion of Heisenberg's report, Pauli made a remark in which he suggested (for the first time "officially") the existence of the neutrino. Another conference participant, Fermi, soon after returning to Rome, proposed a successful theory of β-decay that incorporated Pauli's neutrino and satisfied all the standard conservation laws (see section 4 below). Modelled after quantum electrodynamics (QED) and Dirac's hole theory, Fermi's β-decay theory was a distinctly modern quantum field theory (QFT) of particle creation and annihilation, and it has had an enormous impact on elementary particle physics. Heisenberg's n-p nuclear model, together with the modifications introduced by Majorana, Wigner and others, has led to the modern science of nuclear structure, whereas Fermi's β-decay theory has become, after several modifications, the modern electroweak gauge theory. Beginning in the mid-thirties, new fundamental theories of nuclear forces were proposed, starting again from the nuclear theories of Heisenberg and Fermi [4].

2.2 Heisenberg's model of the nucleus (1932–33)

The modern science of nuclear structure stems from Heisenberg's three-part article of 1932, begun just after Chadwick's announcement of the discovery of the neutron (Heisenberg 1932, 1933). Heisenberg's theory is often described as a neutron–proton model of the nucleus. However, the article is profoundly ambiguous (as seems typical of some of Heisenberg's most seminal work). On the one hand, the constituents (*Bausteine*) of the model nucleus were the neutron and proton, formally regarded as the charged and uncharged states of the particle that we now refer to as the nucleon. (Heisenberg's Hamiltonian function contains only these particles' coordinates and momenta.) On the other hand, the *elementary* particles, in the usual sense of that adjective, were the proton and the electron, while the neutron was regarded as an electron–proton composite. Heisenberg also required additional "loose" electrons, not bound in neutrons. In that fundamental sense, the model was a re-shaping of the e–p model that preceded the neutron's discovery [5].

The power of Heisenberg's model lay in its phenomenological aspect. His article initiated the modern study of nuclear systematics, including the curve of stable isotopes (A versus Z), the calculation of mass defects and some aspects of radioactive decay. Its principal defect was that it required that electrons be present in the nucleus (and *a fortiori* in the neutron itself). This violated basic conservation laws, as well as Heisenberg's own uncertainty principle, and implied (incorrectly) that quantum mechanics failed at distances smaller than the nuclear radius. Exchange forces aside, Heisenberg favoured the presence of electrons in the nucleus on several purely empirical grounds, including β-decay and other cosmic ray and laboratory experiments that we shall not discuss here in detail [6].

The structure of a given nuclear species and its stability or lack of same are determined by the forces that act between the nucleons n and p. In Heisenberg's theory of 1932, the p–p force is taken to be pure Coulomb repulsion, a consequence of the assumption that the protons are elementary particles. He assumed the n–p force to be an exchange force of the type found in the molecular ion H_2^+, and the n–n force to be analogous to the homopolar force found in the molecule H_2. Since the composite neutron cannot be treated by quantum mechanics (according to which it cannot exist!), no fundamental calculation of the n–p or the n–n potential is possible. (For the same reason, the theory cannot deal adequately with β-decay either.) Thus Heisenberg proposed to describe the n–p force by the product of a short range potential function $J(r)$, to be determined empirically, and an operator that changes the nucleon type (either n to p or p to n). This last operator is a part of what became the *isospin formalism*, which is very important in current nuclear and particle physics. In the same spirit, Heisenberg described the n–n force in terms of an ordinary short-range interaction potential function $K(r)$.

In part I of his paper, Heisenberg established this theoretical apparatus and used it to discuss various aspects of nuclear systematics. For example, by assuming that the n–p force dominates, one can understand the approximate equality of the numbers of protons and neutrons in relatively light stable nuclei. In heavy nuclei, on the contrary, Coulomb repulsion among the many protons will cause instability leading to the emission of α-particles. A neutron excess in heavy nuclei leads to β-emission. Decays cause unstable nuclei to approach the stability curve, i.e., the line connecting the stable nuclear species in a plot of A versus Z, often accompanied by the emission of γ-rays.

Parts II and III of Heisenberg's paper extended the treatment of these stability problems, using the method of the self-consistent field (the Fermi–Thomas method) that had been successfully applied in atomic physics. However, each of these parts contains a section discussing the neutron's structure and one on the scattering of γ-rays from the nucleus. These two topics explicitly involve electrons (both "loose" and "bound" electrons) in the nucleus, although these discussions are generally passed over in historical accounts of the Heisenberg nuclear model, both by physicists and by historians of science [7].

In part I also, Heisenberg called for electrons in the nucleus and stressed the compositeness of the neutron, e.g., in the following passage [8]:

> One must realize that there are other physical phenomena for which the neutron can no longer be considered a static structure (*statisches Gebilde*)... To these phenomena belong, e.g., the Meitner–Hupfeld effect, the scattering of γ-rays on the nucleus. Likewise to this class belong all experiments in which the neutrons can be split into protons and electrons; an example of this is the slowing of cosmic ray electrons in passing through nuclei.

In part II, in a section entitled "The properties of the neutron", the author wondered how composite neutrons "with their small mass defect (1 million volts) can survive as a fixed elementary particle (*Elementarbaustein*) in nuclei where the interaction energy is much greater" [9], and he gave this extraordinary answer [10]:

> In the defence of this hypothesis, one can at once adduce that the very existence of the neutron contradicts the laws of quantum mechanics in their present form.

Also the admittedly hypothetical validity of Fermi statistics for neutrons, as well as the failure of the energy theorem for β-decay, proves the inapplicability of present quantum mechanics to the structure of the neutron. However, even if one disregards these properties of the neutron, already the circumstance that the neutron is a structure of approximate extent $\Delta q \approx e^2/mc^2$ means a contradiction to quantum mechanics if the neutron is taken to be a composite of electron and proton.

To paraphrase: the neutron is more complex than the proton, but not in any way describable by quantum mechanics. So when we do certain kinds of nuclear physics, such as systematics — though not when we do, e.g., cosmic ray physics — we should treat the neutron as an elementary particle.

Heisenberg's clearest statement of the advantage of considering most of the nuclear electrons to be bound in neutrons is given in the last section of part III [11]:

> The discovery of the stability of the neutron, not describable by present theory, allows a clean separation of the realms in which quantum mechanics is applicable from those in which it is not, for this stability allows purely quantum mechanical systems to be built up out of protons and neutrons, in which the new kind of features which show up in β-decay do not occasion any difficulty. This possible sharp separation of the quantum mechanical aspects and those new features characteristic for the nucleus seems to get lost if the electrons are considered as independent nuclear constituents.

Up to now we have been calling attention to some problematic aspects of Heisenberg's nuclear theory, but we must emphasize that it was, nevertheless, a major step towards understanding the nucleus as a quantum mechanical system. Heisenberg applied it successfully to nuclear systematics already in

W. Heisenberg (1901–1976) in 1933. Photograph reproduced by permission of Werner-Heisenberg-Archiv.

part I, studying the stability curve. The qualitative conclusions were refined in part II, in which conclusions were drawn concerning the four radioactive series of elements.

In part III, the molecular binding analogy was extended, the n–p exchange force being supplemented with an "electrostatic" (non-exchange) force, as occurs in the molecular ion H_2^+. Heisenberg used the well known Fermi–Thomas method to minimize the energy of the approximate many-nucleon Hamiltonian, with the restriction that the magnitude of the total ρ-spin (the present-day isospin!) was fixed. This yielded an effective mean potential in which the neutrons and protons behave as confined gases of non-interacting particles obeying Fermi–Dirac statistics. The results were seen to justify the conclusions in parts I and II regarding the stability curve, in which similar assumptions had been made concerning the forces.

By October 1933, at the Seventh Solvay Conference (Institut International de Physique 1934; Mehra 1975), Heisenberg had rejected his purely charge-exchange force in favour of a modified force proposed by Ettore Majorana (Majorana 1933a, b) and he also took into account work by Eugene Wigner on the lightest nuclei (Wigner 1933).

2.3 The 1933 Solvay Conference

The theme and title of the Seventh Solvay Conference on physics was chosen to be *The Structure and Properties of Atomic Nuclei*. Taking place after the brilliant discoveries of 1932, its agenda was actually broader than this, and sessions were devoted to the positron and the neutron in addition to nuclear structure. It is also noteworthy that Pauli gave his first official presentation of the neutrino idea and for the first time permitted it to be published [12]. Our discussion here will be confined to those papers bearing on nuclear forces.

We note that, in 1933, there was nothing that could be called a fundamental theory of nuclear forces. The Heisenberg, Majorana and Wigner theories all made use of potentials whose functional form had to be assumed, or fitted from experiment; i.e., they were phenomenological theories. Indeed, there was no general agreement that there existed a specifically *nuclear force*, as opposed to some form of electromagnetic interaction. Certainly no one suggested that there were indeed *two* nuclear forces, one strong and one weak, until Yukawa proposed his meson theory in 1935 [13].

Although, purely logically, one could consider alternative formulations, practically speaking, a fundamental theory for short-range forces must be a quantum field theory [14]. In constructing such a theory, two questions must be addressed [15]:

(a) What are the characteristics of the force that one expects to derive from the fundamental theory, and how can they be inferred from observation and/or experiment?

(b) What are the *constituent objects*, i.e., what particles are involved in the fundamental interaction?

We have avoided the use of "elementary" in connection either with the particles or with the interactions, because that word often connotes the notion of *irreducible*. The interaction between two electrons, e.g., can be regarded as fundamental in QED (quantum electrodynamics), but it is not elementary, insofar as it involves the exchange of photons. Already in the early 1930s, the classification of particles as elementary was becoming problematic because of such processes as pair production.

The most general properties of the nuclear force were known, of course, before 1932. It manifested itself in nuclear binding and in the anomalous (non-Coulombic) large angle scattering of α-particles from light nuclei. Obviously the force was of short range and the attractive binding force was strong enough to overcome the powerful repulsive Coulomb interaction. From the curve of binding energy versus atomic number (*BE* versus *A*), one deduced that for stable nuclear species

(i) *BE* grows linearly with *A*;
(ii) $Z \approx A/2$ for light nuclei and $Z < A/2$ for heavier nuclei.

Property (i) is related to the linear growth of nuclear volume with *A*, sometimes called *saturation*. One knew also that the α-particle was an especially stable structure (also sometimes referred to as *saturation!*).

The discoveries of 1932 helped to complete this general phenomenological picture, especially the discoveries of the neutron and the deuteron. The loosely bound deuteron, besides playing its role as a simple system on which to test hypotheses concerning the n–p interaction, also served as a vehicle by which its contained neutron could be accelerated, which greatly extended the range of possible artificially produced nuclear reactions. At the Solvay Conference, Cockcroft, Rutherford and Lawrence presented results from accelerated deuterons. These were pioneering efforts, not yet productive of firm conclusions about the forces.

George Gamow discussed excited nuclear energy levels and the origin of nuclear γ-rays [16]. In his introduction he gave a clear discussion of what one might learn about nuclear constituents from studying nuclear binding energies (*BE*) We *paraphrase* Gamow as follows:

One generally supposes that two kinds of particles, protons and electrons, are in the nucleus, but that the nucleus also contains stable complexes of these, such as the α-particle and the neutron. Thus the total *BE* has two parts: the internal *BE* of the stable structures and the *BE* of protons, electrons and stable structures with each other. The second part seems to vary in a continuous manner as nuclear particles are added to the nucleus, which rules out certain assumptions. For example if, as it was once assumed, as many α-particles as possible are formed from the protons and electrons, then certain discontinuities in the *BE* curve should be found, but they are not found. So one must assume instead that often α-particles do not form in the nucleus, even when a suitable number of protons and electrons is present. In fact, the behaviour of the *BE* with A suggests that one first forms from the protons and electrons the maximum number of neutrons, and then the maximum number of α-particles [17]. The results regarding the nuclear spin are also in accord with the idea that it is necessary to include neutrons among the nuclear constituents. The saturation property resembles the situation in a liquid drop or in molecules, in which short-range repulsion prevents collapse.

Another part of Gamow's report concerned the anomalous scattering of γ-rays from matter of high Z, which had been reported to produce secondary radiation with components of 0.5 and 1.0 MeV quantum energy [18]. Gamow referred to this as *nuclear fluorescence* and he explained it as follows [19]:

> The γ-quantum of the incident radiation acts on the electron of a nuclear neutron and expels it from the nucleus, producing an artificial β-disintegration. If one finds that the dissociated neutron belongs to a high-lying energy level, the proton that remains, exactly as in the case of a spontaneous β-disintegration, is in an excited state and falls to a lower level, emitting a γ-ray.

Gamow mentioned an alternative explanation proposed by P.M.S. Blackett — namely, that the γ-ray produces an electron–positron pair in the field of the nucleus, and the positron subsequently annihilates with another electron. (Blackett's explanation later proved to be correct!)

Although aware of Pauli's neutrino suggestion, Bohr (following Gamow's talk at the Solvay Conference) called attention to a theory of β-decay that had been advanced by Guido Beck and Kurt Sitte (Beck 1933a,b: Beck and Sitte 1933, 1934a,b) [20]. This theory proposed that an e^+–e^- pair could be created by the strong electric field just outside a heavy nucleus, the e^+ being immediately captured by the nucleus and the electron emitted. Among the fundamental difficulties *not* solved by this theory was the non-conservation of energy and angular momentum. Beck, however, assumed the following attitude towards the neutrino [21]:

> It has been suggested that the [lost mechanical] quantities be ascribed to an unknown particle which it is proposed to call a "neutrino". There is, however, at present no need to assume the real existence of a neutrino and the assumption of its existence would even be an unnecessary complication of the description of the β-decay process.

Chadwick also gave a Solvay report that dealt with nuclear reactions and anomalous scattering of α-rays, and with the properties of the neutron [22]. We mention here only several points about the last item. Using α-particles from a polonium source directed onto a lithium target, Chadwick produced very-low-energy neutrons by the reaction

$$\alpha + {}^7\text{Li} \rightarrow {}^{10}\text{B} + \text{n}.$$

From the existence of this reaction and the isotopic masses, he deduced an upper limit for the mass of the neutron and concluded that "... there seems no doubt that the mass of the neutron should be less than that of the hydrogen atom. That is what we would expect if the neutron results from the intimate union of a proton with an electron." [23]

That sounds like an endorsement of the composite neutron, but Chadwick clearly had some doubts about the matter, for he also presented some arguments in favour of a simple neutron. If composed of a proton and an electron, he asked, why does the hydrogen atom not collapse to the "neutron state"? — and how does one account for its spin and statistics? He continued, rather ambiguously [24]:

> It seems that the assembled facts suggest that the neutron and proton are both elementary particles. However, I shall show later that one can deduce from results concerning the collisions between neutrons and protons some arguments favouring the complexity of the neutron and the proton.

Heisenberg's Solvay report (Heisenberg 1934), which bore the title "General Theoretical Considerations on the Structure of the Nucleus", had three sections dealing, respectively, with principles, hypotheses and applications [25]. From June to October 1933, Heisenberg exchanged about a dozen letters with Pauli, much of the correspondence concerning Heisenberg's preparation of his Solvay report, which he sent to Pauli in July [26]. Topics that were discussed included the properties of the nuclear particles, the possible existence of a neutrino and the exchange character of the nuclear force.

The original manuscript of the report contained the sentence: "At the moment it is not clear whether the statement 'energy conservation is violated in β-decay' represents a valid application of the energy concept." [27] However, this sentence was struck out and replaced by a statement that shows the beginning of a shift in attitude on Heisenberg's part toward the neutrino. Probably he made the replacement after he received a letter of 2 June 1933 from Pauli, which contained this paragraph [28]:

> Concerning nuclear physics I again believe very much in the validity of the energy theorem in β-decay, since still other very penetrating light particles will be emitted. I also believe that the symmetry character of the total system as well as the momentum will always be preserved in all nuclear processes.

In any event, whether because of Pauli's letter or not, Heisenberg replaced the expunged sentence by the following [29]:

> Pauli has discussed the hypothesis that, simultaneously with the β-rays, another very penetrating radiation always leaves the nucleus — perhaps consisting of "neutrinos" having the electron mass — which takes care of energy and angular momentum conservation in the nucleus. On the other hand, Bohr considers it more probable that there is a failure of the energy concept and hence also of the conservation laws in nuclear reactions.

The "principles" section of the Solvay report begins by stating that one of the first tasks of theory in the nuclear domain is to determine "as precisely as possible a limit to the possibility of applying quantum mechanics" [30]. There was evidence from α-decay that quantum mechanics applied to the heavy constituents and Bohr had recently concluded that nuclear mass defects were consistent with the uncertainty principle of quantum mechanics. However, Bohr had used the nuclear model that included electrons and

E. Majorana (1906–1938) with his sisters in 1932. Photograph from Istituto di Fisica Teorica dell' Università and reproduced by permission of AIP Emilio Segrè Visual Archives.

Heisenberg offered two objections to Bohr's work. First, there was no justification to ignore, as Bohr had done, the substantial contributions of the electrons to the energy. Secondly, he said, unlike in the atomic case, "we have no theoretical means to study the forces that act between the various heavy constituents" [31].

Heisenberg then reviewed the other difficulties in principle that arise when electrons are allowed in nuclei. He concluded that one could not apply quantum mechanics to the electron (as opposed to n and p), not even the correspondence principle — or even the electron theory of Lorentz. Indeed: "The statement that electrons act as nuclear constituents possesses no well-defined meaning other than the fact ... that some nuclei emit β-rays" [32].

In the section on "hypotheses" in his Solvay report, Heisenberg introduced Gamow's liquid drop model, which emphasized the α-particle structure of the nucleus; next he discussed whether neutrons were nuclear constituents [33] and then the laws of interaction between pairs of nuclear particles. He explained the approximate equality of neutron and proton numbers for the stable light nuclei (given the repulsive Coulomb force of the protons and the possibly attractive force between neutrons) by assuming that the like-particle forces are dominated by the n–p forces. For the latter he considered both a molecular-type exchange force and an "ordinary" (i.e., non-exchange) force.

An n–p exchange force could be of pure charge-exchange type, as Heisenberg had originally suggested, or, as proposed by Majorana, of a type that exchanges both the charge *and* the spin direction of neutron and proton [34]. Majorana had pointed out, said Heisenberg, that, if one wanted to ensure the saturation requirement without arbitrarily introducing a short-range repulsion between neutron and proton (something "rather difficult to accept"), some type of exchange force was necessary [35]. Furthermore, according to Heisenberg's report, "Majorana legitimately drew the conclusion" that a space-exchange, rather than a charge-exchange, force fits the nuclear systematics best — for, with Heisenberg's force, the deuteron would already be a closed system, rather than, as observed, the α-particle [36].

Heisenberg's report concluded with a recapitulation of nuclear systematics, e.g., the stability curve, on the basis of his own model and that of Majorana, using the statistical (Thomas–Fermi) theory. With that, the subject of nuclear structure physics was fairly launched. One grave problem remained, however, and its solution came very soon after the Solvay Conference of 1933.

2.4 Fermi's theory of β-decay (1933–34)

Among the problems faced by the early nuclear models, one of the most serious was that of explaining β-decay. On the one hand, the most basic laws of physics seemed to be violated if electrons were present in the nucleus. On the other hand, in β-decay they were seen to be leaving it! Moreover, while the nucleus lost a fixed amount of energy in the β-decay transition, the electron carried off a lesser variable energy and no other particle could be observed to carry off the difference. After the advent of quantum mechanics, the other forms of radioactivity found at least a satisfactory qualitative understanding, but the puzzle of β-decay radioactivity only deepened. Then in the early 1930s, the crucial ingredients of a successful approach began to appear: first, the idea of the neutrino; and then, the conviction that electrons were not normally present in the nucleus, but were produced only at the moment when the transition took place.

The idea that β-decay involves the emission of a new particle, now called the neutrino, goes back to Pauli's proposal of December 1930 in a letter to Lise Meitner and Hans Geiger [37]. The first clear statement concerning the creation of the electron in β-decay is contained in a note of Iwanenko of August 1932 [38]. One may well ask why Pauli did not proceed to formulate the theory of β-decay himself. The probable answer is that Pauli, like other leading physicists such as Bohr and Heisenberg, felt that the existing physical theories, including quantum mechanics and relativistic quantum field theory, were not capable of dealing with nuclear phenomena. Thus, the solution to the problem was found instead by Pauli's less hesitant Italian friend, Enrico Fermi [39].

Since the late twenties "it had been felt by Fermi that physicists would be ready in the near future to attack problems of nuclear structure" [40]. He began in 1929 by considering the magnetic moments of nuclei, derived from atomic spectral hyperfine structure (Fermi 1930a, b). A little later, he tried to begin the experimental study of cosmic ray and nuclear physics [41]. During this period, Fermi also familiarized himself with QED, reformulating the theory in a new Hamiltonian scheme, applied it to selected problems and presented lecture courses on it [42].

Fermi learned quite early about Pauli's neutrino hypothesis, when the latter attended the first international conference on nuclear physics, the Rome Convegno di Fisica Nucleare, 11–18 October 1931. Pauli had given a talk in Pasadena on the neutrino (which at that time he called the "neutron") and Fermi asked Samuel Goudsmit to summarize that talk at the Rome meeting (Brown 1978). As Pauli recalled later, Fermi "immediately showed a lively interest for my idea (of β-decay) and a very positive attitude to my new neutral particle" [43]. From this time on, Fermi advocated energy conservation in β-decay (Fermi 1932b). After Chadwick had announced the discovery of the neutron, Fermi invented the presently accepted name for Pauli's particle, the neutrino [44]. At the Seventh Solvay Conference on Physics of October 1933, the last act was presented in the preparation for a theory of β-decay. Both Heisenberg, in his report, and Pauli, in the public discussion, outlined the "new view" of the process, including the neutrino (ν) and the e–ν joint expulsion, ensuring the conservation laws. The task remained, however, to assemble the two new elements into a suitable quantum theory.

On returning home from Brussels, Fermi did the required job during November 1933. His solution had these important aspects.

(i) The problem of β-decay is treated by "second quantization", i.e., a quantum field-theoretical technique using creation and annihilation operators to describe the creation of an e–ν pair [45].

(ii) The e–ν pair acts like a vector field coupled to the charge-changing p–n current, the latter involving the isotopic spin raising and lowering operators (which also occur in Heisenberg's nuclear exchange force).

(iii) Conservation of energy and momentum is enforced by the use of a Hamiltonian scheme, analogous to that of QED, in which the vector potential of the radiation field is coupled to the electron current.

(iv) A coupling constant G is introduced, its magnitude fixed by the rates of β-decay; but no attempt is made to relate G to other interactions (e.g., electromagnetic or strong nuclear).

It is remarkable that the entire approach, including extensive applications, was worked out in less than two months. As Rasetti recalled [46]:

Fermi intended to announce the results of his β-decay theory in a letter to *Nature*, but the manuscript was rejected by the Editor of that journal as containing abstract speculations too remote from physical reality to be of interest to the readers. He then sent a somewhat larger paper to *Ricerca Scientifica* where it was promptly published. The article includes all essential results, showing that the calculations (including the numerical fit) had been completed. The larger papers in *Nuovo Cimento* and *Zeitschrift für Physik* were sent to the respective journals very early in 1934.

Thus the main results, apart from theoretical details, can be found already in the first paper (Fermi 1933, title in English: "Attempt at a theory of the emission of β-rays"). It provided a quantitative theory of β-decay on the basis of known principles of relativistic quantum theory, starting from the assumption that "the total number of electrons and neutrinos in the nucleus is not necessarily constant", and considering "the heavy particles, neutron and proton, as two quantum states connected with the two possible values of an internal coordinate ρ" [47]. The crucial step in the theory consisted of a particular choice of the form of the interaction energy, which assured that in each n → p transition, brought about by the operator Q, a pair is created — an electron (field operator ψ) and a neutrino (field operator ϕ).

Thus Fermi wrote down the following general *Ansatz* for the interaction Hamiltonian H:

$$H = QL(\psi \cdot \phi) + Q^*L^*(\psi^* \cdot \phi^*), \qquad (2.1)$$

with L representing a bilinear expression in the field operators ψ and ϕ, and the stars indicating the Hermitian conjugate. He then restricted the form of L by assuming it to behave under coordinate transformations like the fourth component of a polar four-vector, namely

$$L(\psi \cdot \phi) = G(\psi_2\phi_1 - \psi_1\phi_2 + \psi_3\phi_4 - \psi_4\phi_3), \qquad (2.2)$$

with G a constant expressing the strength of β-decay.

By using the above interaction, Fermi obtained an expression for the inverse of the decay time τ,

$$\tau^{-1} = \text{constant} \times G^2 q F(\eta_0), \qquad (2.3)$$

where q denotes the space integral over the neutron and proton eigenfunctions and $F(\eta_0)$ is a certain function of η_0, the maximum momentum of the electron. The product $\tau F(\eta_0)$ (the *ft* value), in the observed β-decay reactions, assumed values roughly in the range 1–100. The case in which the neutron–proton space integral q vanishes is analogous to a forbidden transition in spectroscopy. Finally, the Fermi constant G can be calculated to be

$$G = 5 \times 10^{-5} \text{ cm}^5 \text{ g s}^{-2}. \qquad (2.4)$$

These results were further refined in Fermi's longer papers (Fermi 1934a, b), which firmly established the foundations of β-decay theory. The next years (and decades) would alter some details of the interaction expressions (2.1) and (2.2), but not the principle of Fermi's description of nuclear β-decay. It was, in fact, so successful that certain physicists even tried to construct a quantum field theory from the Hamiltonian (2.1) to derive the strong nuclear binding forces as a second-order effect, that is through the exchange of electron–neutrino pairs. This unified field theory of strong and weak nuclear forces occupied theoretical physicists until the late thirties, as we shall discuss in the next chapter.

Notes to text

[1]   Heisenberg to Bohr, 20 June 1932: *'Die Grundidee ist: alle prinzipiellen Schwierigkeiten auf das Neutron abzuschieben und im Kern Quantenmechanik zu treiben.'* (Bohr Archives, Copenhagen).

[2]   The view that homopolar binding is an exchange force is not uniformly accepted. Thus, Victor Weisskopf has stated: "The chemical bond is often described as an 'exchange effect'. I believe that such a formulation is misleading. It refers to mathematical terms appearing in the detailed calculation, in which two wave functions appear, differing by an exchange of coordinates. These terms are a consequence of the Pauli principle requiring antisymmetric wave functions. They have no direct physical significance. Electrons are 'exchanged' only in the sense that in the merged molecular quantum state it is no longer possible to assign an electron to one or the other nucleus." (Weisskopf 1985, p 399).

[3]   "Saturation", as used here, implies a nuclear volume that is proportional to A, i.e., constant volume per nucleon.

[4]   The new fundamental theories will be discussed in later chapters.

[5]   According to Miller (1984), "Modern nuclear physics and particle physics began" with Heisenberg's introduction of an exchange force (*Platzwechsel*) that was the analogue of the molecular binding force in the ion H_2^+ (see note 2). However, Heisenberg's *Platzwechsel* (mistranslated by Miller as "migration") was not realizable and hence not visualizable in a quantum mechanical theory, unless the "migrating" electron were spinless and obeyed Bose–Einstein statistics. In fact, the first QFT of nuclear forces was that of the Fermi-field theory discussed in chapter 3 below; the second was the meson theory of Hideki Yukawa.

[6]   See Brown and Moyer (1984).

[7]   For example, David Brink's otherwise excellent *Nuclear Forces* (Brink 1965) provides an English translation of only those parts of Heisenberg's paper not referring to nuclear electrons. Exceptions are Bromberg (1971) and Brown and Moyer (1984). Bromberg's paper is a pioneering account, based on her study in Copenhagen in the Bohr Archive and her interview with Heisenberg in Munich on 16 June 1970 (AHQP). Bromberg sketches the concerns of Bohr and Heisenberg before the neutron discovery, who were led to believe in the dawning of a new dynamical era in physics by the puzzles of nuclear physics, cosmic rays and QED. Although Bromberg's views on Heisenberg's n–p model do not differ substantially from ours, we do not agree with her statements such as that to Heisenberg "the neutron seemed to be elementary as well as complex". His aim was rather to decide to which nuclear phenomena quantum mechanics might be applicable and to which it was not.

[8]   Heisenberg (1932, p 11).

[9]   Heisenberg (1932, p 163). The neutron's mass defect is actually negative; i.e., the free neutron is unstable against β-decay, a fact unknown at the time of Heisenberg's writing.

[10] Heisenberg (1932, p 163).

[11] Heisenberg (1933, p 595). Although the free neutron is actually unstable (not known to Heisenberg at that time), yet it is stable in the nucleus, except in the β-radioactive nuclei. Thus the argument presented is valid.

[12] Institut International de Physique (1934, pp 324–5). See also Brown (1978).

[13] Yukawa (1935). Yukawa introduced separate strong and weak "coupling constants" analogous to electric charge.

[14] By "quantum field theory" we mean to include not only linear theories, but also gauge theories of Yang–Mills type, as well as others that are nonlinear, supersymmetric, relativistic non-local theories, etc., provided that they imply the conservation of electric charge and the relativistic energy-momentum and angular momentum tensors.

[15] These are the first two stages in the pattern proposed by Mituo Taketani in his "methodology of three stages", a general historical analysis that he applies to physics. The search for the characteristic phenomena occurring in a system, the constituents of the system and finally the fundamental theory of interaction within the system, correspond respectively to stages that Taketani calls *phenomenological, substantialistic* and *essentialistic* (Taketani 1971). We shall return to this methodology in chapter 7.

[16] Note 12, pp 231–88.

[17] For the lightest elements the two methods do not differ very much. E.g., for $^9Be_4$, one gets by the first method $2\alpha + 1p + 1e^-$ and by the second method $5p + 4n \Rightarrow 2\alpha + 1n$. However, for $^{208}Pb_{82}$, the first method gives $52\alpha + 22e^-$, whereas the second gives $82p + 126n \Rightarrow 41\alpha + 44n$. These examples were given by Heisenberg in his Solvay Conference report (note 12, p 298).

[18] This is related to the Meitner–Hupfeld effect, discussed in Brown and Moyer (1984).

[19] Note 12, p 259.

[20] Bohr remarked that "even if it does not resolve the fundamental difficulties, it nevertheless deserves our full attention". (Note 12, p 287).

[21] Beck (1933a, p 967).

[22] Note 12, pp 81–120.

[23] Note 12, p 102.

[24] Note 12, p 103.

[25] Note 12, pp 289–344.

[26] Pauli to Heisenberg, 14 July 1933. Letter [384] in [WPSC2], pp 184–7; facsimile on pp 174–83.

[27] W. Heisenberg, *Allgemeine theoretische Überlegungen über den Bau der Atomkerne* (original German manuscript contained in [WHA]), p 27.

[28] Letter [311] in [WPSC2], pp 166–7.

[29] Note 27, p 27.

[30] Note 12, p 289.

[31] Note 12, p 292.

[32] Note 12, pp 292–3. This statement marked a considerable change in viewpoint from that of his three-part paper of 1932.

[33] That is, in greater numbers than as the "remainder" of a dominant α-particle structure.

[34] In note 12, pp 301–2, Heisenberg showed that charge exchange is equivalent to space and spin exchange combined, hence Majorana's exchange is equivalent to space exchange alone (Majorana 1933a, b). According to his Roman colleagues, Majorana's ideas on nuclear forces were well formed before he went to Leipzig to work with Heisenberg on a fellowship in January 1933, but he was reluctant to publish them until he was urged to do so by Heisenberg (Amaldi 1966).

[35] There would be an effective repulsion due to the exclusion principle (as occurs in atomic physics) with an exchange force. E.g., the deuteron having spin 1, a third nucleon of either charge could not have the same wavefunction and exchange charge with either of the first two constituents. With Majorana's force, the closure does not occur for the deuteron, but it does for ^4He.

[36] Note 12, pp 303–4. At present, nuclear saturation is believed to come about mainly due to short range repulsive forces between nucleons.

[37] Pauli to Meitner *et al.*, 4 December 1930, in [WPSC 2], pp 39–40. (English translation in Brown (1978).)

[38] Iwanenko (1932a). He wrote (in translation): "The electrons in nuclei are really quite analogous to the absorbed photons, the emission of the β-electron parallel to the birth of a new particle which, in the state of absorption has no individuality."

[39] Historical accounts of the development of Fermi's theory of β-decay have been given by Pauli (1961) and Rasetti (1962).

[40] E. Segrè, in Fermi (1962, p 328).

[41] In the winter of 1930–31, Fermi started "as a first task the construction and operation of a cloud chamber, with the help of E. Amaldi", according to Rasetti (1962, p 348). However, the workshop in Fermi's Rome institute proved inadequate to the task and the project failed. Rasetti was sent in the autumn of 1931 to Meitner in Berlin to learn to use such items of apparatus as the cloud chamber and electrical counters that eventually were used to detect neutrons. When Rasetti returned, the Rome group built a working cloud chamber and a crystal γ-ray spectrometer.

[42] Fermi began his study of QED in the winter of 1928–29, reading Dirac's fundamental article (Dirac 1927). Fermi submitted his first paper in March 1929 (Fermi 1929). His first lecture course was given in April 1929 at the Institut Henri Poincaré in Paris and he repeated it during the summer of 1930 at the Summer School of Theoretical Physics at Ann Arbor, Michigan. Out of these lectures grew a well-known review article (Fermi 1932a).

[43] Pauli (1961, p 161).

[44] Fermi gave a talk on 7 July 1932 at the *Vème Congrès International d'Electricité* in Paris, when he still used the term *neutron* (*neutrone*) for Pauli's particle. However, he later renamed it in "Rome seminars, in order to distinguish it from the heavy neutron, ('neutrone')". (Pauli 1961, p 362)

[45] Although Fermi was an expert in quantum field theory and had contributed significantly to QED, until this point he had avoided the second quantization method, advanced some six years earlier (Jordan and Klein 1927). According to Franco Rasetti: "Apparently he had some difficulty with the Dirac–Jordan–Klein method of the second quantization of fields, but eventually also mastered that technique and considered a beta-decay theory as a good exercise on the use of creation and destruction operators" (Rasetti 1962, p 539).

[46] Rasetti (1962, p 540).

[47] Fermi (1933, p 492). The quantum number ρ that labelled the proton and neutron states was that of Heisenberg (1932).

Chapter 3

The Fermi-field Theory

3.1 Introduction

In the early 1930s, as the new quantum mechanics was demonstrating its power, opinions were nevertheless divided about the role that it played within the atomic nucleus. Before the discovery of the neutron in 1932 (and even for some time afterwards) electrons were assumed to be present in the nucleus and to take part in nuclear binding. The assumption seemed to preclude any complete quantum mechanical nuclear theory, since the continued presence of even a single electron within the small nuclear volume violated Heisenberg's uncertainty principle. Indeed, one could give *other* good arguments why electrons within the nucleus were incompatible with quantum mechanics, as well as compelling reasons for them to be present, as we have discussed in chapter 2 [1]. It thus appeared inevitable that some aspects of quantum theory would have to be generalized or abandoned.

Soon after the discovery of the neutron, however, a few physicists began to ask whether it might now be possible to make a nuclear theory analogous to non-relativistic atomic and molecular theories, assuming that the neutron could be considered in some respects an elementary particle. This approach was pursued throughout 1932 and 1933, and it was one of the main subjects discussed at Brussels, at the Seventh Solvay Conference on Physics of October 1933, as we have already emphasized in section 3 of the previous chapter.

At this conference Werner Heisenberg and Paul Dirac presented major theoretical reports on the topics that concern us here. During one of the discussion periods, Wolfgang Pauli proposed the existence of another light particle, the neutrino. Enrico Fermi, one of the participants, used Pauli's neutrino soon afterwards to construct a quantum field theory of β-decay, involving an electron–neutrino pair (section 4 of chapter 2). However, even after its publication, this theory had still to compete with a theory advanced by Guido Beck and Kurt Sitte, and incorporating pre-neutrino views, as we describe below in section 2.

After 1934, however, the electron–neutrino pair theory was adapted, especially by Heisenberg, as a fundamental theory of the strong, as well as of the weak, nuclear interactions (section 3). Known as the theory of the "Fermi-field", this unified field theory proved to be flexible enough to adapt

From left to right: E. Fermi (1901–1954), I. Waller (1898–1991) and W.W. Hansen (1909–1949) in 1933. Photograph reproduced by permission of AIP Emilio Segrè Visual Archives.

to such new features as the charge-independence of nuclear forces (section 4). Even after the emergence of Yukawa's meson theory of nuclear forces, some important theoreticians continued considering various versions of the Fermi-field for strong interactions (section 5).

3.2 Nuclear electrons, the neutron and the neutrino (1933–34)

As we noted in chapter 2, the theme of the Seventh Solvay Conference was the atomic nucleus (Institut International 1934). Theorists, including Bohr, Dirac, Fermi, Heisenberg and Pauli were in attendance, as well as prominent nuclear experimentalists from important centres such as Paris, Cambridge and Berlin, and from Berkeley, Leningrad and Rome. There were talks on the nucleus itself, as well as discussions of Pauli's neutrino and the newly discovered elementary particles: the neutron (reported by its discoverer James Chadwick) and the positron (discovered by Carl Anderson and reported at the conference by Dirac).

Heisenberg reported on the theory of nuclear structure, including his own neutron–proton (n–p) model of the nucleus, and on further considerations by Ettore Majorana and Eugene Wigner. The phenomenology initiated by Heisenberg and these authors made use of an effective energy operator (Hamiltonian) that contained the space, spin and charge coordinates of neutrons and protons, but not those of electrons. Heisenberg's new charge coordinate ρ took on the value $+1$ for a neutron and -1 for a proton; otherwise, those heavy nuclear constituents were not distinguished. To treat the approximate n–p symmetry and to describe the transformation of one kind of nucleon into another (to use the modern generic term for the neutron and proton), he introduced what is now called the isospin formalism [2].

Heisenberg's theory of nuclear structure was a great improvement over the earlier electron–proton model, but it suffered from severe conceptual problems that arose because the nucleus still contained electrons, even though their coordinates were suppressed in the phenomenological Hamiltonian. Thus the discovery of the neutron did not immediately solve all the puzzles of nuclear physics. Indeed, further problems had already been anticipated by Pauli, before Heisenberg's theory, in a letter on 29 May 1932 to Lise Meitner [3]:

> Towards the non-Pauli neutrons I am quite *positively* disposed. Although they do not solve by their existence the fundamental problems of nuclear structure (β-decay, wrong statistics), they seem to be very useful in many respects for understanding the nuclear structure. Also I believe that, when one further studies experimentally their actions on other nuclei (e.g. through impacts with protons), there may follow new valuable hints for the theory.

Pauli's positive judgment was confirmed in the discussion period after Heisenberg's presentation at the Solvay Conference. No one objected to Heisenberg's treatment of nuclear systematics, but it was equally clear that fundamental problems remained. Thus most of the participants' remarks concerned Pauli's "β-decay" and "wrong statistics".

The statistics problem showed up in the nucleus $^{14}N_7$ with $N = Z = 7$, which experiment showed to have spin 1 and to obey Bose–Einstein statistics, i.e., it had to be a *boson*. If the neutron were an e–p composite then nitrogen-14, with 7 protons and 7 neutrons, would contain a total of 7 electrons and 14 protons, an odd number; thus it had to be a fermion. Hence, the e–p theory gave it the "wrong statistics".

One of the problems with β-decay was related to Heisenberg's charge-exchange interaction. We can represent the latter schematically as

$$n + p' \rightarrow (p + e^-) + p' \rightarrow p + (e^- + p') \rightarrow p + n', \qquad (3.1)$$

where the primed and unprimed quantities refer to different nucleons in interaction with each other. The intermediate steps n → p + e⁻ and e⁻ + p′ → n′ also represent, respectively, β-decay and inverse β-decay reactions. The basic reactions that are assumed in this scheme conserve electric charge, but violate several other conservation laws, including that of angular momentum.

Although many physicists (including Bohr) were willing to accept these violations of the standard theory (quantum mechanics plus relativity theory), Pauli wrote to Heisenberg before the Solvay Conference, warning him that he would "publicly confess" his belief in the conservation laws. He stated: "I shall insist in Brussels that a *neutron can never decompose... into an electron and a proton.*" [4]

At the conference, Pauli suggested that the physical quantities that seemed to be lost in β-decay were in fact carried off with "the emission of a very penetrating radiation of neutral particles, which has not been observed yet". He continued [5]:

> With regard to the properties of these neutral particles, we first learn from atomic weights (of radioactive elements) that their mass cannot be much larger than that of the electron. In order to distinguish them from the heavy neutrons, Fermi proposed the name "neutrino". It is possible that the neutrino proper mass is equal to zero, so that it would have to propagate with the velocity of light, like photons. Nevertheless, its penetrating power would be far greater than that of a photon with the same energy. It seems to me admissible that neutrinos possess a spin $\frac{1}{2}$ and that they obey Fermi statistics, in spite of the fact that experiments do not provide us with any direct proof of this hypothesis.

This statement was followed by remarks by Chadwick, Bohr, Meitner and Francis Perrin concerning the possibility of experiments to detect Pauli's neutrinos. Dirac did not directly refer to neutrinos but said that he thought it would be quite all right if the heavier nuclei (say from chlorine onward) were allowed to contain a small number of electrons, because then the nuclei would still be "essentially constituted of protons and neutrons", so that Heisenberg's analysis would still be applicable.

To a question from Rudolf Peierls about why he thought the neutron was a "truly elementary particle" and not simply one that "never disintegrates into a proton and electron" Heisenberg replied that that was "just an impression". In the light of the previous discussions, this was a rather confusing exchange, which Bohr ended by saying: "It seems to me that one cannot unambiguously attach a meaning to the distinction between elementary and complex particles." [6]

As Pauli noted later, Fermi had named the neutrino, distinguishing it from the heavier neutron (in Italian, *neutrone*) of Chadwick. Fermi had learned about Pauli's proposal much earlier; at the Rome Congress on Nuclear Physics in October 1931, he had held "many private discussions" with Pauli, in which they agreed that they found unacceptable Bohr's idea (put forward at the Congress and also earlier) that energy was conserved only statistically on the average, not necessarily in each elementary act [7]. When Fermi returned to Rome from the 1933 Solvay Conference, he set about to make a field theory of β-decay that obeyed all known conservation laws.

In December 1933, two months after the Solvay Conference, Fermi published his new theory in a short paper in *Ricerca Scientifica* (Fermi (1933), see chapter 2, section 4). In its basic interaction, a neutron n transforms into a proton p, an electron e$^-$, and a neutrino $\bar{\nu}$ (an antineutrino in the present language):

$$n \rightarrow p + e^- + \bar{\nu}. \tag{3.2}$$

The following month, Fermi published a longer version of his paper in Italian and in German [8]. In his theory all the usual conservation laws are obeyed, the differences in energy, momentum, angular momentum and charge between the parent nucleus and the daughter nucleus being carried off by the electron and the neutrino.

The total number of heavy particles, protons plus neutrons, does not change; they are described by non-relativistic Schrödinger functions. In contrast, the light particles, the electron and the neutrino (the latter possibly massless), are treated by the relativistic Dirac theory. Since an electron and a neutrino are produced in the decay, the number of light particles changes; a quantum field theory using the so-called creation and annihilation operators is needed to describe the process. The method is often called *second quantization* when applied to the Schrödinger (or Dirac) wavefunction "field" (which itself already refers to a quantized system) [9]. In modern terms, Fermi's theory has two charge-changing currents (n–p and ν–e$^-$) interacting at a given point — the four-fermion contact interaction. The n–p charge-changing exchange current is identical with that in Heisenberg's nuclear-force model; following Heisenberg, Fermi formulated it using isospin operators. They treated the heavy particles in a non-relativistic approximation, since their kinetic energies were small compared with their rest mass energies and also because, as Fermi remarked, "the relativistic wave equation for the heavy particles is unknown" [10].

The light-particle current, on the other hand, must be treated relativistically. The 16 operator quantities obtained by multiplying each of the four Dirac field components of the electron by each of the four Dirac field components of the neutrino can be combined to give five relativistic covariants: scalar, vector, tensor, pseudovector and pseudoscalar. The most general form possible for the interaction involved all five of these "currents", but Fermi decided to explore only the four-vector current. For that case the e–ν current behaved formally like the quantum of the four-vector potential of the electromagnetic field, i.e., as a kind of photon. Fermi was especially well equipped to study that kind of theory, having given courses on quantum electrodynamics at the Institut Henri Poincaré in Paris in 1929 and at the summer session of the University of Michigan at Ann Arbor in 1930, which courses gave rise to a well-known review article (Fermi 1932a).

Fermi's long comprehensive article on β-decay included detailed applications. For example, he discussed the shape of the electron energy distribution in β-decay and the mean lifetimes of the decays, treating cases in which the nuclear transition associated with the decay was "allowed" or, alternatively, "forbidden" (in the terminology of atomic transitions). Furthermore, he related the shape of the electron distribution near its energy end-point to the rest mass of the neutrino, citing the experimental curves that indicated a very light, perhaps massless, neutrino. The success of Fermi's theory eventually caused most physicists to accept the existence of Pauli's neutrino long before it was detected (which did not occur until the 1950s) [11].

However, the Fermi theory was not immediately accepted. A competing theory, proposed a bit earlier by Guido Beck and Kurt Sitte, also claimed good agreement with experiment. These authors preferred assuming the breakdown of the conservation laws, rather than postulating the existence of a practically unobservable neutrino [12]. In their theory, Beck and Sitte supposed that the strong nuclear potential created a virtual electron–positron pair, the positron being absorbed by the nucleus (which increased its charge by one unit), while the electron escaped. Properties of the positron in addition to its charge, such as spin, magnetic moment and energy, were to be absorbed by the nucleus — in effect, those properties were "lost" [13]. In the Pauli–Fermi theory the neutrino carried off the balancing physical quantities.

In October 1934, another international physics conference was held in London, dealing with nuclear physics and the solid state of matter. In the session on β-decay, Beck's paper (1935) still advocated the Beck–Sitte theory of β-decay and criticized Fermi's. Fermi also addressed the conference, not on β-decay, but on the subject of artificial radioactivity produced by neutron bombardment [14].

Today, when conservation laws are regarded as among the safest foundations of theoretical physics, it may be puzzling that a theory like Beck and Sitte's, not conserving energy and angular momentum, could be competitive with Fermi's, built in analogy with quantum electrodynamics (even though it involved an almost undetectable particle, the neutrino). Much later, Heisenberg referred to this situation as one "of very unclear thinking". He recalled that, just after the publication of his papers on nuclear structure in 1932, "my assistant at that time, Guido Beck ... protested very strongly, and said it was a scandal because I really claimed that there were no electrons in the nucleus, because electrons come out". Heisenberg further recalled that Victor Weisskopf in early 1934 "explained to me that these two papers on the nucleus were very much below my standard, since one can't actually believe that there are no electrons in the nucleus" [15]. (The younger theorists had actually misread Heisenberg's papers, because his nuclei did contain electrons, though most of them were bound in neutrons.)

Heisenberg actually excluded electrons from the nucleus only under the influence of Fermi's successful β-decay theory of late 1933, that is, a year after he wrote his three-part paper on nuclear structure [16]. By March 1934, he could write to Bohr [17]:

The neutrino will perhaps have only the degree of reality of light quanta, but this suffices to carry out a description in the spirit of conservation laws. Perhaps you feel that my newfound love for the neutrino represents a mental confusion, but I must confess that the experiment of Joliot has especially strengthened that love.

Bohr, who had been an advocate of the Beck–Sitte theory, wrote to Pauli a little later: "I am...quite prepared that we do have here a really new situation which might be synonymous with the existence of neutrinos." [18]

3.3 The origins of the Fermi-field theory of nuclear forces (1934)

Pauli, at the Swiss Federal Institute of Technology (ETH), learned about Fermi's new β-decay theory through Felix Bloch, who was spending part of a fellowship year in Rome. He either heard it directly or via Gregor Wentzel of the University of Zurich, to whom Bloch had written [19]:

Fermi has made a beautiful theory of β-decay emission, introducing the neutrino, which so simply reproduces the empirical facts that I believe in it strongly. The mass of the neutrino should be essentially zero, or in any case much smaller than that of the electron.

Pauli was pleased and, describing Fermi's theory in a letter to Heisenberg, declared, "Das wäre also Wasser auf unsere Mühle!" (That would be water to drive our mill!) [20]. Heisenberg's reply stated that he had read Fermi's short report in Italian and that it also pleased him enormously [21].

Barely another week had passed before Heisenberg again wrote to Pauli, saying that, if there really were an interaction that produced an electron–neutrino pair, then "in the second approximation it should give rise to a force between neutron and proton", just as the possibility of producing a photon leads to the Coulomb force [22]. Heisenberg sketched out an approximate evaluation of the force and showed that Fermi's assumed interaction yielded a charge-exchange force of the form that he had proposed in his three-part paper on the n–p nuclear model, whereas a minor modification of Fermi's interaction gave Majorana's exchange force.

Neglecting the contribution to the force from distances smaller than the proton's Compton wavelength $h/(Mc)$ (M being the proton mass), he found that the exchange energy of a neutron–proton pair was proportional to the inverse fifth power of the distance separating them. Also, using the Fermi constant obtained from β-decay, Heisenberg found the exchange force to be too small by many orders of magnitude. "However", he said, "that may not be a misfortune, considering the sloppiness (*Schlampigkeit*) of the calculation". Thus began the saga of the Fermi-field, which for several years would be the dominant fundamental theory of nuclear forces [23].

The value estimated by Heisenberg for the exchange energy was

$$J(r) \approx mc^2 (10^{-14}/r)^5, \tag{3.3}$$

where mc^2 is the rest energy of the electron (about 0.5 MeV) and r is the distance in centimetres between neutron and proton. If one inserts a value of r equal to the range of nuclear forces, about 2×10^{-13} cm, $J(r)$ is too small by at least a factor of 10^6. However, Heisenberg noted that the effective value of the exchange force depended upon the behaviour of $J(r)$ for small r and that, for r less than the proton's Compton wavelength $h/(Mc)$, the perturbation calculation no longer made sense.

In addition to corresponding with Pauli about the electron–neutrino pair as the possible carrier of an exchange force, Heisenberg wrote to Fermi, who replied on 30 January and 6 February 1934 [24]. Fermi wrote that he had tried something similar and also had found that "the interaction which arises in second order between neutron and proton has the right form, but is quantitatively much too small". However, Fermi conjectured that his β-decay interaction might be analogous to the interaction of matter with the radiation field in the form originally proposed by Dirac, who had neglected the scalar and longitudinal components of the four-vector potential. In the electromagnetic case, this gives only a small relativistic correction to the much larger Coulomb interaction.

On the other hand, Fermi argued that he did not see how the coupling constant estimated from β-decay could be much larger. Perhaps a factor of ten was possible; that might come about if the overlap integral between the wavefunction of the initial neutron and the final proton were much smaller than he had assumed, but Fermi thought that unlikely. He was more concerned about whether the neutrino had a rest mass, and if so whether it would obey Dirac's hole theory, which he had answered in the negative in his β-decay article [25]. He mentioned the possibility that two isobars differing by one unit of charge, such as ^{115}Sn and ^{115}In, could *both* be stable against β-decay, provided the neutrino had a non-zero rest mass. In the letter of 6 February 1934 to Heisenberg, Fermi explained that point in more detail.

Fermi's theoretical assistant in Rome, Gian Carlo Wick, was more encouraging when he wrote to Heisenberg in April 1934 [26]:

> Very honoured Professor! In your work on nuclear theory you have remarked that for radiative effects of the nucleus, besides the proton-dipole [moment], one must also take into account the electrons virtually present in the neutrons.

To this the footnote was added: "I don't know what position you are taking on this question after the discovery of absorption through pair production." [27] Wick agreed that, if there really were exchange currents in the nucleus, the proton dipole moment alone would not constitute a conserved current and here radiative calculations without exchange currents must fail [28].

Wick went on to report a "rather crazy idea" (*eine ziemlich verrückte Betrachtung*). Perhaps the smallness of the Fermi constant did not lead to a hopeless situation for nuclear forces, since the Fermi form might not be exact; furthermore, the exchange force involved much shorter wavelengths of the electrons than did β-emission. The discrepancy could thus be blamed on the extrapolation — and, besides, quantum mechanics might not apply in this

From left to right: G.C. Wick (1909–1992), F.D. Rasetti (born 1901) and E. Amaldi (1908–1989) in Rome, 1935. Photograph reproduced by permission of AIP Emilio Segrè Visual Archives.

region. Wick roughly estimated the probability of a virtual dissociation (either $n \rightarrow p + e^- + \bar{\nu}$ or $p \rightarrow n + e^+ + \nu$) to be of order unity, which he thought might explain the anomalous value of the proton magnetic moment [29]. "However", he added, "please don't think that I believe all this!".

As Wick was writing to Heisenberg, the latter was preparing a set of lectures to be given at the Cavendish Laboratory in Cambridge and wrote to Bohr that, in working on the lectures, he had become aware that he was "really very happy about the development of nuclear physics". He briefly sketched the point of view that he would develop in the first of his four Cambridge lectures [30]. He stressed that, as for the case of light quanta in atomic collisions, *"light quanta or electrons or positrons are never knocked out"* directly in nuclear collisions, but appear only after the collision has taken place.

That analogy was meant as preparation for the third Cambridge lecture, in which Heisenberg described the Fermi β-decay theory and its connection with "the law of force between neutrons and protons". Heisenberg obtained a value for the exchange potential energy that was "too small by a factor 10^{10}". However, he mentioned the "analogy of radiation theory", in which radiative effects are small compared with static effects and suggested that "the process of Fermi [is] possibly not yet the primary process".

Others picked up the theme of a possible connection between Heisenberg's exchange force and Fermi's theory of β-decay: Igor Tamm in Moscow, Dimitri Iwanenko in Leningrad, Arnold Nordsieck in Ann Arbor, Michigan and Hideki Yukawa in Osaka. Tamm and Iwanenko had the idea independently of each other, but subsequently discussed it together and then sent separate letters to *Nature*, which were published on 30 June 1934 [31]. Their results agreed essentially with that which Heisenberg had already written in his letter to Pauli of 18 January 1934. They found an exchange energy that decreased with the inverse fifth power of the distance and of such a magnitude that to obtain about 1 MeV of binding energy would require a mean distance of 10^{-15} cm, about 100 times smaller than the range of nuclear forces. The same conclusion, but with further reservations about the mathematical validity of the calculation, was drawn by Arnold Nordsieck, a student of Robert Oppenheimer (Nordsieck 1934).

Yukawa had tried unsuccessfully since 1932 to make a fundamental theory yielding Heisenberg's exchange force, treating the neutron–proton exchange current as the source of an electron field (*without* an accompanying neutrino) obeying Dirac's wave equation (Brown 1985, 1986). He said later that when he read the letters of Tamm and Iwanenko in *Nature*, he was "heartened by the negative result, and it opened my eyes" [32]. Within a few months, he realized that a new particle (the meson) was needed as the quantum of the nuclear force, one of suitable mass and coupling strength, in order to give both the right range and exchange energy. Yukawa read his paper on the new theory on 17 November 1934 at a meeting in Tokyo and it was published in English two months later (Yukawa 1935). His theory was not taken seriously until particles thought to be mesons were identified in the cosmic rays in 1937 (Neddermeyer and Anderson 1937; Street and Stevenson 1937a, b; Nishina *et al.* 1937).

We have already referred to the London Conference of October 1934, at which the Beck–Sitte neutrino-less β-decay theory was presented. At the same session, C.D. Ellis and George Gamow described the empirical situation. The first discussant, Hans A. Bethe, said that he and Rudolf E. Peierls had been considering alternative forms of Fermi's interaction, which would result in an electron energy spectrum showing fewer electrons (or positrons, in the case of β^+-decay) with low energies [33]. (While agreeing that his interaction could be altered if necessary, Fermi doubted the shape of the experimental spectra for small electron energies, which he ascribed to technical difficulties of observation.) Bethe and Peierls proposed to introduce derivatives to suppress the β-decay interaction, thus requiring a larger Fermi constant to fit the experiments. That would make it possible to use the Fermi interaction to provide the needed nuclear exchange force, as Heisenberg had suggested.

The proposal of Bethe and Peierls pleased Heisenberg. He communicated it to Pauli at the end of October 1934 and claimed that with it the proton's anomalous magnetic moment would have the right order of magnitude [34]. However, Pauli was not happy with this suggestion: "The present situation in theoretical physics is this, that a subtraction physics of electrons and positrons stands face to face with a nuclear physics of arbitrary (*unbestimmten*) functions." [35] To Pauli it appeared that the only effect of Bethe's suggestion was to shift the arbitrariness from Heisenberg's and Majorana's exchange potentials to the form of the four-fermion interaction Hamiltonian. Moreover, he declared that Fermi's interaction had an entirely formal character, since it did not follow from any general theoretical principle.

In the next exchange of letters, Heisenberg maintained that the exact form of Fermi's interaction would probably soon be known from careful measurement of the β-decay spectrum and that the reduction of nuclear forces to the Fermi interaction was a great step forward in principle; for example, it applied to the magnetic moments of the proton and the neutron [36]. Pauli was not convinced. Why, he asked, did Heisenberg not include negative proton states (not considered as *anti*protons) in his calculation of the magnetic moment and why did he assume that the heavy particles obeyed the Dirac equation? "This [latter] assumption", he wrote, "is to me not very sympathetic, because then the old comedy with the negative energy states is repeated for the heavy particles." [37]

3.4 The Fermi-field theory and the charge independence of nuclear forces (1935–37)

In February 1935, Heisenberg sent a contribution to a *Festschrift* for the Dutch physicist Pieter Zeeman. It was entitled "Remarks on the Theory of the Atomic Nucleus" and reviewed the status of the Fermi-field theory, using published work and Heisenberg's own mostly unpublished contributions (Heisenberg 1935). Heisenberg's approach was cautious, perhaps in response to objections that Pauli had raised in their correspondence. He carefully traced out the analogy between the descriptions of atomic electron shells and the nucleus, as shown in table 1 [38].

Emphasizing that the Fermi theory, like Maxwell's, was a *local field theory*, Heisenberg explained that "the Fermi theory allows in principle the mathematical execution of the idea that the existence of exchange forces follows from the possibility of β-decay" [39]. Referring to the fact that Tamm and Iwanenko had, assuming the "special [form of] interaction energy chosen

Table 1

	atomic shells	atomic nucleus	
elementary constituents	nucleus electron	protons, neutrons	
particles emitted in transitions	light quanta	electrons positrons neutrinos	light quanta
corresponding field	Maxwell field	Fermi field	Maxwell field
first approximation to the interaction	Coulomb forces	exchange forces	Coulomb forces

by Fermi", found the exchange forces to be too small, he described the suggestion of Bethe and Peierls [40] to include derivatives in the interaction. Depending upon whether they included one or two derivatives, Bethe and Peierls obtained forces varying as r^{-7} or r^{-9}, instead of the r^{-5} obtained by Tamm and Iwanenko from the original Fermi interaction [41].

Heisenberg pointed out that the heavy particles would have an infinite self-energy arising from their interaction with the Fermi-field, as in the case for a charged particle interacting with the Maxwell field. Because of the higher power of the singularity at the origin (e.g., r^{-5} instead of r^{-1}), the divergence became far worse for the Fermi-field. One way to get a finite result was to introduce a proton or neutron "radius". By the same method, one calculated an "addition" to the Dirac magnetic moment, i.e., the anomalous magnetic moment, which is nearly equal and opposite for the neutron and the proton because of the symmetry of the theory. As Heisenberg noted, that work had been done by Wick [42].

During 1935 and 1936, Fermi's original form for the β-decay interaction came under increasing attack on the ground that it did not sufficiently suppress the low-energy decay electrons. (It turned out later that this effect was caused by self-absorption of low-energy electrons in the β-ray source. Hence it was an experimental artefact, as Fermi had maintained from the beginning.) This seeming defect in Fermi's theory was seized upon by the supporters of the Beck–Sitte theory, by those favouring a breakdown of conservation laws and by those trying to construct a theory of nuclear exchange forces and magnetic moments. E.J. Konopinski and G.E. Uhlenbeck took up the suggestion of Bethe and Peierls to introduce one or more derivatives into Fermi's interaction and compared it with experiment [43]. Their theory fitted the observed spectral shapes well, but the authors concluded also that it did not change the order of magnitude of Fermi's coupling constant. Thus, as had been pointed out by Tamm, Iwanenko and Nordsieck for the original Fermi interaction, it was not easy to obtain nuclear forces of the observed strength.

From left to right: G. Uhlenbeck (1900–1988), E. Konopinski (1911–1990) and Julian Schwinger (1918–1994). Photograph reproduced by permission of AIP Emilio Segrè Visual Archives (Uhlenbeck Collection).

Yukawa's meson theory, which afterwards became the real competitor to Heisenberg's theory, had barely appeared in print when Heisenberg sent in his article to Zeeman's *Festschrift*, very probably unaware of the existence of Yukawa's theory [44]. Yukawa's theory was referred to neither by Carl Friedrich von Weizsäcker in his book *Die Atomkerne* (the preface to which was written in September 1936) nor by Gamow in the second edition of his book on nuclear physics (Weizsäcker 1937, Gamow 1937). Bethe and his collaborators, in their influential review articles, which are often collectively called "Bethe's Bible" (Bethe and Bacher 1936, Bethe 1937, Livingston and Bethe 1937), did not mention Yukawa either.

In these review articles, the only fundamental theory of nuclear forces considered is that of the Fermi-field, assumed to have the Konopinski–Uhlenbeck form [45]. Although, like the other authors, Bethe and Bacher found it impossible to fit both the strength and the range of the forces, they held that "the general idea of a connection between β-emission and nuclear forces is so attractive that one would be very reluctant to give it up" [46]. They discussed three ways in which the situation might be remedied, roughly stated: (a) β-emission could be analogous to a "radiative" effect, whereas binding forces are analogous to a stronger "Coulombic" effect; (b) one might introduce even more derivatives; and (c) one could modify the behaviour of very-high-energy electrons.

Bethe and Bacher also proposed another modification in the Konopinski–Uhlenbeck interaction, namely, to change the spin-dependence so as to obtain exchange forces of the Majorana rather than the Heisenberg type. They further asserted that *if* the theory were altered to give the right n–p force, it would give p–p and n–n forces that were "not much smaller, . . . in agreement with the conclusions from nuclear binding energies" [47]. The last point began in 1936 to assume a major role in thinking about nuclear forces, as we turn now to discuss this development.

The first theories of nuclear forces (those of Heisenberg, Majorana, Wigner, etc.) assumed the dominance of an attractive n–p force. The reasons for doing so are not very clear except in Heisenberg's case: having assumed that the p–p force was purely Coulombic, his n–n force had to be negligibly weak in order to explain the equality of the neutron and proton numbers in light nuclei. In trying to account for the large ratio of the binding energies of the α-particle and the deuteron, Wigner had said [48]:

> This [disparity] would rather indicate an attraction between the neutrons or between the protons, which is very unlikely on the basis of the previous discussion. The purpose of the subsequent calculation is to see how far it is possible to explain the large mass defect of He without such an assumption, or even to reconcile it with the existence of some repulsive forces between the different neutrons and also the different protons.

However, by late 1935 evidence was accumulating for a "short range attractive interaction between two protons with anti-parallel spins" (Present 1936). This evidence came from several sources: p–p scattering experiments, the binding energies of the hydrogen and helium isotopes, and the stability conditions of heavier nuclei of even A and Z [49].

The main experimental contributions were due to American physicists, but Wolfgang Pauli was the first to transfer the news to Europe, writing in February 1936 from Princeton to his colleague Gregor Wentzel in Zürich [50]:

> I just returned from the New York meeting, where we not only have amused ourselves very well, but where I also learned much physics. There were above all Tuve's new experiments on the proton–proton scattering — the first reliable ones. That is, he uses counters instead of the Wilson [cloud] chamber, to detect protons; hence the statistical fluctuation errors — which have made everything irregular and uncertain in the published results of White — are eliminated. The outcome is: *one needs additional attractive forces between two protons, which have the same order of magnitude as the forces between proton and neutron.* This has now been completely substantiated. From the mass defects of the light nuclei one can conclude that such forces must also exist between two neutrons. (Concerning the relevant computations I received a long private lecture by Bethe in New York, thus I am again quite up to date.) — About the experiments on slow neutrons, Breit and Wigner have now demonstrated definitely that they are not interesting for the theoreticians, and Bethe agreed with this.

All these empirical results led to the perception that appreciable and substantially equal like-particle forces existed in nuclei. It was then a

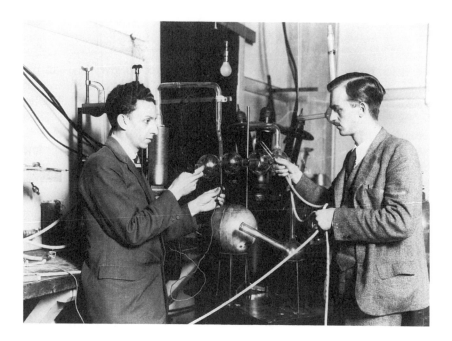

E. Gavriola (left) and M. Tuve (1901–1982). Photograph from The Smithsonian Institute and reproduced by permission of AIP Emilio Segrè Visual Archives.

relatively small step, but one of major conceptual importance, for theorists to postulate the equality of n–p, n–n and p–p forces (in the same spin and orbital states), a principle known as the *charge-independence* of nuclear forces [51].

Strict, or even approximate, charge independence was difficult to incorporate in the Fermi-field theory of nuclear forces. Although this theory did predict n–n and p–p forces in addition to the n–p exchange force, the like-particle forces had a somewhat different origin from the n–p force. Whereas the latter arose from the exchange of an e–ν or an e$^+$–ν pair (technically, it was of second order in perturbation theory), the like-particle forces required the exchange of *two* pairs (fourth order in perturbation theory) [52]. Thus there was no reason to expect the like- and unlike-particle forces to have the same order of magnitude.

Wentzel first tried to modify the theory, so that all three types of nucleon pairs would interact only in the fourth order [53]. To do so he made the basic β-decay interaction a process of second order, for this purpose inventing a second kind of heavy nucleon having zero spin ("singlet" neutron N and "singlet" proton P). The ordinary β-decay, for example, became a two-step process:

$$n \rightarrow N + \bar{\nu} \rightarrow p + e^- + \bar{\nu} \qquad (3.4)$$

or

$$n \rightarrow P + e^- \rightarrow p + \bar{\nu} + e^-. \tag{3.5}$$

Soon thereafter, however, Wentzel dropped the idea of heavy nucleons, at least for the time being. Retaining the Fermi picture of β-decay, he now tried to make charge-independent nuclear forces all appear in the second order by introducing additional lepton pairs, namely, $\bar{\nu}\nu$ and e^+e^- [54]. These "neutral currents" would be hard to observe directly, since they would be emitted in nuclear processes in which the nuclear charge was unchanged; in such processes γ-rays would also be emitted, with much higher probability. Upon their emission, the γ-rays could be converted into e^+e^- pairs, thus masking any "weak" neutral current.

Since the neutral currents would give n–p as well as p–p and n–n forces, whereas the charged currents (e^- $\bar{\nu}$ and $e^+\nu$) would contribute *only* to the n–p force, Wentzel concluded that the forces would not be charge-independent unless the charge currents were negligible. In that case, saturation of the forces would require some other mechanism [55].

Nicholas Kemmer carried Wentzel's idea a step further in a path-breaking work that is the first true charge-independent field theory of nuclear forces (Kemmer 1937). Kemmer had been working under Pauli's direction at Zürich during 1936 and was spending the autumn of that year on a fellowship at Imperial College, London. Pauli wrote to Kemmer, urging him to study the issue of the *Physical Review* for 1 November, which contained the papers on charge-independence. Pauli closed with the remark, "It could well be that it would provide sensible problems for calculation." [56]

Kemmer's work was a field-theoretical extension of a phenomenological analysis of charge-independence by B. Cassen and E.U. Condon, which made use of the isospin operators introduced by Heisenberg in 1932 and used by Fermi in 1934 to describe "exchange currents" of the nucleons. The American authors showed that, if τ_1 and τ_2 were, respectively, the isospin operators of the two nucleons, an interaction that involved only the scalar product of those (vector) operators would be charge-independent [57]. (We now say that the interaction is "invariant under rotations in charge space".)

Kemmer started with the observations that an intermediate field (or set of fields) to transmit the nuclear forces was necessary in a relativistic treatment of the problem and that Fermi's electron–neutrino field was the only real candidate. However, he cautioned [58]:

> Any modification of Fermi's theory that leads to forces of the correct magnitude can however be shown to be incompatible with the saturation conditions required in the theory of heavy nuclei. In deriving these results the Fermi interaction should therefore ... be regarded merely as a "model" suitable for our purpose, without being necessarily generally correct.

To use isotopic spin in this problem, one must attribute it not only to the nucleon currents but also to the intermediate field. Kemmer did so, letting the pair consisting of a positron and a neutrino have the same isospin assignments, respectively, as the pair proton and neutron [59]. Thus, as in Wentzel's theory, neutral currents, e^+e^- and $\bar{\nu}\nu$ appeared. Imposing the requirement that the nuclear forces be charge-independent fixed the ratio of the coupling strengths of the various currents. Contrary to the assertion of Wentzel (who had overlooked the possibility of interference of amplitudes), a fully charge-independent field theory was possible [60]. Spin-dependence and saturation conditions, as given by Gregory Breit and Eugene Feenberg (1936), were also accommodated.

3.5 Conclusion: the Fermi-field and related theories (1938–41)

As we know, although Hideki Yukawa published his meson theory of nuclear forces in early 1935, it received its first international notice only in 1937, when particles of mass roughly 200 times that of the electron, with positive and negative unit charges, were first identified in the cosmic rays (Neddermeyer and Anderson 1937). The earliest reference to Yukawa's paper occurred in a rather ambiguous letter of J.R. Oppenheimer and R. Serber to *Physical Review* (1937). A week later, the Swiss physicist E.C.G. Stueckelberg sent a letter to the same journal, remarking: "Independently of Yukawa the writer arrived at the same conclusion." [61]

A large number of papers concerning the meson theory began to flow into the journals during 1938, from Yukawa and his collaborators in Japan and from European theorists. It is not our purpose to discuss them here, except to point out that some meson calculations were modelled on those already done with the Fermi-field, whereas others entered entirely new territory. An example of a new process was the calculation of the meson's mean lifetime [62]. Among the Fermi-field concepts adapted to meson theory were the calculations of anomalous magnetic moments and the formulation of charge-independent nuclear forces [63]. A charge-independent field theory requires a neutral current, which in a meson theory implies the existence of a neutral meson (Kemmer 1938c; Yukawa *et al.* 1938b).

The lack of observation of such a neutral meson (detected first in 1950) was, however, only one of the many disparities between meson theory and the observed properties of cosmic-ray particles. The latter, called *mesotrons* during the period under discussion, had about the right mass but otherwise did not behave as Yukawa's mesons should have done. No nuclear scattering was observed, and very little nuclear absorption, so that mesotrons could penetrate into deep tunnels and mines. The mass and the mean lifetime also did not seem quite right [64]. Thus, some physicists continued the Fermi-field approach [65].

In particular, three papers of Critchfield, Teller and Wigner were based upon the suggestion (Gamow and Teller 1937) that one should make the neutral current e^+e^- dominate the nuclear force, since its most natural form of coupling would be charge-independent. In an attempt to derive the nuclear saturation property, Critchfield and Teller (1938) introduced a special form of interaction energy H' between the nucleons and the electrons, such that $H' \gg E_{kin}$, where E_{kin} is the kinetic energy of the emitted pair (that is, the opposite of the assumption that would allow perturbation theory to be used).

When someone (probably Wigner) noticed that the Critchfield and Teller interaction violated inversion invariance (parity), they adopted a new form, one that still allowed saturation of the forces provided that the interaction between heavy and light particles was bounded below (Wigner *et al.* 1939). Perhaps in concession to the meson theories then coming into vogue, the paper contained the following statement [66]:

> We wish to mention, finally, that we consider the mathematical formalism as the most permanent part of the present paper. It is possible that this can be taken over into a new theory the physical foundation of which is quite different.

None-the-less, despite this disclaimer, Critchfield extended the theory by making the basic interaction dependent on the spins of heavy particles. Thus he was able to treat the spin-dependence of the forces, the anomalous magnetic moment of the nucleons and the quadrupole moment of the deuteron (Critchfield 1939).

We close by mentioning a class of theories that combined features of Fermi-field theory and Yukawa theory. As noted above, Bethe suggested that only neutral mesons be used, in order to obtain charge-independent nuclear forces [65]. Robert E. Marshak started a new development, which he called heavy-electron pair theory (Marshak 1940). He described his quanta as follows [67]:

> The heavy electrons are assumed to be identical with electrons in every respect ("hole" theory, Fermi statistics, etc.) except that their rest mass is taken equal to the cosmic-ray meson mass. [Furthermore] ... the range is directly connected with the rest mass of the heavy electron pair field (in contrast to the Gamow–Teller pair theory). At small r, the potential goes as $1/r^5$ so that one has to cut off in the same way as in the original electron–neutrino theory. The advantage of the heavy electron pair theory is that it deals with particles which can be identified with the cosmic-ray meson.

The last sentence of Marshak's paper (other than an acknowledgment) reads, "This paper has established that the existence of unobserved neutral mesons is not indispensable to a theory of nuclear forces." Marshak and others used the "heavy electron pair theory" throughout the 1940s [68].

There are multiple ironies here. Although the "cosmic ray mesons" (mesotrons or muons) turned out to have spin-$\frac{1}{2}$ and no neutral companion, as Marshak assumed, they play no significant role in nuclear forces, nuclear magnetic moments or β-decay. The muons appear in the cosmic rays mainly as decay products of the Yukawa meson, the pion, as shown first by the nuclear emulsion group of Bristol in 1947. There is indeed a neutral *pion*, first detected at Berkeley in 1950 [69]. Furthermore, when Bethe and Marshak constructed a theory proposing the existence of two types of mesons, one of spin-$\frac{1}{2}$ and one of spin 0, decaying in cascade, one strongly and one weakly interacting, they incorrectly assigned to the spin-$\frac{1}{2}$ meson the strong-interaction role [70]. By 1950, however, the Yukawa theory had entirely replaced the Fermi-field theory of nuclear forces.

Notes to text

[1] See also Stuewer (1983) and Brown and Rechenberg (1988).
[2] The isospin operators used by Heisenberg (and later by Fermi) were 2×2 matrices of the same form as the Pauli spin matrices. They were used to make "projection operators" that selected either the neutron or the proton state of the nucleon and to make "raising" and "lowering" operators that changed one type of nucleon into the other.
[3] [WPSC2], letter [291], pp 113–4. The reference to "non-Pauli neutrons" meant those of Chadwick. Pauli himself had proposed a different particle that he called the "neutron", but it is now called the "neutrino".
[4] Pauli to Heisenberg, 14 July 1933. [WPSC2], letter [314], note 8, pp 184–7, especially p 184 (original emphasis).
[5] Institut International (1934, p 325).
[6] Institut International (1934, p 331).
[7] Discussions between Fermi and Pauli in Rome are cited by Pauli (1961). In his Faraday lecture of 1931, Bohr said that "we have no argument, either empirical or theoretical, for upholding the energy principle in the case of β-ray disintegrations" (Bohr 1932, especially p 383). He kept this opinion and, according to Pauli: "Only in 1936 did he accept completely the validity of the energy theorem in β-decay and the neutrino, when Fermi's theory had already developed successfully". (Pauli 1961, p 163).
[8] Fermi (1933, 1934a, b). These papers are numbered 76, 80a, and 80b in Fermi (1962).
[9] Jordan and Klein (1927), Jordan and Wigner (1928). The second paper, concerning the second quantization method for fermions, was not referred to by Fermi, who quoted instead Heisenberg (1931), which applied the Jordan–Wigner method to atomic shells. Perhaps, as suggested by Rasetti (in Fermi 1962, vol. I, p 539), he had had some difficulty in following the Jordan–Wigner paper.
[10] Fermi (1934b), p 166.
[11] See Pauli (1961). Bohr's full acceptance of energy conservation in β-decay is in Bohr (1936).
[12] Bohr, e.g., urged the 1933 Solvay Conference participants to give the Beck–Sitte theory their careful attention. (See Institut International (1934, p 287–8).)
[13] G. Beck and K. Sitte (1933, 1934b); and G. Beck (1933a).
[14] See International Conference (1935). Beck did not actually attend this London conference and his paper was probably read by Sitte (private communication by Guido Beck to L.M. Brown).
[15] The quotes are from the transcript of the interview of Joan Bromberg with Werner Heisenberg, dated 16 June 1970 (deposited in [WHA] and [AHQP], especially pp 5–6.
[16] Brown and Rechenberg (1988, section 3).
[17] Heisenberg to Bohr, 13 March 1934 (Bohr-Archives, Copenhagen): "*Die Neutrinos werden vielleicht nur den Realitätsgrad der Lichtquanten besitzen, aber das reicht doch hin, um mit ihnen eine Beschreibung im Sinne der Erhaltungssätze durchzuführen. — Vielleicht empfindest Du meine neue Liebe zu den Neutrinos als seelische Verwirrung, aber ich muß gestehen, daß mich insbesondere das Experiment von Joliot sehr in dieser Liebe bestärkt hat.*" The paper of F.

Joliot addressed here is Curie and Joliot (1934), presented at the meeting of 15 January to the Paris Academy.

[18] Bohr to Pauli, 15 March 1934, reproduced in [WPSC], pp 307–10, especially p 310. Nevertheless, Bohr sent off the corrections of the Solvay Conference discussions containing his advocacy of the Beck–Sitte theory at about the same time. However, Bohr in 1936 accepted the neutrino (Bohr 1936).

[19] Felix Bloch to Gregor Wentzel, 24 December 1933; in [AHQP]. In [WPSC2] on p 245, von Meyenn states that Pauli "would have learned about it through Felix Bloch directly from Rome".

[20] Pauli to Heisenberg, 7 January 1934, [WPSC2], p 248.

[21] Heisenberg to Pauli, 12 January 1934, [WPSC2], p 249.

[22] Heisenberg to Pauli, 18 January 1934, [WPSC2], pp 250–3. The idea was not really a new one for Heisenberg. He had already written to Pauli in July 1933: "From the standpoint of your theory one would have to say always: [neutron] decay is into an electron, proton and neutrino. Therefore, exchange forces would be present". (Heisenberg to Pauli, 17 July 1933; [WPSC2], p 195).

[23] For an earlier clear notice of this historical fact see Barut (1982).

[24] Fermi to Heisenberg, 30 January 1934 and 6 February 1934 ([WHA] and [AHQP]). We have not been able to locate Heisenberg's letter to Fermi which initiated this correspondence.

[25] Fermi (1934b, p 162). Soon after Fermi's β-decay paper was published, artificial radioactivity involving positron decay was discovered by the Joliot-Curies (Curie and Joliot 1934). To apply Fermi's theory to this new process, the idea of neutrino "holes" (i.e., the idea of a neutrino and an antineutrino) was introduced by Wick (1934).

[26] Wick to Heisenberg, 16 April 1934 [WHA]; the letter bears no year, but it is almost certainly 1934.

[27] When he wrote his three-part paper on nuclear forces, Heisenberg was very concerned about high-energy radiative interactions with the nucleus, whose strength is so great that Heisenberg wrote, in part III (1933, p 595): " ...α-particles in the nucleus must be built of protons and electrons (not protons and neutrons) and the electrons bound in α-particles, in spite of their large binding energy, must contribute to γ-ray scattering at least as much as do the free nuclear electrons." Another explanation for large radiative effects was soon given in terms of the "shower" processes: *Bremsstrahlung*, pair production and pair annihilation. For more on this question, see Brown and Moyer (1984).

[28] Wick's well-taken point was considered explicitly (and independently) by Siegert (1937). The difficulty of defining a current "does not occur if one tries to explain the mechanism of the exchange by that suggested in Fermi's theory of the β-decay", according to Arnold J.F. Siegert and "the general features will remain in any theory in which the exchange is explained by the creation and reabsorption of light particles". In July 1937, when the paper was submitted, one such theory was Yukawa's meson theory, to which Siegert's conclusions do in fact apply.

[29] A charged particle obeying Dirac's relativistic wave equation has the magnetic moment $e\hbar/(2Mc)$, where M is the particle's mass. (A neutral Dirac particle should have zero moment.) The proton was found to have an "anomalous" value, about three times larger. This result was published by Frisch and Stern (1933) and Estermann and Stern (1933a). See Wick (1935).

[30] Heisenberg to Bohr, 17 April 1934. Heisenberg left for Cambridge on 21 April 1934. The unpublished sketch of the Cambridge lectures bears the title, "*Disposition der Vorträge in Cambridge*" [WHA].

[31] Tamm (1934) and Iwanenko (1934). Iwanenko had suggested soon after learning of Chadwick's neutron that nuclear electrons were "*all* packed in α-particles or neutrons" (Iwanenko 1932a,b).

[32] Yukawa (1982, p 201).
[33] International Conference (1935, p 66 ff). Bethe was repeating a suggestion that he had made in September 1934 at a conference in Copenhagen on nuclear physics.
[34] Heisenberg to Pauli, 28 October 1934 [WPSC2], pp 354–6.
[35] Pauli to Heisenberg, 1 November 1934, [WPSC2], pp 357–8. According to von Meyenn, the discussion of Dirac's hole theory between Heisenberg and Pauli reached its peak in 1934 with the problems of the infinite quantities that had to be removed by dubious "subtraction physics" ([WPSC2], p 245).
[36] Heisenberg to Pauli, 4 November 1934, [WPSC2], pp 358–60, especially p 362.
[37] Pauli to Heisenberg, 7 November 1934, [WPSC2], pp 360–3.
[38] The table was reproduced in Weizsäcker (1937, p 29). Weizsäcker added to it a bottom row that gave the "corresponding description of particle emission in transitions", namely, "Dirac radiation theory" under "Coulomb forces" and "Fermi theory of β-decay" under "exchange forces".
[39] Heisenberg (1935, p 112).
[40] See International Conference (1935).
[41] The perturbation calculation leading to r^{-5} was given in Weizsäcker (1937, pp 188–91). He obtained for the second order exchange energy, neglecting the electron and neutrino masses, and using Fermi's interaction:

$$J(r) = -\left(\frac{g^2}{ch^6}\right) \int d^3 p_s \int d^3 p_t \frac{e^{i(p_s-p_t)\cdot r/h}}{p_s+p_t}\left(1 - \frac{p_s\cdot p_t}{p_s p_t}\right)$$

It is obvious on dimensional grounds that $J(r)$ is proportional to $1/r^5$. If the β-decay interaction has one (or two) derivatives, then two (or four) powers of momentum will appear in the numerator of the integrand, whence a behaviour as r^{-7} (or r^{-9}).
[42] See note 29. The proton's anomalous moment and the neutron's moment (which is *all* anomalous) appeared to be nearly equal and opposite from a comparison of the measured moments of the proton and the deuteron.
[43] Konopinski and Uhlenbeck (1935). They seemed not to have known that Bethe and Peierls had offered the same suggestion.
[44] The first mention of Yukawa's theory in the Pauli letters is in that from Pauli to Heisenberg, 22 February 1938, [WPSC], p 552.
[45] Konopinski and Uhlenbeck (1935). Bethe and Bacher (1936) attributed the first suggestion of a connection between Fermi β-decay and nuclear forces to Heisenberg's unpublished lectures at the Cavendish Laboratory in 1934.
[46] Bethe and Bacher (1936, p 203).
[47] Bethe and Bacher (1936, p 204).
[48] Wigner (1933, p 253).
[49] Present (1936) and Bethe and Bacher (1936, sections 10 and 21).
[50] Pauli to Wentzel, 24 February 1936, [WPSC2], pp 440–1, especially p 441.
[51] The main p–p results reported were by White (1936) and Tuve et al. (1936). Analysis of the p–p scattering results was performed by Breit et al. (1936). The extension to charge independence was made by Breit and Feenberg (1936). Finally, the formulation of the charge-independent interaction in terms of isospin was done by Cassen and Condon (1936).
[52] For example, each of a pair of protons could emit an $e^+-\nu$ pair and turn into a neutron. Then each of the "intermediate" neutrons could absorb the other's $e^+-\nu$ pair and become again a proton.
[53] Wentzel (1937a,b). This theory and the other unusual "pair" theories have been discussed by Mukherji (1974).
[54] Wentzel (1937c). Heavy nucleons, the so-called nucleon isobars, appeared later in Wentzel's work on the strong-coupling meson theory; see Wentzel (1940).

[55] A purely neutral current hypothesis was advanced just at the same time, in January 1937 (Gamow and Teller 1937). Noting that the then accepted Konopinski–Uhlenbeck modification of Fermi's interaction still gave nuclear forces too small by a factor 10^{12}, Gamow and Teller observed that a process involving the exchange of an e^+–e^- pair could be assigned the right strength arbitrarily and would be charge-independent.

[56] [WPSC2], p 490. The incident is also discussed in Kemmer (1983a).

[57] Cassen and Condon (1936). The first use of the term *isotopic spin* (now usually shortened to *isospin*) was apparently by Wigner (1937).

[58] Kemmer (1937, p 908).

[59] That is, τ_3 has the value $+1$ for the proton and the positron, the value -1 for the neutron and the neutrino. The nucleon assignments are in agreement with the current ones, but are opposite from the original assignments of Heisenberg.

[60] In a footnote (number 10) Kemmer (1937) states that his result "has been evolved in a discussion with Professor G. Wentzel who has kindly permitted the writer to point out that the statement contained in his paper...needs correction".

[61] Stueckelberg (1937a). Stueckelberg had proposed a "unitary theory" in which electrons, neutrinos and nucleons were aspects of a single "matter field" (Stueckelberg 1936a, b). This theory had only one effective coupling strength, as opposed to Yukawa's strong and weak coupling constants, so it was closer in spirit to the Fermi-field theory.

[62] This was calculated, following a suggestion of Bhabha, in Yukawa and Sakata (1939b). See Yukawa *et al.* (1938a).

[63] Just as a Fermi-type β-decay interaction could be formulated in terms of currents of different space-time transformation properties (such as scalars or vectors), so Yukawa's meson could have different forms. Kemmer developed four of the five possible relativistic theories, omitting the symmetric tensor field (Kemmer 1938b).

[64] For these discrepancies see, e.g., Hayakawa (1983).

[65] Other than Kemmer's formulation, another way to achieve charge-independence of nuclear forces in meson theory was to use *only* neutral mesons. This was considered by Bethe (1940a,b). This approach separated the nuclear force problem from the behaviour of the cosmic ray "mesotrons".

[66] Wigner *et al.* (1939, p 533).

[67] Marshak (1940, abstract).

[68] Critchfield and Lamb (1940); Marshak and Weisskopf (1941). Other works up to 1949 are cited in Mukherji (1974).

[69] Bjorklund *et al.* (1950); Steinberger *et al.* (1950). In the same year, neutral pions were detected in the cosmic rays by Carlson *et al.* (1950).

[70] Marshak and Bethe (1947). A two-meson theory having the correct assignment of spins had been proposed in Japan some years earlier (Sakata and Inoue 1942, 1946).

Chapter 4

Cosmic Rays, Quantum Field Theories and Nuclear Forces

4.1 Introduction

During the 1930s, the cosmic rays constituted the most important source of data on the high-energy behaviour both of quantum electrodynamics and of nuclear forces. In the laboratory, physicists could not achieve particle energies above a few mega-electronvolts, so the mysterious extraterrestrial radiation — whose composition was only partially understood — had to serve as a poor substitute for a scientific laboratory ruled by controllable conditions. This lent an extra urgency to the cosmic ray physicists, who were already motivated by an adventurous curiosity. Thus the French cosmic-ray specialist Pierre Auger spoke of his colleagues as playing the roles of "mountaineers, mine workers, divers and air riders". Calling the decade between 1925 and 1935 a "heroic period", he continued [1]:

> We have seen that, according to Hess [the discoverer of cosmic rays], this radiation must possess an extraordinarily large penetration power...The dimensions for the necessary shielding were so big that hardly other hindrances than those erected by nature can satisfy these conditions. The water masses of the lakes and the sea, the earth's crust and the atmosphere provide as powerful shieldings as one might just wish to have. The different places for carrying out measurements ridicule any description. Thus also the amusing story occurred that I once heard a Russian physicist tell in a French lecture: "I have measured cosmic radiation in the sea and in high mountain ranges; I have measured it on the ground of lakes and in the highest atmosphere, in rock-salt and carbon mines and in deepest caverns. Finally I have measured it 'en fer' [which means 'in hell']". Of course, he wanted to say "dans le fer", i.e., "in iron".

The "divers" included researchers like the German Erich Regener, who sank his "*Bodensee-Bombe*" in Lake Constance; among the "mountaineers" were the Dutch Paul Ehrenfest, Jr., who carried apparatus up to an altitude of 3500 m; the Hungarian Jenö Barnóthy and the Japanese Yataro Sekido descended into mines and tunnels; the Swiss Auguste Piccard and others ascended to the skies in balloons and airplanes to make their observations.

Measuring devices were constantly improved in those "heroic days". Cloud chambers were built with powerful magnets to measure charged particle momentum, by Carl Anderson in California, Patrick Blackett in London and Manchester, and Louis Leprince-Ringuet in Bellevue, France. Complex telescope arrangements of counters were built by the Italian Bruno Rossi and high-pressure ionization chambers in Germany by Gerhard Hoffmann. Thus the extraterrestrial radiation gradually yielded its secrets and physicists sorting out the various cosmic-ray phenomena in the atmosphere found that the less-penetrating component was associated with electromagnetic cascade showers and that the more-penetrating component contained a new elementary particle, the mesotron. However, there remained puzzling features of the cosmic rays that left adequate room for other explications [2].

The wealth of cosmic-ray observations in the 1930s demanded interpretation by the theories available to describe the high-energy behaviour of matter, namely quantum electrodynamics and theories of nuclear forces, both of "fundamental" and of "phenomenological" types. In chapter 2, we have given our definition of these terms, together with our opinion that *practically speaking, a fundamental theory for short-range forces must be a quantum field theory*. We described there how the "discovery" of three new particles in 1932–3 (the neutron, the positron and the neutrino) established a coherent basis for the quantum mechanical treatment of strong and weak nuclear forces by allowing the banishment of electrons from the nucleus. (See also Stuewer (1983).) Enrico Fermi's theory of β-decay provided a key development. The Fermi-field theory, extended by Werner Heisenberg and others to give the first quantum field theory of the strong nuclear forces, had serious difficulties, but it also possessed the strong appeal of economy, being a unified theory of weak and strong nuclear interactions with a single coupling parameter (chapter 3).

Towards the end of 1934, Hideki Yukawa, in Osaka, Japan, isolated from the mainstream of developments in Europe and America, invented a new theory of nuclear forces based upon the exchange of a conjectured massive spinless charged particle, which he called the *U-quantum* and which is now called the *meson* (Yukawa 1935). Yukawa's meson theory has become the paradigm theory of elementary particles; variants of it are used repeatedly for a wide range of basic phenomena. We shall retrace the steps by which Yukawa arrived at his pathbreaking result and its further development in part B below.

Although Yukawa's work was crucial for a fundamental quantum field theory of nuclear forces, physicists did not notice it for two and a half years (until June 1937), and even then did not greet it with great enthusiasm. In the present chapter on the interpretation of cosmic-ray phenomena, we shall therefore follow the mainstream of physics during this period, stressing the connections between nuclear and cosmic ray physics from 1934 to 1938 including the high-energy applications of the Fermi-field theory until the adoption of the meson theory of nuclear forces. We leave for later treatment the application of meson theory to cosmic rays.

Before embarking upon our programme, we would like to mention some of the secondary literature from which we have benefited and also to compare our treatment with earlier historical studies. A valuable survey of meson theory was given by Visvapriya Mukherji, based upon published works and letters exchanged between the author and scientific contributors (Mukherji 1974). More recent works, partially overlapping our own, have been given by David Cassidy and Peter Galison (Cassidy 1981; Galison 1983a).

Galison's paper dealt with identifying the composition of the cosmic rays observed in the atmosphere and with the doubts that were raised regarding QED as applied to the high energy interactions of the light cosmic-ray particles. His account ends with the vindication of QED and the identification of the "mesotron" (now called the *muon*), a new particle of mass intermediate between that of the electron and that of the proton [3]. Galison especially presented the motivations and experimental techniques of one of the discovery groups, that of J.C. Street. Among the historical questions that he raised in connection with the muon is that of the "moment of discovery" or "when does an experiment end?", a question that Galison has pursued in his work on other discoveries.

Cassidy, in his article, tried to identify "two different programmatic approaches to field theory and its applications". One group, including Walter Heitler and Robert Oppenheimer, Cassidy called "cosmic-ray physicists"; he described them as making only conservative adjustments to existing theories, i.e., the minimum required to fit the data. The other group that he called "field theorists", which included Heisenberg, Dirac and Pauli, according to Cassidy, "embraced" rather than "avoided" an apparent breakdown of QED at high energies, in seeking clues to formulate a new and revolutionary field theory.

Cassidy focused his attention on a class of cosmic ray events, referred to by Heisenberg as "explosive showers", for which he sought a non-electromagnetic origin. In Cassidy's view, the field theorists made explosive showers the touchstone of quantum field theory, concluding that they required the introduction of a new universal constant having the dimension of a length, which would be analogous to the constants c, the velocity of light, and h, Planck's constant.

Our consideration of the Fermi-field theory as applied to nuclear forces and to cosmic-ray interactions overlaps a little with that of Galison's paper, and somewhat more with Cassidy's and Mukherji's. (However, we believe that Cassidy has overstated his distinction between field theorists and cosmic-ray physicists.) We differ in focusing attention on the triple confluence of nuclear, cosmic ray and field theory physics that took place in the 1930s, which led directly to modern particle physics [4]. We treat in this chapter, as we have already stressed, only the first phase of this development, one which was completed essentially in 1938, when meson field theory replaced the Fermi-field theory of nuclear forces.

4.2 The cosmic rays and nuclear interaction (1934–6)

The diversity of the secondary phenomena produced by the high-energy particles of the cosmic rays made them the natural testing ground for the limits of validity of fundamental field theories: QED, the Fermi-field theory of nuclear forces and later also the meson field theories. "In fact", said P.M.S. Blackett at an international physics conference in London in October 1934 [5],

> ... it is only through the study of cosmic rays that we can hope at present to learn about the properties of very energetic radiations. But since the experimental phenomena are both complicated and hardly at all under the experimenter's control, it is by no means easy to find their correct interpretation. For to do this implies the analysis of obviously complex radiation into simpler constituents and then the decision as to the nature and properties of the various radiations.

Blackett stated the problem of cosmic rays as it appeared in the mid-1930s: to separate and classify the constituents of the cosmic rays, and then to study their separate interactions. However, the process required a kind of bootstrap, in that one of the theories whose validity was to be tested, namely QED, had to be used to interpret the test results. It thus became a

P.M.S. Blackett (1897–1974) in 1948. Photograph by Lotte Meitner-Graf and reproduced by permission of AIP Emilio Segrè Visual Archives.

question of whether it was possible to find a self-consistent interpretation of a wide range of cosmic phenomena.

At the London conference of 1934, nearly all the main centres of cosmic-ray research were represented. A partial list of speakers included: P. Auger and L. Leprince-Ringuet (France); G. Hoffmann (Germany); P.M.S. Blackett (Great Britain); B. Rossi (Italy); and C.D. Anderson, S.H. Neddermeyer, A.H. Compton and R. Millikan (USA) [6]. The papers gave a good overview of the state of knowledge of cosmic rays at that time. Absorption measurements made since the 1920s showed that the rays in the atmosphere consisted of a highly absorbable "soft" component and a penetrating "hard" component. The part of the soft component visible in the cloud chamber consisted of groups of a few up to a few hundred positively and negatively charged particles, usually forming a cone, and apparently originating in dense matter (i.e., not in the chamber gas but in the chamber walls or in shielding material). The hard component consisted mainly of single tracks of particles with low ionizing power. Both components existed at mountain altitude and persisted down to sea level. That is, the soft component was diminished through absorption, but was also regenerated as the rays passed through the atmosphere [7].

In 1934, two views were held regarding the nature of the *primary* cosmic radiation, i.e., the radiation falling upon the earth from space, in contrast to its secondaries produced in the atmosphere. Millikan (with his associates from Caltech, Anderson and Neddermeyer) maintained at London that the primaries were high-energy γ-rays, "born" in space in the course of formation of atomic nuclei, a process that he called "atom building". He acknowledged that these primary photons would interact to produce secondary electrons and it was to these secondaries that he attributed the observed latitudinal distribution of cosmic rays brought about by the earth's magnetic field. The other speakers at the London conference, however, interpreted this same distribution as proof that the cosmic ray primaries were charged particles, not photons.

Just as two views were held regarding the cosmic ray primaries, so there were two interpretations of the shower phenomena. (Recall Blackett's quoted remark that "the experimental phenomena are complicated and hardly at all under the experimenter's control".) The view of the majority was that the following process takes place: a high energy electron is deflected in the field of a nucleus, emitting a high-energy photon (the *Bremsstrahlung* process); the photon then creates an electron–positron pair in the field of a nucleus (the pair production process). Each of these secondaries produces another photon by *Bremsstrahlung* — and so a cascade results in which the energy, through a sequence of individual events, is eventually divided among many light particles, which then lose their energy by ionization. A minority held the view that some of these showers were explosive events, in which many particles were simultaneously produced by the encounter of a fast particle with a nucleus.

During 1933 and 1934 the first QED cross-sections for the radiative processes which contribute to the cascade showers were calculated, namely *Bremsstrahlung* and pair production [8]. Heitler's book on the quantum theory of radiation later characterized the result for the energy loss of electrons as surprising: "We obtain, therefore, the striking result, that for *energies higher than a certain limit, the energy loss is almost entirely due to emission of radiation and reaches a value which is much higher than that due to ionization.* For lead this limit lies at $20mc^2$ and for water at $250mc^2$." [9] (Here mc^2 is the rest energy of the electron, 0.51 MeV.)

Commenting specifically on the results presented at the London conference by Anderson and Neddermeyer, who had studied the energy lost by particles in lead plates in a cloud chamber with a magnetic field, Bethe remarked that "the radiative energy loss seems far smaller than that predicted by theory" [10]. This apparent disagreement persisted until mid-1936, when Anderson and Neddermeyer reported larger radiative energy losses, which "do not so far indicate a breakdown of the theoretical formula" [11].

That QED would show a breakdown at some larger energy was anything but unexpected. The group of theorists that Cassidy called "field theorists" were hoping to use the prospective breakdown as a clue to how to modify QED so as to avoid the so-called "divergences", that is, infinite self-mass and infinite charge, predicted by the theory. The divergences came from the inclusion of virtual (energy non-conserving) processes of high energy entering in the calculations. Bohr's correspondence principle provided no guidance in these calculations; indeed classical electrodynamics had analogous problems. However, even Cassidy's other group, the "cosmic-ray physicists", were convinced that QED would go wrong at some not terribly high, and certainly observable, energy. Speaking about the Oppenheimer school, Robert Serber said: "Oppie at first disbelieved at mc^2 ..., then retreated to $137mc^2$, but could hardly write a paper without a lament." [12] On the other hand, there was no reason to expect a failure of QED (at least in lower order perturbation theory) at the energies being considered, as E.J. Williams and C.F. von Weizsäcker pointed out; for they showed that, viewed in a suitable coordinate reference frame, energies of no more than a few mega-electronvolts were involved—for which QED gave correct results, as shown by laboratory experiments! (Weizsäcker 1934; Williams 1935). Even after those results, according to Serber, Oppenheimer "struggled to maintain his lack of faith" [13]. We shall not further pursue this "failed revolution" against QED, which forms a major theme of Galison (1983a).

Before returning to the resolution of the problem of soft and penetrating components and the identification of the cosmic ray *mesotron*, later called the *muon*, we consider the interpretation of showers as explosions initiated by strong nuclear interaction, an interpretation proposed first by Heisenberg in 1936. The abstract of his paper reads: "The Fermi theory of β-decay leads to a qualitative explanation of 'shower' formation, whose consequences will be discussed in detail" [14]. The idea, as Heisenberg had communicated by letter two weeks earlier to Pauli, was the following [15]. The basic quantized electromagnetic interaction couples a photon, described by the four-vector field A_μ to a four-vector current qj_μ^{el}, where q is the electric charge. Thus, the interaction energy H'_{el} is given by the relativistic scalar product

$$H'_{el} = qj_\mu^{el} A_\mu. \tag{4.1}$$

The probability of deflection of a high-energy electron (of charge $-e$) in the field of a nucleus of charge Ze, accompanied by the emission of n photons, is proportional to the product of $[Ze^2/(\hbar c)]^2$ for the deflection and a factor $e^2/(\hbar c)$ with $\hbar = (h/2\pi)$ for each emitted photon, or

$$\text{Probability} \propto [Ze^2/(\hbar c)]^2 [e^2/(\hbar c)]^n. \tag{4.2}$$

The quantity $e^2/(\hbar c)$, the fine-structure constant, has a value of nearly 1/137, so that single-photon emission $n = 1$ is strongly favoured (as postulated by the cascade shower theorists). The calculation that leads to this result uses the perturbation method, whose validity depends upon the smallness of $e^2/(\hbar c)$ relative to unity.

Similarly we can write the Fermi β-decay interaction in the form of a relativistic scalar product,

$$H'_F = f j_\mu^{np} j_\mu^{e\nu}, \tag{4.3}$$

where j_μ^{np}, the neutron–proton current, and $j_\mu^{e\nu}$, the electron–neutrino current, are analogous to j_μ^{el}, the electromagnetic current. The constant $f = G_F/(\hbar c)$, where G_F is the Fermi constant. The electromagnetic interaction of two currents can also be written in a form analogous to equation (4.3) if A_μ in equation (4.1) is expressed in terms of the current which produces it. The difference is that, in the electromagnetic case, the interaction is delayed (being carried over a distance by the photon field A_μ), whereas the β-decay interaction was assumed by Fermi to be instantaneous (point-like). In his theory, the analogue of $e^2/(\hbar c)$ is f^2; however, f^2 has the dimensions of the fourth power of a length. This has as a consequence that the validity of the perturbation treatment of the Fermi-field is limited by the requirement that the dimensionless quantity f/λ^2 be small relative to unity, where λ is the de Broglie wavelength of the emitted electron or neutrino. Hence there is a characteristic length $\lambda_0 = \sqrt{f}$, such that $f/\lambda_0^2 = 1$; for wavelengths smaller than λ_0 (or for energies larger than $\hbar c/\lambda_0$), the perturbation method cannot be applied to the Fermi-field. Collisions involving multiple production of e–ν pairs may therefore not be small. Heisenberg saw this as a possible mechanism for what he called "explosive showers".

Since the perturbation method fails for the Fermi theory at high energy, Heisenberg thought that conclusions about its predicted self-energy being infinite, etc., could be false. As he wrote to Pauli [16]:

The non-convergence of the self-energy for high momenta in the Fermi theory is no argument at all against this theory; but so long as one has not constructed a new mathematical procedure to handle the high-energy region, it is only an argument that the physicists doing the calculations lack mathematical understanding; thus all arguments about infinite energies in the Fermi theory are nonsense.

He added that it was most important to find out how theories of the Fermi type behave qualitatively at high energy. On dimensional grounds (i.e., due to the small *dimensionless* fine-structure constant), he argued that QED cannot behave otherwise at high than at low energies, whereas on the other hand the formalism of wave quantization might be changed as much by introducing the Fermi constant as had physics in general by the introduction of the dimensional constants c and h.

Although one may easily see why Heisenberg's advocacy of the Fermi-field theory of nuclear forces led him to expect explosive showers to occur in the cosmic rays [17], it is more difficult to understand why he did not consider that at least the majority of the showers observed were produced by the electromagnetic processes of *Bremsstrahlung* and pair production, after Bethe and Heitler had shown that their cross-sections were large enough [18].

A possible reason is that Heisenberg did not believe in the validity of QED, on which the Bethe–Heitler cross-sections depended, at the high energies of the cosmic rays, in spite of the relativistic considerations of von Weizsäcker and E.J. Williams! [19]. A second possibility may be given: as typically observed in a cloud chamber, much of the multiplication by cascade occurred within the chamber wall or in a plate of heavy metal, so that the cone of emerging shower particles had what appeared to be a point-like vertex. Hence it could be interpreted as a single explosive process, a process excluded by perturbative QED. Additionally, a part of the radiation succeeded in penetrating hundreds of metres of water equivalent (Regener 1937; Euler 1937). According to the theory of explosive showers, this highly penetrating radiation would consist of "slow" neutrinos produced in the explosive showers. Recall that the Fermi-field theory predicted that high-energy electrons and neutrinos would have strong nuclear interactions, whereas low-energy electrons and neutrinos would interact only weakly.

As a third reason, Heisenberg argued that, on the assumption that the primary cosmic rays were protons, their Fermi-field interactions would create shower-producing particles, i.e., electrons and photons, whereas their origin was unexplained by the cascade theory [20]. Fourthly, large bursts of ionization ("*Stöße*") had been observed in ionization chambers, beginning in 1928, and these appeared not to be merely large cascade showers (Hoffmann 1935). We shall see below that, even after the bulk of showers had been explained as cascades, these other considerations, and additional observations as well, kept alive the idea of explosive showers.

4.3 The neutrino theory of light (1934–37)

Before continuing with cosmic ray problems, we shall deal briefly with a different topic that played a role in the theoretical discussions of that time:

the *neutrino theory of light*. This theory attempted to establish a relationship between the nuclear phenomena (e.g., β-decay) and electromagnetic phenomena. Although it did not lead to experimentally substantiated conclusions, it stimulated the fantasy of many theoreticians who hoped for a unified picture of all forces in nature.

Soon after the 1933 Solvay Conference, at which Wolfgang Pauli finally made his neutrino hypothesis public, Louis de Broglie, another participant in the conference, sought to apply the neutrino concept in electrodynamics, suggesting that the light quantum is composed of a neutrino–antineutrino pair [21]. His brief paper was immediately noticed and studied by Pauli, who wrote to Heisenberg [22]:

> In the *Comptes rendus* of 8 January 1934...there has appeared a quite interesting note of de Broglie, in which he discusses the point of view of composing the photon from two neutrinos. The main problem — as it seems to me — is to formulate reasonably the *interaction terms of neutrinos and electrons* in the Hamilton function. One cannot grasp *a priori* how the particular neutrino pairs that stick together and build up the photon occur much more easily than any two neutrinos having different directions of momenta and different energies.

In the following months Pauli and Heisenberg eagerly discussed the problems of de Broglie's neutrino theory of light, of which Pauli thought

L. de Broglie (1892–1987). Photograph from Burndy Library and reproduced by permission of AIP Emilio Segrè Visual Archives.

"very highly" [23], without forgetting about the remaining difficulties, such as the problem of how the *static* electromagnetic field could be represented by a product of two spinor fields [24]. Heisenberg raised a still more fundamental problem, namely, the self-energy difficulty in quantum electrodynamics [25]. In spite of these difficulties, they both kept their interest in the theory for a number of years [26].

We sketch here the rich work on the neutrino theory of light between 1934 and 1937, which Max Born called in 1936 "the most important contribution during the last years to the fundamental conception of theoretical physics" [27]. During the year 1934, Louis de Broglie wrote a set of papers on the subject, in which he derived several results about the equation of motion and the spin of the photon (Broglie 1934c and Broglie and Winter 1934). By the end of the year, other theoreticians started to submit papers, including G. Wentzel (Zürich), P. Jordan (Rostock), R. Kronig (Groningen), O. Scherzer (Munich), M. Born and N.S. Nagendra Nath (Bangalore), V. Fock (Leningrad), E.C.G. Stueckelberg (Geneva) and A. Sokolov (Tomsk). Indeed, a worldwide enterprise emerged, filling many pages of the scientific journals in the mid-1930s.

Gregor Wentzel showed in a paper, which was received in October 1934 by the *Zeitschrift für Physik*, "that the fundamental equations of electrodynamics can be derived from a formal scheme, in which the electromagnetic fields enter as operators representing the creation and destruction of *pairs of corpuscles*" [28]. Nearly simultaneously, Pascual Jordan carried out a similar programme [29]. He especially focused on the "*kernel (Kernfrage)* of the problem, i.e., the emergence of *Bose statistics* for light quanta if one begins with *Fermi statistics* for the fundamental objects [neutrinos]" [30]. In a one-dimensional model, he succeeded in demonstrating that the wavefunction of the composite photon possesses the commutation relations of Bose particles [31].

Jordan's paper by no means finished the problem. Thus Ralph Kronig entered the field, contributing three papers submitted in March, June and August 1935 to the Dutch journal *Physica* (Kronig 1935). He established a relationship between the number of light quanta and the number of neutrinos, on the one hand, and between their energies, on the other hand. Both relationships were of a statistical nature, i.e., to a given number of neutrinos and antineutrinos there corresponded only a certain probability for a given number of photons, and the total neutrino–antineutrino pair energy would not convert directly into the energy of the associated light quanta. Then Otto Scherzer, a student of Sommerfeld, sought to derive the known optical phenomena by assuming a special form for the Hamiltonian describing the interaction of neutrino pairs with matter (Scherzer 1935). Jordan simplified Kronig's calculations in the one-dimensional neutrino–photon model (Jordan 1936a). Jordan and Kronig then collaborated to prove their results also in three dimensions (Jordan and Kronig 1936). Later in 1936 they each tried, separately, to remove some of the non-physical assumptions that they had made earlier [32].

At the end of 1936, Vladimir Fock raised objections against the Jordan–Kronig theory (Fock 1936). He argued, first of all, that the light-quantum field cannot satisfy a linear differential equation since it depends quadratically on the neutrino field. Secondly, Jordan's expression for the photon operator,

$$\sqrt{\nu} b(\nu) = \int_{-\infty}^{+\infty} \gamma^+(a)\gamma(a+\nu)\, da \qquad (4.4)$$

(with γ and γ^+ denoting neutrino and antineutrino operators, respectively) would commute with its conjugate operator. To prove this point, Fock wrote the operator as

$$\sqrt{\nu} b(\nu) = \int_{-\infty}^{+\infty} e^{-i\nu x} \psi^+(x)\psi(x)\, dx \qquad (4.5)$$

and claimed that its action in configuration space corresponded just to a multiplication by a number, $\sum_{k=1}^{n} e^{-i\nu x_k}$ (with x_k denoting the coordinate of the kth object). Hence he expected that "no consistent neutrino theory of light ... can be constructed" [33].

Fock's conclusions were attacked immediately by Ernst Carl Gerlach Stueckelberg who stressed that "a configuration space representation is only possible if the number of neutrinos n is finite" [34]. With an infinite number of neutrinos being involved in the construction of one photon, however, the commutation relation of Jordan, i.e.,

$$b^+(\nu)b(\nu) - b(\nu)b^+(\nu) = -1 \qquad (4.6)$$

would indeed follow for $\nu > 0$. Similar objections were put forward by N.S. Nagendra Nath [35] and A. Sokolov (Sokolov 1937). Still Fock stuck to his criticism of the neutrino theory of light, raising further arguments against it [36].

In a new paper published in April 1937, Jordan defended his theory and worked out further results concerning the relationship between the wave amplitudes of the neutrino and the photon fields (Jordan 1937). He found that the neutrino field $\psi(x)$ (in a one-space-dimensional model) can be considered as a "fusion" of two partial systems, one, $\Omega(x)$, being the light field; the other, $\chi(x)$, the "pure" neutrino field. In particular, he concluded that the "amplitude $\chi(x)$ must replace the amplitude $\psi(x)$ used in Fermi's theory of β-decay" [37].

Clearly, if the neutrino played a role both in β-decay (perhaps also in nuclear interaction, as assumed by the Fermi-field theory) and in electromagnetic interactions, it had to possess a rather peculiar coupling with matter. Thus Ralph Kronig remarked in a letter to *Nature* dated 31 December 1935 that "the interaction energy must be such that the neutrino field is excited by the [β-] disintegration process of the radiationless type", and he noted that "the forms hitherto proposed for it have not the character required above" [38]. Not much progress in the problem of neutrino coupling to matter was achieved in the late 1930s; instead the statistics problem raised by Fock in 1936 drew new attention in 1938, but without being settled [39].

4.4 Electromagnetic shower theory and the interpretation of the hard component (1937)

4.4.1 *Electromagnetic cascade showers*

That standard QED could predict the main part of the shower phenomena was shown by Carlson and Oppenheimer (1937) in a paper received by the *Physical Review* on 8 December 1936 and by Bhabha and Heitler (1937) in a paper received by the *Proceedings of the Royal Society* three days later. Bhabha and Heitler quoted Anderson and Neddermeyer (1936) to claim that there were "no *direct* measurements of energy loss by fast electrons which conclusively prove a breakdown of the theory" and thus "it is reasonable *as a working hypothesis* to assume the theoretical formulae for energy loss and pair creation to be valid for all energies, however high" [40]. Carlson and Oppenheimer took substantially the same position.

The cascade process that they assumed can be illustrated as follows. An electron of high energy, say 10^{12} eV, passes through matter and is deflected in the field of a nucleus, emitting a photon (*Bremsstrahlung*); for example, the photon may take half of the energy. Both electron and photon continue to travel at small angles to the original direction, which is typical of high-energy processes. The photon then materializes as an electron–positron pair, again contained within a small angle. The resulting high-energy electron and positron make more photons by *Bremsstrahlung* and these in turn make more pairs. As Bhabha and Heitler stated [41]:

> Firstly, after passing through a plate of some heavy substance of a suitable thickness we can show that the original electron may emerge accompanied by a large number of electrons of large energy, which would all appear to come from a small region in the plate. Such a phenomenon would resemble the showers observed in experiments on cosmic radiation. In other words, some large cascade showers can be mistaken for "explosive" showers.

Bhabha and Heitler also claimed that their quantitative predictions for the number of electrons having energies above a given large value and found at a certain penetration depth agree well with experiments. They went on to state:

> Secondly, ... the effective "absorption coefficient" calculated from the tail end of these [absorption] curves has a value which is much less than the smallest absorption coefficient for hard γ-radiation. Similar curves have been wrongly used to prove the existence of a radiation much more penetrating than any possible theoretically.

This second point says that arguments that the theory breaks down in predicting lower electron penetrability than observed are falsified by the statistical nature of the shower. Hence they continued: "Though it is true that no electron of any reasonable energy has any chance of penetrating the whole atmosphere, there is a large probability that one of its secondaries arrives at the bottom of the atmosphere with a comparatively high energy" [42].

Carlson and Oppenheimer agreed in the main with the Bhabha–Heitler paper (which they had seen in manuscript form) but they had one reservation, namely [43]:

> Their results differ from ours primarily because of their neglect of ionization losses; apart from this the agreement between their values and ours is excellent. We do not agree with their conclusion that these calculations make it possible to ascribe the greater part of sea-level cosmic radiation to degraded electrons and photons of high initial energy.

4.4.2 *Penetrating showers*

Although both pairs of cascade shower theorists felt that they had accounted for a good part of the cosmic ray phenomena, they were aware that the cosmic rays still held mysteries [44]. For example, Bhabha and Heitler could invoke the cascade process to explain small absorption effectively, but, they continued [45]:

Left, J.R. Oppenheimer (1904–1967) (photograph from Lawrence Berkeley Laboratory) and right, J.F. Carlson (1899–1954) about 1933. Photographs reproduced by permission of AIP Emilio Segrè Visual Archives.

> We must mention . . . that the absorption coefficient calculated from Regener's curve under 100 m of water has a value which is about a hundred times smaller even than the values quoted above . . . We must conclude, either that the extremely hard radiation which penetrates 250 m of water consists of particles of protonic mass, or that the quantum theory of radiation breaks down for radiation of the highest energies if it consists of electrons.

Carlson and Oppenheimer made similar remarks about small absorption. Since 20 cm of lead should absorb practically all showers if the primary energies are less than 10^{11} eV, then [46]:

> One can conclude, either that the theoretical estimates of the probability of these processes are inapplicable in the domain of cosmic ray energies, or that the actual penetration of these rays has to be ascribed to the presence of a component other than electrons or photons . . . ; and if these are not electrons, they are particles not previously known to physics.

Most of the sea-level cosmic radiation, we now realize, consists of "particles not previously known", called muons, with mass about 200 times that of the electron. Positive and negative, they are, to all intents and purposes, "heavy electrons" in their properties. Although they were given that name early on (as well as other names: mesotron, yukon, meson, etc.), they were at first ascribed richer properties than the electron has, namely, participation in the strong, as well as in the weak and electromagnetic, interactions. One reason for this is that they were mistakenly considered to be candidates for the U-quanta of Hideki Yukawa's meson theory. Because Galison has dealt at length with the muon discovery, especially the role played by the Jabez Street group of Harvard University, we shall merely add a few comments [47].

Heitler has explained why it was so difficult to identify the muon [48]. Highly relativistic particles of the same charge all have essentially the same ionizing power, whatever their masses; in contrast, the radiative energy loss is proportional to the inverse square of the mass. Thus, for example, protons radiate with a probability nearly four million times less than electrons of the same energy. By measuring the curvature of the cloud chamber track formed by a charged particle in the presence of a magnetic field, one obtains the particle's momentum and, by assuming a mass, its energy. Measurements of energy lost in metal plates in a cloud chamber, assuming that all the particles observed were positive or negative electrons, led to a peculiar result. Electrons of energy less than about 200 MeV lost energy as theory predicted [49]. However, at higher energies the "electron" energy loss dropped to a very small value, far lower than theory predicted (Blackett and Wilson 1937; J.G. Wilson 1938). Thus, Heitler concluded, either QED was breaking down (for a long time the preferred explanation) or "the particles with the small energy loss *cannot be electrons*" [50].

Just about the time that the cascade papers were written, Anderson and Neddermeyer and Street and Stevenson were concluding that the non-radiating particles were not protons (plus and minus!) but new particles, the mesotrons. The rough mass measurements made in 1937 confirmed their judgment [47].

The cascade showers could thus account for the soft component of cosmic rays, while the mesotrons could be the penetrating component; but was that all there was? What of the expectation, based on the widely accepted Fermi-field theory, that multiple particle production by nuclear interaction (involving not only protons and neutrons, but also electrons and neutrinos) should occur at high energies? Bhabha and Heitler referred to Heisenberg's explosive shower paper of 1936 as "elegant" and said that perhaps it must be used to explain the largest showers.

On the other hand, Carlson and Oppenheimer took a harsh view of Heisenberg's explanation of large showers and bursts "as highly multiple elementary processes", stating that it was "without cogent experimental foundation", and that "in fact it rests on an abusive extension of the formalism of the theory of the electron neutrino field" [51]. One might guess from this emphatic language that they were rejecting the Fermi-field approach to strong interactions, but nothing could be farther from the truth!

Also the existence of two types of shower was not questioned. Carlson and Oppenheimer wrote [51]:

> Cloud chamber observations have shown the existence of two fairly well differentiated types of shower. In one of these, and by far the more common, only electrons, positrons and γ-rays appear to take part... In the other and rarer type of shower, transverse momenta of the order of 100 MeV are common; the shower is usually not collimated at all; heavy recoil particles are frequently seen; and the total number of particles is small.

They then stated that their cascade shower analysis was believable only

> if [one] admits the presence of another component to which the analysis is not at all applicable, and of other types of elementary processes, which essentially involve the heavy particles and their coupling with electrons, and which find no place in this treatment.

The other component was, of course, the penetrating component; the other elementary processes which couple heavy and light particles were those involving the Fermi interaction.

To avoid Heisenberg's "abusive extension of the formalism" of the Fermi field and to explain a supposed excess of small showers below thick absorbers, Oppenheimer and his associates proposed a "modified and 'cut-off' Fermi coupling convenient for the treatment both of nuclear forces and of high energy disintegrations" [51]. To paraphrase: the usual way to avoid divergences is to reduce the coupling strength at high energies, as they point out, and this cut-off procedure also limits Heisenberg's multiple processes. For low energies, one wants a weak Fermi coupling, to account for long β-decay lifetimes. For nuclear forces, the coupling should be strong at intermediate energies. What could be simpler than to multiply the usual Fermi interaction by a function of the energy of interaction, an $f(E)$ of the desired shape? They did so and claimed a reasonable range for the nuclear force [52].

4.4.3 *Heisenberg's response to the cascade papers*

By now it has become evident to the reader that Blackett's remark (at the 1934 London conference) that it was not easy to find a correct interpretation of cosmic ray data was a vast understatement [5]. One reason for this difficulty arose from different theorists trying to account for different sets of phenomena. In a letter to Pauli, Heisenberg wrote of cosmic ray studies (Barnóthy and Forró 1937) performed in mines in Hungary [53]:

> Barnóthy and Forró in Budapest sent me a work on cosmic rays, which investigates shower formation at large depths (730 m water equivalent). There it is shown that there is a non-ionizing shower producing radiation of absorption coefficient $\mu = 2.1 \times 10^{-5}$ cm^2 g^{-1} (corresponding to a cross-section of approximately 10^{-28} cm^2). Light quanta can scarcely have such penetrability (according to Bethe, certainly not). Neutrons also surely not, so Barnóthy and Forró thus conclude that we are dealing here with neutrinos. That seems to me quite convincing.

A few weeks later, Heisenberg reported to Bohr that [54]:

> for very energetic electrons the Bethe–Heitler radiation will always be most frequent, as Oppenheimer has stressed; but for protons and neutrinos the most important stopping process will be the Fermi showers. It will be especially simple at great depths ... There the whole of the cosmic rays will consist of slow neutrinos that make small showers.

By 21 January 1937, Heisenberg then wrote to Pauli that he had received the cascade shower paper by Bhabha and Heitler, that it seemed to be "very good" and that

> it looks as though practically all processes above sea-level could be explained by electromagnetic theory alone ... The true [*echten*] showers would be very few in comparison to the cascades.

At sea-level, explosive showers would be no more than a thousandth part of all showers, Heisenberg agreed. But there would be an important difference at great depths, where Bhabha and Heitler would need to explain the observed showers by neutrons and protons "which practically do not radiate". After a short calculation regarding the number of neutrinos expected on the basis of his shower theory, Heisenberg says the best experimental check would be to study the electron energies at great depth: "According to my reckoning these energies should on average be smaller than at the surface, but greater than by the proton theory." [55]

Later, in mid-1937, Heisenberg wrote to Bohr in much the same spirit [56]:

> It looks as though many showers can only originate via cascades; however, it seems certain ... that the explosive type of shower does occur with [a cross-section] of approximately 10^{-27} cm^2, i.e., a thousand times less frequently than the cascades, and that "bursts" have mainly this origin [57]. So far the experiments agree well with our earlier programme ... An interesting novelty,

W. Heisenberg (1901–1979) discussing his shower theory in 1936. Photograph reproduced by permission of AIP Emilio Segrè Visual Archives.

which Anderson claims to have found, is the unstable heavy electrons. I...do not know how well-founded it is experimentally. At any rate, Blackett has strong arguments against the idea that all penetrating radiation consists of heavy electrons. He claims that most particles of about 6×10^8 eV are penetrating in Pb but produce cascades in Cu and Al; he therefore believes in a failure of the Bethe–Heitler formula for large energies, with the limiting energy being different for different elements. About this question, which can only be decided by experiment, I have no definite opinion.

From these selections we can see that Heisenberg was actively interested in the cosmic ray phenomena, that he carefully read the experimental papers and thus that he has equal claim to be placed with Oppenheimer and Heitler in Cassidy's group of "cosmic-ray physicists" [58]. (It would be equally appropriate to include Oppenheimer among Cassidy's "field theorists".) Like many others trying to solve puzzles, physicists are usually willing to use any information and any insights that they can gather—without strong prior commitment.

However, that is not to say that they cannot be strongly influenced by their earlier experiences—whether of success or failure. Thus it was part of the "collective consciousness" of field theorists that infinite self-energies and radiative reactions already arose in Lorentz's classical electron theory if one assumed a point electron. A miraculous solution to this problem did not emerge in QED, but instead even more intractable infinities were added, connected with the electric charge (polarization of the vacuum). Thus, as we have mentioned earlier, a breakdown of electrodynamics was anticipated at the so-called classical electron radius $r_0 = e^2/(mc^2) \approx 10^{-13}$ cm (which happens to be also about the size of the range of nuclear forces). Expressing $1/r_0$ in energy dimensions i.e., multiplying by $\hbar c$, one finds the expected breakdown energy to be $137mc^2$, where mc^2 is the rest energy of the electron, about 0.5 MeV.

In their path-breaking systematic formulation of QED in 1929 and 1930, Heisenberg and Pauli had become aware of the importance of these problems. At that time they had considered the introduction of a lattice-world (*Gitterwelt*) of cells of volume r_0^3, which would give a finite electron self-energy. However, because this concept was difficult to reconcile with relativistic invariance, they soon rejected it (Heisenberg 1930). Nevertheless, Heisenberg kept returning to the idea (for decades) that a new universal constant, a length of order r_0, would play a role in a future convergent quantum field theory.

4.5 A fundamental length and cosmic-ray bursts (1937–39)

4.5.1 *A new fundamental length*

As we pointed out in section 2, the Fermi coupling constant G_F itself provides a critical length, namely $\lambda_0 = \sqrt{G_F}/(\hbar c)$, such that interactions characterized by distances smaller than λ_0 lead to multiple, rather than single, production of particles (if the theory is extended uncritically to high energies). For example, a proton with de Broglie wavelength $\lambda < \lambda_0$, incident upon a nucleus at rest, should produce multiple electron–neutrino pairs. The value of G_F, obtained by comparing the original Fermi theory with β-decay lifetimes, yields $\lambda_0 \approx 10^{-4} r_0$, but with a suitable (Konopinski–Uhlenbeck) modification, one can get λ_0 to be of the order of r_0 (Heisenberg 1936). Heisenberg's motive in 1936 for having $\lambda_0 \approx r_0$ was that with this larger value he might be able to explain showers. However, when the cascade theories proved successful, that motive was largely, though not entirely, removed.

We stress that there exists no necessary logical connection between the *critical* length λ_0, at which multiple production overtakes single production, and the conjectured *universal* length at which quantum field theories fail, which would therefore characterize (and possibly emerge from) a future convergent theory. Nevertheless, it appeared natural to guess that there might be such a connection [59].

To see whether a nonlinear field theory which contains a length parameter and for which, consequently, perturbation theory fails above some energy, might be convergent when given appropriate mathematical treatment, Heisenberg and Pauli in 1936 began to consider some simplified model theories. They resorted again, as they had in 1930, to a "lattice-world" picture, applied this time to a relativistic quantum field theory of spinless particles with self-interaction [60]. After an exchange of letters on such models, however, Pauli was inclined to drop the idea of a universal length. It did not seem to solve the problem of divergence at high energy, it made special difficulties for relativistic invariance and "seemed to be no longer supported by experiment" [61].

With regard to the last point, Pauli declared that he had been influenced by Bhabha, who had visited him in Zürich during the Easter holidays. Agreeing that there was, in fact, besides the cascades, a second kind of shower, the Indian physicist had suggested that they might be proton-initiated nuclear disintegrations, rather than multiple Fermi-field processes. That the former type of nuclear process took place was shown soon afterwards by two Viennese physicists, who pioneered the observation of cosmic rays by means of nuclear emulsion [62]. They observed so-called "stars", i.e., tracks of heavy particles emerging from a point. Heisenberg analysed this nuclear process using then standard methods; he found general agreement with experiment and concluded that there was no reason for such collisions to result in light particles. However, he still maintained at the end of the paper that "the discussion of the Hoffmann bursts on the basis of cascade theory still makes the existence of explosive showers very probable" [63].

In the following year, 1938, physicists gave up the Fermi-field explanation of nuclear forces in favour of Yukawa meson theory. Even then, Heisenberg still clung to the idea of a universal length that would limit the validity of quantum field theory and at the same time lead to its finiteness. He reviewed his arguments for such a future theory in an article dedicated to the eightieth birthday of Max Planck and in another article on the limitations of quantum field theory (Heisenberg 1938a,b). Another year later, he would develop a new explosive-shower theory in which the vector meson field replaced the Fermi-field (Heisenberg 1939b).

4.5.2 *The problem of cosmic ray bursts*

One reason for Heisenberg's insistence on the possibility "that very energetic particles create *in a single event* a very large number of secondary particles" [64] was the observation of so-called bursts (*Stöße*) in cosmic rays. The first indication of the existence of the bursts appeared in early 1928, when Gerhard Hoffmann and F. Lindholm systematically monitored the intensity of cosmic radiation on the Swiss mountain Muottas Muraigl near the town of Davos (Hoffmann and Lindholm 1928). Making use of a high pressure ionization chamber (filled with CO_2 at 40 atmospheres) to increase the ionization effect and applying a voltage compensation method, they achieved a sensitivity of 1–2%, and thus detected "fluctuations of the radiation that exceed 1% and occur independently of the [external] air pressure". Hoffmann continued to study this phenomenon over many years, gradually improving his experimental technique. In the 1930s "spontaneous bursts" or "Hoffmann bursts", as they were called, were observed not only by Hoffmann and his students but by many others.

Regarding the origin of bursts, Steinke and Schindler (1932) noticed in 1932 that "the spontaneous bursts are created by cosmic radiation in the shielding material". An American group concluded in 1934 that "the production of bursts must either be due to a very soft component of the cosmic radiation or else some mechanism of their [local] production... must be supposed" (Montgomery and Montgomery 1935a); these authors also observed that the presence of nearby material (such as water above the chamber) enhances the bursts (Montgomery and Montgomery 1935b). W. Messerschmidt detected maxima in the size and frequency distribution curves as a function of the atomic number of the shielding material; specifically "the burst frequency drops in lead already behind a shielding thickness of 5 cm, whereas it rises in aluminium up to 30 cm" (Messerschmidt 1936).

Jørgen Bøggild's Copenhagen Ph.D. thesis covered a wide range of problems connected with bursts. He concluded that his results could be explained well by the Bhabha–Heitler cascade shower theory [65]. The theorist Hans Euler discussed Bøggild's data at the Bad Kreuznach meeting of the German Physical Society in September 1937, during a similar conclusion regarding most of the bursts. However, he argued that, besides the normal cascade showers, there must be "still another type of process which accounts for bursts that can be absorbed only in iron of several metres thickness". He continued [66]:

> Whether these processes, which the large bursts create behind thick layers [of shielding], also include Heisenberg's multiple processes, or whether they are entirely caused by a hard [i.e., penetrating] particle splitting off, once in a while, a soft electron or photon, which in turn gives rise by multiplication to the burst—information on this problem can only be obtained from bursts behind very, very *thin layers*.

Thus Euler was still considering the possibility of "explosive showers" as well as the possibility that the penetration of thick shields might involve the "heavy electron" that had recently been reported by Anderson and Neddermeyer and others.

Through his *Habilitation* thesis in 1938 and subsequent work, Heisenberg's student Hans Euler became the leading theoretical expert on Hoffmann bursts. In his first analysis, he separated the bursts into two nearly equal groups, one of which he ascribed to cascade showers (Euler 1938a). Concerning the others, which he said were dominant both below very thin and below very thick absorbers, he concluded that these "non-cascade bursts are created in an explosive manner" as Heisenberg had proposed in 1936 (Euler 1938b). About a year later, in June 1939, Euler and the Norwegian theorist Harald Wergeland (also a student of Heisenberg) reached only a slightly different conclusion after making a new kind of analysis of the large air showers (an analysis that had been suggested by the experimental cosmic ray physicists Lajos Janóssy, Pierre Auger and their collaborators). Euler and Wergeland concluded [67]:

It follows that air showers may explain the zero effect of the Hoffmann bursts [i.e., bursts in unshielded ionization chambers] and via multiplication a certain part of the bursts in the case of thin layers. However, the Hoffmann bursts in the case of medium and thick layers (>15 cm of iron, 20 cm of lead) cannot be understood on the basis of multiplying process of air showers, air electrons, or Bhabha's ionization electrons, because they are larger in light than in heavy material; *these bursts contribute therefore, also in future, an argument in favour of the existence of explosion-like showers.*

After the Second World War, refined methods of ionization chamber construction allowed observers to distinguish between the various processes initiating bursts (Montgomery and Montgomery 1947; Bridge *et al.* 1948; Rossi 1949). The following were found to be responsible for Hoffmann bursts [68]:

(1) *In unshielded chambers*: air showers; nuclear collisions within the chamber wall which lead to the emission of strongly ionizing particles; at extremely high altitudes, heavy nuclear fragments passing through the chamber.

(2) *In chambers shielded or covered by material*: cascade showers arising in the material; nuclear collisions in the surrounding material, yielding strongly ionizing fragments which reach the inside of the chamber or producing muons, which create showers either by *Bremsstrahlung* or by ejection of an electron.

Thus, by the early fifties, any association with Heisenberg's favoured explosive showers had completely disappeared, whereas other processes of high-energy physics helped to provide sources for all types of Hoffmann bursts.

We have presented in some detail the story of the Hoffmann bursts, not only because they represented cosmic-ray phenomena widely discussed in the 1930s, but also because the explanations offered for them were connected with the latest ideas of that time on nuclear forces, especially at high energies. Cosmic ray phenomena continued to play an important role in the formation and testing of various theoretical ideas throughout the 1940s and well into the 1950s, until advanced high-energy accelerators replaced the cosmic rays as sources of high-energy particles.

Notes to text

[1] Auger (1946, p 22). See also Auger (1983).
[2] For detailed recollections of the "heroic age" of cosmic rays, see Sekido and Elliot (1985).
[3] See Brown (1989) and references therein to earlier studies of Yukawa's work; also see Darrigol (1988).
[4] This is the point of view expressed already in chapter 2.
[5] Blackett (1935, especially p 199).
[6] International Conference (1935, p 3).
[7] Excellent detailed studies of the cosmic rays at mountain altitude and sea-level,

the establishment of the existence of hard and soft components, and comparisons of cloud chamber and counter experiments were carried out by French workers. As examples see Auger *et al.* (1936) and Leprince-Ringuet and Crussard (1937). See also articles by Pierre V. Auger and by Louis Leprince-Ringuet in Brown and Hoddeson (1983) and Colloque International (1982).

[8]   Bethe and Heitler (1934). An indispensable reference on all radiative processes is Heitler (1936).

[9]   Heitler (1936, p 174) (original emphasis). The ionization energy loss had been evaluated from quantum mechanics, especially in Bethe (1932) and Bloch (1933). However, the behaviour for highly relativistic energies was first given correctly by Fermi (1940). The need for this substantial modification of the high-energy ionization was apparently not noticed earlier.

[10] Bethe, in International Conference (1935, p 250).

[11] Anderson and Neddermeyer (1936). In this paper the authors' observations were based on a subset of particles that could be clearly identified as electrons.

[12] Serber (1983, especially p 207). See also Brown and Hoddeson (1983, introduction, pp 13–6).

[13] Serber (1983, p 210).

[14] Heisenberg (1936, p 533). Cassidy (1981) has contrasted Heisenberg's shower theory with the cascade shower theory as an illustration of the differing aims of the field theorists and the cosmic-ray theorists. However, it would appear that the contrast is not so much with "theoretical" versus "practical" objectives, as with which "facts" appeared more urgently to require explanation. For example, Heisenberg was concerned about the *origin* of the shower-producing radiation (which was not clear until about 1950) while, on the other hand, Bethe, Heitler and Oppenheimer were mainly concerned — as were also Heisenberg and Pauli — about the divergent behaviour of QED.

[15] Heisenberg to Pauli, 26 May 1936, [WPSC2], pp 445–6. Heisenberg emphasized, "It appears to me thus that one can understand the existence of cosmic ray showers directly from the Fermi theory".

[16] Heisenberg to Pauli, 30 May 1936, [WPSC2], p 446.

[17] In chapter 3 we have already indicated how widespread the attraction of the unified Fermi theory of nuclear forces was. The attitude of Bethe and Bacher was that "one would be very reluctant to give it up". (Bethe and Bacher 1936, especially p 203). To get a reasonable fit both to β-decay and to the strong nuclear force, it was necessary to modify Fermi's original form of interaction, introducing derivatives of the field operators, as in Konopinski and Uhlenbeck (1935). Now Heisenberg (1936) did the same to get a bigger cross-section for shower production. Most other authors treating nuclear forces at this time espoused the Fermi-field: cf. Weizsäcker (1937) and Gamow (1937).

[18] Bethe and Heitler (1934). Earlier calculations that neglected screening of the nuclear field by atomic electrons had given even larger cross-sections, but were rightly suspected to be false.

[19] Although Anderson and Neddermeyer had found by June 1936 that their earlier reported measurements indicating a breakdown of the radiation formulae had resulted from a bias in the counter triggering of their cloud chamber (Anderson and Neddermeyer 1936), this did not put to rest the idea that QED would break down at some higher energy. Nor were the Weizsäcker–Williams results (Weizsäcker 1934; Williams 1935) conclusive on this point. Both Bohr and Oppenheimer suggested that the Fourier analysis of the nuclear electric field in the rest frame of the electron, which was the essential point of the Weizsäcker–Williams treatment, might not be correct for the higher frequency components that were responsible for larger radiative losses. (In fact, in the 1950s one recognized that nuclei and nucleons have smaller high-frequency components than do point-like sources and thus radiate less.) As late as July 1937,

Heisenberg wrote to Bohr that Blackett had "strong arguments against the idea that all penetrating radiation consists of heavy electrons". Blackett claimed that most particles of about 6×10^8 eV are penetrating in lead, but make cascades in Cu and Al; hence he "believes in a failure of Bethe–Heitler for large energies". (Heisenberg to Bohr, 5 July 1937, Niels Bohr Archives).

[20] The presently accepted main source of the shower-producing radiation is nuclear collision resulting in neutral pions, discovered in 1950, which decay into two γ-rays.

[21] Broglie (1934a,b). Louis de Broglie had previously already contemplated the idea of a composite photon, suggesting that electron–positron pairs might constitute the electromagnetic field (Broglie 1932a, b, 1933).

[22] Pauli to Heisenberg, 19 January 1934, [WPSC2], pp 253–4.

[23] Pauli to Joliot, 26 January 1934, [WPSC2] p 265.

[24] Pauli to Heisenberg, 30 January 1934: "De Broglie only demonstrates how this works for electromagnetic fields." [WPSC2], p 271.

[25] Heisenberg to Pauli, 5 February 1934, [WPSC2], p 273.

[26] See, e.g., W. Pauli to R. Kronig, 5 April 1935, [WPSC2], p 385.

[27] Born and Nagendra Nath (1936, especially p 318).

[28] Wentzel (1934, especially p 337).

[29] Jordan (1935). Jordan pointed out that he had earlier suspected the photon to be a composite object (see p 464); and he referred to Jordan (1928, especially p 207).

[30] Jordan (1935, pp 464–5). Besides the problem of the commutation relations, the author studied the problem of obtaining the correct black-body radiation distribution as well as the absorption of a neutrino-pair and the Raman effect of single neutrinos.

[31] Jordan claimed explicitly that de Broglie's original idea of assuming a photon of energy $h\nu$ to consist of a neutrino–antineutrino pair, with each partner contributing the energy $\frac{1}{2} h\nu$, would not solve his *Kernfrage*. Pauli had drawn his attention to that point.

[32] Jordan (1936b); Kronig (1936a). Jordan and Kronig had earlier not included all degrees of freedom of the photon but had assumed the latter to be a scalar object; also they had not considered a fully relativistically invariant treatment.

[33] Fock (1936, p 1012).

[34] Stueckelberg (1937b, especially p 198).

[35] Nagendra Nath (1937). Nagendra Nath referred to a proof of the Jordan commutation relation given in Born and Nagendra Nath (1936). See also Pryce (1937).

[36] Fock (1937). He claimed, e.g., that the operators $b(\nu)$ in equation (5) do not form a complete set and are not uniquely determined; in addition, only neutrino–antineutrino pairs whose partners have strictly parallel momenta would contribute to form photons, but their probability should be negligible.

[37] Jordan (1937, p 121).

[38] Kronig (1936b). Kronig (1935) had shown the possibility of constructing a suitable neutrino–matter interaction that would satisfy the necessary requirements.

[39] Pryce (1938a, b). Pryce claimed that the correct Bose–Einstein commutation rules for photons could not be achieved unless one supposed them to be composed of neutrinos having non-zero mass. Such massive neutrinos were considered, e.g., by Sokolov (1938), in order to obtain Coulomb's law in some approximation from the neutrino theory of light. It should also be mentioned here that Born and Nagendra Nath (1936) considered scalar neutrinos; for such a theory the result of Pryce would not hold.

[40] Bhabha and Heitler (1937), original emphasis.

[41] Bhabha and Heitler (1937, p 434).

[42] Bhabha and Heitler (1937, p 454).
[43] Carlson and Oppenheimer (1937, p 222).
[44] The main experimental results cited by Carlson and Oppenheimer were in Geiger (1935) (wrongly cited by them); also in Pfotzer (1936). Bhabha and Heitler also cited Pfotzer, Auger *et al.* (1936) and others.
[45] Bhabha and Heitler (1937, p 455).
[46] Carlson and Oppenheimer (1937, p 220).
[47] Three significant short papers in connection with the identification of the muon as the penetrating component of cosmic rays are Neddermeyer and Anderson (1937), Street and Stevenson (1937a, b), and Nishina *et al.* (1937). All three groups used counter-triggered cloud chambers. The last two papers roughly estimate the mass; the third paper was received two months before, but published a month later than the second one.
[48] Heitler (1936, second edition 1944, pp 228–9). In most other respects the second and first editions are printed from the same plates, due to wartime conditions at the time of the second edition. (Hereafter we shall refer to the muon during this period by the name proposed in Anderson and Neddermeyer (1938), namely the *mesotron.*)
[49] Nordheim *et al.* (1937, especially p 1039).
[50] Heitler (1936, p 229) (original emphasis).
[51] Carlson and Oppenheimer (1937, p 221).
[52] In the β-decay energy range, their formulation is equivalent to the proposal in Konopinski and Uhlenbeck (1935). See also our discussion in chapter 3, section 4.
[53] Heisenberg to Pauli, 18 December 1936 [WPSC2], p 491. Pauli's reply (21 December 1936) asked Heisenberg to send the cited work for his study. Obviously Pauli was interested in this possibility of demonstrating the existence of neutrinos.
[54] Heisenberg to Bohr, 11 January 1937 (Niels Bohr Archives).
[55] Heisenberg to Pauli, 21 January 1937 [WPSC2], pp 503–4. By "the proton theory", Heisenberg probably referred to the theory that the penetrating radiation consisted of fast primary protons and negative protons, not *necessarily antiprotons*. Recall that this was still earlier than the mesotron discovery. There are many references to the proton theory in the literature, but we shall not discuss them here. For a work sent at nearly this time, but after the Bhabha–Heitler work, see Bhabha (1937).
[56] Heisenberg to Bohr, 5 July 1937 (Niels Bohr Archives). It is not evident how Heisenberg concluded that the heavy electron is unstable.
[57] At this point Heisenberg inserts the names Street, Stevenson and Fussel. It is most likely that the references are to two abstracts of papers at an American Physical Society meeting in April 1937: Street and Stevenson (1937a); Fussel (1937).
[58] Pauli reacted differently; in thanking Heisenberg for sending the manuscript of Barnóthy and Forró, he said that "it will now be studied by the experimental physicists". Pauli to Heisenberg, 19 January 1937 [WPSC], p 502.
[59] A connection might have been inferred from analogy with the Born–Infeld convergent nonlinear theory of electrodynamics, in which the nonlinearity introduced a fundamental length which served as a cut-off (Born 1933; Born and Infeld 1934). Heisenberg pointed out that this length-dependent theory had also the consequence that "in the collision of two very energetic light quanta a shower of longer nonlinear light quanta results" ([WPSC2], p 537).
[60] The Lagrangian *without* self-interaction was that of the linear theory of scalar particles, which Pauli called the "anti-Dirac" theory, because it has no negative energy states (Pauli and Weisskopf 1934).
[61] Pauli to Heisenberg, 2 May 1937 [WPSC2], p 520.

[62] Blau and Wambacher (1937). For cloud chamber observations, see Brode and Starr (1938).
[63] Heisenberg (1937, especially p 384).
[64] Heisenberg (1936, p 533) (our emphasis).
[65] Bøggild (1937).
[66] Euler (1937, especially p 524).
[67] Euler and H. Wergeland (1939, especially p 485) (our emphasis).
[68] Meyer (1953, especially p 105).

Part B

Yukawa's Heavy Quantum and the Mesotron

The history of fundamental theories of nuclear forces begins with Heisenberg's theory of 1932 (chapter 2). In this first fundamental nuclear theory, the force that binds the nucleus together results from the *exchange of an electron* between a neutron and a proton, and the same basic interaction is responsible for nuclear β-decay. That forces arise through the exchange of particles (quanta) is the idea underlying the relativistic quantum theory of fields; in the present context, we call such theories "fundamental". The archetype of quantum field theories is quantum electrodynamics (QED), in which the exchanged particles are massless light quanta.

Heisenberg's electron-exchange model, if taken literally, violated the conservation laws of energy and angular momentum. In 1934, he replaced the electron emitted and absorbed (in the proton–neutron interaction) by an electron–neutrino pair, taking the elementary interaction to be that of Fermi's β-decay theory. Heisenberg's new fundamental theory of nuclear forces thus became the "Fermi-field theory", a theory that respected the conservation laws (chapter 3). Both of Heisenberg's exchange theories were "unified" in that the same elementary interaction was responsible both for the strong nuclear binding force and for the weak β-decay type of nuclear force. Both theories also shared the well-known difficulties of quantum field theory (QFT): namely, they predicted infinite self-charge and self-mass, in obvious contradiction to observation. Already present in QED, these so-called divergences were even more intractable in the nuclear field theories.

In spite of their divergent behaviour at high energy — from which the infinities arose — the fundamental nuclear field theories served as useful guides to phenomenological treatments, in which high energy cut-offs were introduced in integrals to avoid the divergences; also other parameters were fitted to the observed nuclear forces. These phenomenological models yielded reasonable approximate descriptions of low-energy phenomena such as nuclear binding, nuclear reactions and nuclear decays. However, to test the relativistic field-theoretical aspects of the fundamental theories involving, for example, particle creation and annihilation, physicists needed data that came from the *high-energy* interactions of fast cosmic-ray particles (chapter 4). When cosmic-ray observations of the mid-1930s and new experiments in the nuclear laboratory were confronted with the fundamental nuclear theories, some issues were clarified, but several puzzling features remained still without adequate interpretation. Moreover, the main difficulty of the Fermi-field theory was never really resolved, namely: how to reconcile the elementary weak coupling with the strong nuclear binding force, in both range and strength.

Meanwhile, Hideki Yukawa tried to derive a proper quantum field theory of Heisenberg's electron-exchange model of nuclear forces, a theory that Heisenberg had adopted in 1932 only tentatively and, for the most part, had used only phenomenologically (in the belief that quantum field theory needed a thoroughgoing reform before it could be applied within regions as small as the nucleus). Yukawa's first Heisenberg-like nuclear theory and a later attempt at a fundamental treatment of the Fermi-field theory were also failures. However, he learned from these failures that the range of force in a field theory is the inverse mass of the exchanged quantum m (measured in an appropriate unit: the Compton wave length of the particle $\hbar/(mc)$). As a result, Yukawa invented "heavy quanta", charged particles of positive and negative electronic charge with mass about 200 times that of the electron, in order to obtain forces having the empirical nuclear force range (see chapter 5 below).

Yukawa's quantum field theory, like the other fundamental nuclear-force theories proposed earlier, was also 'unified'; this had the consequence that the free *U-quanta*, as he called the new particles, would decay spontaneously with a mean lifetime of about a microsecond. This fact, together with the large mass of the *U-quantum* (excluding its free creation in the nuclear laboratory), would account for its not having been observed earlier. Indeed, in his first paper, Yukawa pointed out that only in the cosmic rays would enough collision energy be available to produce free *U*-quanta.

By mid-1937, several cosmic ray groups agreed that such particles existed. Now called pions, at first they were called "heavy electrons", then given various other names, including "mesotron", the name that we shall provisionally adopt here. Yukawa's theory was published in a respected journal of wide circulation and written in more than adequate English, but it had been neglected (even in Japan) for more than two years. Now the new particles were hailed as being identical with Yukawa's "heavy quanta" and they aroused world-wide interest in his theory (chapter 6).

In 1938 the meson theory effectively replaced the previous Fermi-field theory as the most accepted fundamental theory of nuclear forces (although other theories were still being considered). Besides Yukawa and his collaborators at Osaka University, European theorists picked up the idea of heavy quanta and worked out a detailed version based upon exchanged particles of spin 1, the vector-meson theory. This scheme appeared to provide a satisfactory treatment of much of the available data on nuclear structure and reactions, as well as of cosmic-ray phenomena (chapter 7).

Chapter 5

The Origin of Yukawa's Meson Theory

5.1 Introduction

Science, physics in particular, did not exist in Japan before the Meiji restoration (1868). What is perhaps the main reason for this was pointed out by Shigeru Nakayama [1]:

> The Platonic conviction that eternal patterns underlie the flux of nature is so central to the Western tradition that it might seem no science is possible without it. Nevertheless, although Chinese science assumed that regularities were there for the finding, they believed that the ultimate texture of reality was too subtle to be fully measured or comprehended by empirical investigation. Japanese paid even less attention to the general while showing an even keener curiosity about the particular and evanescent. In keeping with the orientation toward regularity in the early West, phenomena that could not be explained by contemporary theory, such as comets and novae, were classified as anomalous and given scant attention. In the history-oriented East, extraordinary phenomena were keenly observed and carefully recorded.

In order to learn Western science and technology, Japanese were sent after the Meiji restoration to Europe and the United States, and many foreign scientists were invited to teach at the newly established Japanese institutions. By the turn of the century, the country possessed some scholars who could compete with their Western counterparts. Among these early Japanese physicists some outstanding names are Kotaro Honda, a pioneer in metallurgy and magnetism; Hantaro Nagaoka, who invented a "Saturnian" atom, a precursor of Rutherford's nuclear atom; and Jun Ishiwara, who contributed to relativity and quantum theory [2].

The introduction of modern atomic physics, including quantum mechanics, was again aided by exchanges with the West. In the twenties, for example, Takeo Shimizu worked in Ernest Rutherford's Cavendish Laboratory and Toshio Takamine learned spectroscopy at Bohr's Copenhagen institute. By far the most important propagator of advanced atomic and nuclear science in Japan was Yoshio Nishina, who spent eight years in Europe, on leave from the Institute of Physical and Chemical Research (Riken) in Tokyo [3]. He worked at Cambridge, Göttingen and especially Copenhagen, where he carried out important theoretical research (Yagi 1990).

When Nishina returned to Tokyo in 1928 he was not able to obtain a position at the prestigious Tokyo Imperial University; a year later the world-wide economic depression hit Japan and its academic job market. Nishina gave some lectures on quantum mechanics, e.g., in May 1931 at Kyoto Imperial University. In July of the same year he started a major laboratory of nuclear physics at Riken, which, together with the laboratory of Masaharu Nishikawa, built several nuclear accelerators (a Cockcroft–Walton machine and two cyclotrons) [4]. To head the theoretical group, Nishina invited Sin-itiro Tomonaga from Kyoto.

Important impulses radiated from Riken that gave Japanese physics a modern face. That Tokyo institution also supported physics at Osaka University, when it founded its Science Faculty in 1932. Seichi Kikuchi moved there from Riken in 1933 to establish a new nuclear physics laboratory. He also attracted another Kyoto theorist, Hideki Yukawa, who soon originated a new and original theory of nuclear forces.

The importance of Yukawa's theory extended far beyond nuclear physics, becoming paradigmatic for the whole of elementary particle physics, which entered a new phase in the early thirties. In this chapter we shall trace the intellectual trail that the young Japanese physicist followed to arrive at his goal, making use of many documents discovered among Yukawa's papers at Kyoto University [5]. The meson theory was the result of a powerful creative act. It incorporated a number of ideas that are commonplace today, but which were novel and surprising sixty years ago. These are some of the original features proposed by Yukawa.

(a) The nuclear binding force is transmitted by the exchange of massive charged particles (the heavy quanta).
(b) The range of the force is inversely proportional to the mass of the quantum.
(c) There are *two* nuclear forces, one strong and one weak, and thus two coupling constants.
(d) The weak interaction is mediated by the exchange of the heavy quanta; that is, they serve as charged intermediate bosons for β-decay.
(e) As a result of assumption (d), the heavy quanta are unstable, decaying via the weak interaction.

When Yukawa won the Nobel Prize in Physics in 1949 "for his prediction of mesons on the basis of theoretical work on nuclear forces", thus becoming Japan's first Nobel Laureate, the whole nation rejoiced. Coming at the time when Japan was just beginning to emerge from the terrible destruction of the Second World War, this international recognition of one of her citizens provided a tremendous encouragement to the whole Japanese people, but especially to the scientists among them. By any measure, Yukawa ranks with the world's outstanding physicists of the twentieth century. The method that he introduced to treat the nuclear force problem, namely the exchange of massive quanta, has been used repeatedly to encompass an ever-widening

H. Yukawa (1906–1979) in 1939. Photograph reproduced by permission of Yukawa Hall Archival Library.

range of physical phenomena that were unimaginable when he first proposed the meson theory.

Yukawa occupies a position in Japanese physics that is central in several aspects. Before him there were Japanese (perhaps half a dozen) who had made significant contributions to physics and he had brilliant contemporaries, including his classmate, Sin-itiro Tomonaga (a future Nobel Prize winner). Both Yukawa and Tomonaga had as their mentor Yoshio Nishina, who had graduated from Tokyo Imperial University in 1916. Yukawa and his students and collaborators, including Shoichi Sakata, Mituo Taketani and Minoru Kobayashi, all graduates of Kyoto Imperial University, formed a major school of theoretical elementary particle physics in Kyoto and Osaka, matched in Japan only by that of Nishina and Tomonaga in Tokyo [6].

5.2 Yukawa takes up the problem of nuclear forces (up to 1933)

Yukawa was born as Hideki Ogawa in Tokyo on 23 January 1907, the fifth child and third son of Takuji and Koyuki Ogawa, who would eventually have two more sons. His father, Takuji, was called from the Bureau of Geological Survey in Tokyo in 1908 to be Professor of Geography at Kyoto Imperial University, and it was in Kyoto that Hideki grew up and went to school. Although at first not especially interested in physics and mathematics, but rather in history and literature, in the later high school years he began to read books on atomic problems in English and German [7]. In 1926 he entered the programme in physics at Kyoto Imperial University, together with his high-school classmate Tomonaga.

Soon afterwards, the two young students learned from a lecture by Hantaro Nagaoka on "The Past and Present of Physics" about the quantum-mechanical revolution in Europe that had begun with Heisenberg's seminal paper of 1925. They read Max Born's *Mechanics of the Atom* (of 1925, in German) and they studied the papers of Schrödinger in the journals as they arrived. In his third (and final) university year, Yukawa did experimental research in spectroscopy under the direction of Masamichi Kimura, but then decided to do his final thesis on Dirac's relativistic electron theory.

There were no courses on the new quantum mechanics, but occasional lectures were given by well-known visiting European physicists, such as Arnold Sommerfeld in 1928. Dirac and Heisenberg, invited by Nishina, gave eight lectures in Tokyo in 1929 (attended by Tomonaga, but not by Yukawa); later, the two Europeans lectured in Kyoto (Brown and Rechenberg 1987). Nishina also gave lectures occasionally in Kyoto, which had a strong influence on Yukawa, because they "were not only explanations of quantum mechanics, for he carried with him the spirit of Copenhagen, the spirit of that leading group of theoretical physicists with Niels Bohr at its centre" [8].

Yukawa graduated from Kyoto Imperial University in 1929, the first year of the Great Depression, an almost impossible time to find a job. He therefore continued to live with his family in Kyoto. For three years, together with Tomonaga, Yukawa worked as an unpaid assistant in the theoretical physics research room of Professor Kajuro Tamaki. Intensely ambitious, he was afraid that he had arrived upon the physics scene too late to be able to make a fundamental contribution to quantum theory. Much later, Yukawa wrote in his autobiography: "As I was desperately trying to reach the front line, the new quantum physics kept moving forward at a great pace." [9]

After his graduation, Yukawa considered two outstanding problems facing physics: the structure of the nucleus and the questions raised by Dirac's relativistic electron theory of 1928, especially the puzzle of its negative energy states (for which there appeared to be no meaning according to the theory of relativity). With these problems in mind, he began to look into hyperfine structure, a small but observable splitting of atomic spectral lines. This effect was believed to be due to the action of the magnetic field of the nucleus on the extra-nuclear atomic electrons, the main influence being on those electrons which penetrated nearest to the nucleus and thus experienced the strongest part of the magnetic field. At the same time, because of the strong attractive electric field they would experience near the nucleus, those electrons would attain velocities near that of light. Thus the problems associated with electrons in or near the nucleus and the problems of Dirac's electron theory would be present at the same time. Yukawa hoped they would shed light upon each other.

As it happened, Enrico Fermi in Rome was thinking along the same lines [10]. He was obviously a more experienced theorist; also, he did not have to convince a mentor, as Yukawa did Professor Tamaki. When Yukawa handed the latter his manuscript, he put it in a drawer and said that he would study it later. Meanwhile Fermi published his treatment of the problem (Fermi 1930a,b), which discouraged Yukawa from pursuing it any further. Instead, he began to study a major two-part paper on quantum electrodynamics (Heisenberg and Pauli 1929, 1930), the first to quantize all four components of the electromagnetic field, in interaction with Dirac electrons. Yukawa later referred to this paper as a partial "settling of accounts" in the balance sheet of the quantum theory [11].

The account that needed settling was the wave–particle duality. The quantum of action had been introduced into physics by Planck in 1900 in his treatment of the spectrum of black-body radiation. In 1905, Einstein showed that Planck's theory implied that light and other electromagnetic radiations, typically wavelike, had a microstructure that was particle-like. Since waves and particles are conceptual opposites, Einstein's quantum hypothesis gave rise to what was called the wave–particle paradox. After Bohr's successful treatment of the hydrogen spectrum in 1913, Planck's quantum of action became recognized as setting the scale of quantum effects in matter as well as radiation [12]. In the 1920s, the mystery was only deepened by Louis de Broglie's conjecture that electrons have a wave-like nature in addition to their particle properties, and by the experimental confirmation of these "matter waves" [13]. By 1929, after the new quantum mechanics of Heisenberg and Schrödinger had resulted in a broad understanding of atomic phenomena, with the wave–particle duality incorporated in the Copenhagen interpretation, Yukawa felt that the time had come to make a fresh attack on resolving the wave–particle paradox.

Dirac had already in 1927 introduced methods that took into account the quantum nature of the electromagnetic field in calculating, for example, the spontaneous decay rates of atomic states (determining the intensity of the lines in atomic spectra). However, the 1929 paper by Heisenberg and Pauli provided the first *fully relativistic* quantum theory of the electromagnetic field. Whereas Dirac had quantized only the transverse degrees of freedom of the field of free radiation, but left the static Coulomb potential in its classical form, Heisenberg and Pauli found a way to quantize both transverse and longitudinal components of the electromagnetic vector potential, as required in a fundamental relativistic theory. However, they were discouraged to find that the theory predicted infinite electron mass and other infinities that contradicted observation. Tracing the source of those infinities and curing them became a major task of theoretical physics in the 1930s, a task that was accomplished (at least provisionally) only in the 1940s by the so-called renormalization programme, in which Yukawa's former classmate Tomonaga played a leading role.

Disappointed in losing his race with Fermi on the hyperfine structure problem, Yukawa applied his energy fully to quantum field theory, hoping to complete the partial "settling of accounts" that Heisenberg and Pauli had begun. In his autobiography he vividly describes the frustration of his attack on the infinities of quantum electrodynamics [14]:

> Each day I would destroy the ideas that I had created that day. By the time I crossed the Kamo River on my way home in the evening, I was in a state of desperation. Even the mountains of Kyoto, which usually consoled me, were melancholy in the evening sun... Finally, I gave up that demon hunting and began to think that I should search for an easier problem.

These demons were to haunt Yukawa throughout the rest of his career. Even after the renormalization theory of QED in the 1940s, he continued to regard the forging of a finite quantum field theory as the main problem of theoretical physics.

In 1932, however, Yukawa resolved to think of other matters. He saw Chadwick's announcement of the neutron and soon afterwards read part I of Heisenberg's paper "On the Structure of Atomic Nuclei" (Heisenberg 1932). This re-kindled his interest in the nuclear force problem, which he still thought (as in his earlier study of hyperfine structure) was closely related to the problems of relativistic field theory. An unpublished document of early 1933 by Yukawa is entitled "On the Problem of Nuclear Electrons. I". It is in English and begins: "The problems of the atomic nucleus, especially the problems of nuclear electrons, are so intimately related with the problems of the relativistic formulation of quantum mechanics that when they are solved, if they ever be solved at all, they will be solved together." [15]

Recognizing the importance of Heisenberg's contribution, he prepared a summary in Japanese of the first two parts of Heisenberg's paper, shortly after they became available. This became his first publication (Yukawa 1933a) [16]. In the introduction to his account, Yukawa discussed the strengths and weaknesses of Heisenberg's model. Placing particular stress on issues of principle, he wrote [17]:

> In this paper Heisenberg ignored the difficult problems of electrons within the nucleus and, under the assumption that all nuclei consist of protons and neutrons only, considered what conclusions can be drawn from the present quantum mechanics. This essentially means that he transferred the problem of the electrons in the nucleus to the problem of the make-up of the neutron itself, but it is also true that the limit to which the present quantum mechanics can be applied to the atomic nucleus is widened by this approach. Though Heisenberg does not present a definite view on whether neutrons should be seen as separate entities or as a combination of a proton and an electron, this problem, like the β-decay problem... cannot be resolved with today's theory. And unless these problems are resolved, one cannot say whether the view that electrons have no independent existence in the nucleus is correct.

For the next year and a half, Yukawa pondered these problems, until he came up with an original solution.

5.3 A new fundamental theory of nuclear forces (1933–34)

During 1933, Yukawa argued (in several unpublished documents) that Heisenberg's programme could be carried one step further towards a fundamental theory of nuclear forces and β-decay. To do so, it was necessary to treat the neutron as a truly elementary particle because, he said, considering its structure would lie "outside the applicability of present quantum mechanics" [18]. He attempted to formulate the Heisenberg charge exchange force in analogy to QED. The exchange of an electron between a neutron and a proton would produce the nuclear force, just as photon exchange produces the electromagnetic force in QED; in contrast, Heisenberg used the analogy of molecular exchange forces.

In classical electrodynamics, the electric and magnetic fields can be derived from a "vector potential" A_μ ($\mu = 0,\ 1,\ 2,\ 3$), a relativistic four-vector obeying the wave equation

$$\Box A_\mu = j_\mu \tag{5.1}$$

Here, $\Box$ is the d'Alembertian operator, and j_μ is the four-vector charge-current density, regarded as the "source" of A_μ. In a region free of charges and currents, the right-hand side of equation (5.1) is zero and the equation describes the free electromagnetic field. Equation (5.1) holds also in QED, with A_μ the *quantized* electromagnetic field and j_μ the appropriate *quantized* charge-current density (for example, that of the Dirac electron–positron field).

Yukawa now started from the Dirac relativistic electron equation, which he treated as the "free-field equation" of the "classical electron field"

$$\mathbf{D}\psi = 0 \tag{5.2}$$

where $\mathbf{D}$ is the 4×4 matrix Dirac differential operator (including the electromagnetic potentials) and ψ is the four-component Dirac relativistic spinor wave function of the electron. He then modified this equation by introducing on its right-hand side a source term $\mathbf{J}$, depending on the neutron and proton (regarded as the neutral and charged states, respectively, of the nucleon):

$$\mathbf{D}\psi = \mathbf{J}. \tag{5.3}$$

$\mathbf{J}$ in equation (5.3) has a form somewhat analogous to the Dirac j_μ, but it is significantly more complicated. Yukawa constructed it out of an eight-component nucleon spinor and its adjoint; in $\mathbf{J}$ appear Dirac matrices acting upon the spinor, as well as isospin matrices that change a neutron into a proton, or vice versa. j_μ (in equation (5.1)) transforms as a relativistic four-vector, the "current" J must transform as a relativistic spinor (like ψ). That was an essentially new and possibly problematic aspect, as Yukawa realized. In QED, the same j_μ that is responsible for the exchange of photons between charges, accounting for electromagnetic forces, acts also as the source of free photons. However, if the nuclear current J (producing an electron) were to account for β-decay as well as nuclear forces, then β-decay would fail to conserve energy and angular momentum, as in any theory in which the electron emerges unaccompanied from the nucleus [19]. That was also true in Heisenberg's theory and, indeed, Yukawa was aiming to construct a fundamental theory that would justify Heisenberg's phenomenological approach.

Equation (5.2) is the usual Dirac equation, whose Green function solution was known. Yukawa set out to construct the analogous solution of equation (5.3), namely one depending upon J. He found the following form for the Green function: a Dirac operator acts upon an integral that for vanishing electron mass is entirely analogous to the retarded potential due to the source J [20]. Explicitly, for zero electron mass, the integral is

$$\int \mathrm{d}r' J(r,\ t - |r' - r|/c)/|r' - r|. \tag{5.4}$$

Here r' is the source point, r the field point, t the time and c the velocity of light. However, for an electron of mass m, there is an additional exponential factor in the integral, with an oscillating character.

In his manuscript, Yukawa stated:

> Substituting this solution into [the Hamiltonian] H, we find the interaction energy that corresponds to Heisenberg's "*Platzwechsel*" interaction. Its exact form and the retardation effect can be seen, and a factor
>
> $$\exp(i\rho_3 mc|r' - r|/\hbar) \tag{5.5}$$
>
> [ρ_3 is a Dirac matrix] appears as a kind of phase factor, so that the term corresponding to Heisenberg's [*Platzwechselintegral*] $J(r)$ has a form like the Coulomb field and does not decrease sufficiently with distance.

Yukawa referred here to the most striking difference between the nuclear force and either electromagnetism or gravitation, namely its short range of influence. Although the small length $\hbar/(mc)$ (the electron's Compton wavelength divided by 2π) appears in the exponent in equation (5.5) and sets the scale of oscillation of the phase factor, it does not limit the range of the force.

However, in preparing the quantized field version of the theory based on equation (5.4), Yukawa thought for a time that he had the kind of solution he was seeking. On 3 April 1933, he gave his first talk to a meeting of the Physico-Mathematical Society of Japan; in the published abstract, he claimed to have obtained the nuclear charge exchange force of range $\hbar/(mc)$ (Yukawa 1933b). The mass m is that of the electron, giving a range still about 200 times too large. Nevertheless, had the result been mathematically correct, it would have been highly suggestive, for a heavier "electron" would have given a shorter range. Yukawa later recalled that, when he gave the talk at Sendai, Yoshio Nishina actually proposed to him that he introduce a "Bose electron". However, in the unpublished manuscript that he read at Sendai, he withdrew the promising result, saying:

> The practical calculation does not yield the sought result that the interaction term decreases rapidly as the distance becomes larger than $\hbar/(mc)$, unlike what I wrote in the abstract of this talk.

On the blank reverse side of the previous page is written, "mistaken conjecture" [21].

Yukawa was trying to derive Heisenberg's nuclear theory from deeper principles. However, that theory contained the questionable feature that the emission of an electron by a neutron violated the conservation of angular momentum, whether the neutron was treated as fundamental or composite. It should be noted that the young Japanese physicist was unaware of Pauli's neutrino proposal or Fermi's β-decay theory until, early in 1934, one of his colleagues at Osaka University called his attention to Fermi's publication (Fermi 1934b).

At first Yukawa thought that Fermi had once again beaten him to the punch (as he had in the case of the hyperfine structure of spectral lines), for Yukawa had assumed with Heisenberg that the β-decay interaction was also responsible for the strong nuclear force. The success of Fermi's β-decay theory gave rise to the suggestion (as we discussed in chapter 3) that Heisenberg's charge-exchange nuclear force might be carried by the exchange of a *pair* of particles, namely an electron and a neutrino, rather than by a single electron. In this way it would preserve all conservation laws. Yukawa independently arrived at the same idea and began calculations along this line (as did Heisenberg and others). Before he finished, the first published results using this approach were given in independent letters to *Nature* by two physicists from the USSR, Igor Tamm and Dmitri Iwanenko (Tamm 1934; Iwanenko 1934). Both authors concluded that the Fermi-field, as the electron–neutrino exchange was called, could explain *either* the short range *or* the large strength of the nuclear binding force, but not both properties simultaneously.

When Yukawa read the results of Tamm and Iwanenko in the fall of 1934, his reasoning and intuition came together, and, as he recalled in his autobiography [22]:

> I was heartened by the negative result [of Tamm and Iwanenko] and it opened my eyes, so that I thought: let me not look for the particle that belongs to the field of nuclear force among the known particles, including the new neutrino... When I began to think in this manner, I had almost reached my goal... The crucial point came to me one night in October... My new insight was that [the range of the force] and the mass of the new particle that I was seeking are inversely related to each other. Why had I not noticed that before? The next morning I tackled the problem of the mass of the new particle and found it to be about two hundred times that of the electron.

During the weeks that followed this discovery, Yukawa became very busy in implementing it, and his progress is documented by sets of calculations, drafts of talks and drafts of articles. These are deposited in the Yukawa Hall Archival Library in Kyoto and we are fortunate that they are generally dated by the month, some even by the day [23]. The contents reveal the topics that the young Osaka theorist—he had become a lecturer at the newly established Science Faculty at Osaka University in 1933—treated in order to test his new ideas on the nuclear forces.

The first four sets of calculations dealt with neutron–proton collisions; the next two sets compute the binding energy of the deuteron. Then, in the following two sets, he solved the time-independent wave equation obeyed by the postulated new particle with mass m. These calculations, dated October 1934, were probably carried out before Yukawa composed his early drafts of talks and papers. He wrote the earliest of these drafts (dated 27 October) in Japanese, unlike the later ones, and it may contain the contents of Yukawa's first presentation of the theory in Osaka [24].

The draft of the talk began with the application of the Fermi-field theory to neutron–proton interactions. Pointing out that the result is much too small, the author then introduced his new heavy quantum field, in analogy to the electromagnetic field, consisting of a four-vector with scalar and vector potentials, respectively U and B. He then wrote the static n–p interaction, using only the scalar part (corresponding to the Coulomb interaction, see equation (5.6) below). In general, U obeys the time-dependent equation with a nucleon source term (see equation (5.10) below). Finally, he considered β-decay and showed that it followed from coupling the U-field to a different source, consisting of an electron–neutrino pair.

The Osaka lecture contained all the features of the theory of nuclear forces that Yukawa presented at the Regular Meeting of the Physico-Mathematical Society in Tokyo on 17 November 1934. The programme of the meeting listed seven topics, from "Meromorphic Functions" to "The Polarity of Thunder Clouds", the last one being "On the Interaction of Elementary Particles". Yukawa was allotted ten minutes to present his talk, whereas the "thunder clouds" got forty minutes [25]. However, he wrote a full improved version in English and sent it to the *Proceedings of PMSJ*, where it appeared in the January–February issue of 1935 (Yukawa 1935) [26].

5.4 The U-quantum and the cosmic-ray "mesotron" (1934)

Yukawa's meson theory article, entitled "On the Interaction of Elementary Particles I", aimed at nothing less than a quantum field theory that would account in a unified way for the nuclear binding force and the nuclear β-decay interaction. That is, he set out, in the light of the insight gained by his reading the analysis of Tamm and Iwanenko, to modify the theories both of Heisenberg and of Fermi so that they could be combined. The model for such a theory was obviously and explicitly QED; following that model closely had important heuristic and pedagogical advantages. Yukawa's paper appears transparent and appealing to us now, and it remains something of a puzzle why it was entirely ignored for two years both in Japan and in the West. His work began to be noticed only in 1937, when a particle whose mass closely fitted the requirements of meson theory was detected in the cosmic rays (see chapter 6 below).

In the introduction to his paper, Yukawa mentioned the charge-exchange theory of Heisenberg and the unsuccessful attempts of the Russian theorists, Tamm and Iwanenko, to obtain the nuclear force from the Fermi interaction (see chapter 3), which resulted in an "interaction energy . . . much too small to account for the binding energies of neutrons and protons in the nucleus" (p 48) [27]. Therefore, he went on (p 48):

> To remove this defect, it seems natural to modify the theory of Heisenberg and Fermi in the following way. The transition of a heavy particle from neutron state to proton state is not always accompanied by the emission of light particles, i.e., a neutrino and an electron, but the energy liberated by the transition is taken up sometimes by another heavy particle, which in turn will be transformed from proton state into neutron state. If the probability of occurrence of the latter process is much larger than that of the former, the interaction between the neutron and the proton will be much larger than in the case of Fermi, whereas the probability of emission of light particles is not affected essentially.

Yukawa then remarked that the "interaction between the elementary particles can be described by a field of force, just as the interaction between the charged particles is described by the electromagnetic field". Furthermore (p 49): "In the quantum theory, this [nuclear force] field should be accompanied by a new quantum, just as the electromagnetic field is accompanied by the photon." Thus Yukawa exploited the electromagnetic analogy from the outset. Beginning with the classical theory, he generalized d'Alembert's equation for the scalar potential of electromagnetism, having the well-known spherically symmetric static solution $1/r$ for a point source. He pointed out that a potential of finite range $1/\lambda$, namely

$$U(r) = \pm(g^2/r) \exp(-\lambda r) \tag{5.6}$$

is a spherically symmetric static solution of the generalized wave equation

$$(\Box - \lambda^2)U = 0, \tag{5.7}$$

which he assumed to be the "correct equation for U in vacuum".

However: "In the presence of the heavy particles, the U-field interacts with them and causes the transition from neutron state to proton state". Yukawa then (p 50) introduced a source term J' on the right-hand side of this equation, something like the J in equation (5.3) above. This time the particle to be created has the transformation character of U, like that of the electromagnetic scalar potential, not a spinor. (We shall refer to Yukawa's U-particle, which has no spin, as a "scalar" particle, even though it does not transform as a Lorentz scalar.)

The new source term, J', was constructed of nucleon spinors in such a way that the emission of a negatively charged meson corresponded to the transition *neutron* to *proton* (n→p), whereas the emission of a positively charged meson corresponded to the transition *proton* to *neutron* (p→n). Explicitly, Yukawa used the Heisenberg isospin matrices τ_1, τ_2 and τ_3, which are 2 × 2 matrices formally identical to the Pauli spin matrices σ_1, σ_2 and σ_3, to write

$$J' = -4\pi g \tilde{\Psi} \frac{\tau_1 - i\tau_2}{2} \Psi, \tag{5.8}$$

where Ψ denotes the wavefunction of the heavy particles, being a function of time and spin, as well as of τ_3', which takes the value either 1 or -1. Evidently the U-field must carry off positive charge for the p→n transition. In the inverse n→p transition, the field must carry off negative charge, represented by the complex-conjugate of the U-field, namely the $\tilde{U}$-field, which obeys the complex-conjugate wave equation.

At this point Yukawa noted that the U-field analogue of the electromagnetic *vector* potential would be disregarded in the rest of the paper, since "there's no correct relativistic theory for the heavy particles" (p 50). He then carried out a calculation using "the simple non-relativistic wave equation neglecting spin", i.e., the Schrödinger equation, to obtain the effective potential of interaction between two heavy particles and thus the effective Hamiltonian for the system. What he found was that (p 51):

> This Hamiltonian is equivalent to Heisenberg's Hamiltonian... if we take for the "*Platzwechselintegral*" $J(r) = -g^2 e^{-\lambda r}$, except that the interaction between the neutrons and the electrostatic repulsion between the protons are not taken into account.

Unlike the case in Heisenberg's paper (1932), the negative sign was here chosen for the "*Platzwechselintegral*", so that the spin of the deuteron would be 1, in agreement with experiment. (This "choice" was later found to be inconsistent with the original type of meson theory, leading subsequently to the consideration of other types.) The new parameter g, having the dimension of electric charge, and the new inverse length λ, were to be "determined by comparison with experiment" (p 52).

Yukawa next considered the nature of the U-quanta, to be obtained by quantizing the classical U-field "on the line similar to that of the electromagnetic field" (p 52). Making the usual identifications

$$p_x = -i\hbar \partial/\partial x, \text{ etc.,} \qquad W = i\hbar \partial/\partial t,$$

and letting $m_U c = \lambda \hbar$, the free-field equation, equation (5.7), became (pp 52–3, our emphasis):

$$(p_x^2 + p_y^2 + p_z^2 - W^2/c^2 + m_U c^2)U = 0, \tag{5.9}$$

so that the quantum accompanying the field has the proper mass $m_U = \lambda \hbar/c$. Assuming $\lambda = 5 \times 10^{12}$ cm^{-1}, we obtain for m_U a value 2×10^2 times as large

DEPARTMENT OF PHYSICS
OSAKA IMPERIAL UNIVERSITY.

DATE

NO. 9

field. According to the From the The law of conservation of the electric charge demands that the quantum should have + of — e the charge $+e$ or $-e$. The quantized field U corresponds to the operator which increases the number of the negatively charged quanta by one and decreases the number of the positively charged quanta by one. The field $\bar{U}$, the complex conjugate of U, which does not commute with U, corresponds to the inverse operator.

Next, denoting

$$p_x = -i\hbar \frac{\partial}{\partial x} \text{ , etc } \quad W = i\hbar \frac{\partial}{\partial t} \text{ ,}$$

$$m_U c = \lambda \hbar \text{ ,}$$

The wave equation for U in free space can be written in the form

$$\left\{ p_x^2 + p_y^2 + p_z^2 - \frac{W^2}{c^2} + m_U^2 c^2 \right\} U = 0 , \quad (12)$$

so that the quantum accompanying the field has the proper mass $m_U = \frac{\lambda \hbar}{c}$. Assuming, for example, $\lambda = 5 \times 10^{12} cm^{-1}$ we obtain for m_U a value 2×10^2 times as large as electron mass. Thus the result is rather surprising and the existence of such quantum with large mass and positive or negative charge has never

H. Yukawa: manuscript of the first paper showing the mass-range formula (November 1934). Reproduced by permission of Yukawa Hall Archival Library.

as the electron mass. *As such a quantum with large mass and positive or negative charge has never been found by the experiment, the above theory seems to be on a wrong line.*

However, Yukawa then showed that "in the ordinary nuclear transformation" the meson cannot be produced. He did this by considering the equation determining the production of the meson field, namely

$$(\Box - \lambda^2)U = J', \quad (5.10)$$

with J' given by equation (5.8). Yukawa sought the solution analogous to that given in equation (5.4) and he obtained a similar result — the important difference being that the exponential factor, which appeared when the mass m of the U-particle was non-zero, had the form

$$\exp(-\mu|r' - r|), \tag{5.11}$$

with $\mu = (\lambda^2 - \omega^2/c^2)^{1/2}$, $\hbar\omega = W_N - W_P$ being the difference in the energies of the neutron and proton before and after the emission of a U-particle, including their rest mass energies. Thus, provided that the energy difference is less than $\lambda c\hbar$, the exponential is a decreasing function of $|r' - r|$, instead of the oscillating one found earlier in connection with the electron emission theory. Upon quantizing the U-field, Yukawa showed immediately that the quantity λ is $mc/\hbar$ so that the range of the force is inversely proportional to the mass of the U-quantum. That is the insight that crystallized the meson theory in Yukawa's mind on that sleepless October night in 1934! While the range–mass relation is so commonplace today as to seem self-evident to most physicists, that was not the case in the 1930s. Four years later, the well-known Italian physicist Gian Carlo Wick took the trouble to present an *anschaulich* explanation of Yukawa's result (Wick 1938).

Yukawa also assumed that the U-field (or equivalently, the U-quantum) was coupled to an additional charge-changing current, constructed analogously to J', but with the electron and neutrino replacing the proton and neutron. Thus he made the meson an "intermediate boson" carrying the weak as well as the strong interaction. The very notions of strong and weak nuclear interaction were not part of the general thinking before Yukawa constructed his meson theory. On the contrary, as we have seen, Heisenberg tried to make the β-decay and the nuclear charge exchange forces the same, characterizing them both by a single coupling strength. After Fermi's β-decay paper appeared, the same single-coupling idea was carried over into the so-called Fermi-field approach to nuclear forces. However, Yukawa introduced a large constant g and a small one g', so that the decay amplitude of a neutron, say, into a proton, an electron and a neutrino was proportional to g'. If one substituted for Fermi's coupling constant the quantity $g_F = (4\pi/\lambda^2)gg'$, as Yukawa proved, his theory of nuclear β-decay was effectively equivalent to Fermi's.

There was, however, an important additional consequence of Yukawa's assumption that the meson had a β-decay interaction. It implied that the meson itself would decay radioactively if it were produced in a free state (e.g., outside the nucleus). Curiously, this fact was ignored by Yukawa and his associates until 1938, when that important prediction of the theory was pointed out by the Indian physicist Homi J. Bhabha (1938a). The short mean lifetime of the meson (about 10^{-8} s) is the main reason why such particles are not found in abundance on the earth.

The question of why the particles were not commonly observed was raised at Yukawa's first presentation of the theory, which took place at one of the informal luncheon meetings of the nuclear physics group at Osaka Imperial University, which Yukawa regularly attended. Yukawa responded that an energy sufficient to create a meson was required, i.e., at least its rest energy of about 100 MeV, and such energies were not available (in those days) in the nuclear physics laboratory. The only terrestrial furnace in which such a particle could be forged was the atmospheric cosmic rays; hence Yukawa suggested to his experimentalist colleagues at Osaka that they should look for *U*-quanta in cosmic-ray cloud chamber photographs.

5.5 The meaning of the meson

One of Yukawa's first students wrote that the meson theory "opened up a new fundamental view of Nature" and that "this event might be considered a miracle in the history of Japanese physics. Through all of his works and thoughts, we are impressed by the simplicity of approach, the unfailing intuition and the creativity of a great master, which are deep-rooted in Yukawa's culture." [28] Here the meson theory is said to enlarge the class of elementary particles, providing a new view of nature; it is also called a miracle for Japanese physics, in which native cultural traits may have played a significant role. There may be other "meanings" for the meson, but we shall deal with the subject mainly under these rubrics.

In 1974 Yukawa described the natural philosophy that prevailed in 1929 when he graduated from Kyoto University [29]:

At this period the atomic nucleus was inconsistency itself, quite inexplicable. And why?—because our concept of elementary particle was too narrow. There was no such word in Japanese and we used the English word—it meant proton and electron. From somewhere had come a divine message forbidding us to think about any other particle. To think outside these limits (except for the photon) was to be arrogant, not to fear the wrath of the gods. It was because the concept that matter continues forever had been a tradition since the times of Democritus and Epicurus. To think about the creation of particles other than photons was suspect and there was a strong inhibition of such thoughts that was almost unconscious.

The unwillingness to consider the possibility that there might be more elementary constituent particles than two was sometimes explicitly stated [30]. When Pauli proposed the neutrino, he waited three years to publish his idea—and did so only as a discussion remark. When Dirac was "led to a new kind of particle caused by a hole in the distribution of negative energy states" and the theory predicted that the particle would have positive electronic charge, but electronic mass, he nevertheless called it the *proton* [31]:

I just didn't dare to postulate a new particle at that stage, because the whole climate of opinion was against new particles. So I thought that this hole would have to be a proton. I was well aware that there was an enormous mass difference between the proton and the electron, but I thought that in some way the Coulomb force between the electrons in the sea might lead to the appearance of a different rest mass for the proton. So I published my paper on this subject as a theory of electrons and protons.

In chapter 1, we discussed how physicists had been conditioned, since at least the early part of the century, to look for a breakdown of electrodynamics at H.A. Lorentz's "classical electron radius" of $e^2/(mc^2)$, or about 10^{-13} cm. Later, forewarned by the revolutionary new quantum mechanics of the 1920s, they were prepared to find the explanation of the new phenomena of nuclear physics in a breakdown of existing dynamics, rather than in the postulation of new particles and/or new forces. Bohr's insistence on the failure of energy conservation, or rather its statistical character, represents a notable example of this tendency. In the same spirit, Heisenberg maintained his belief in a small universal length, even after the new cosmic ray particles had been announced by Anderson and Neddermeyer, writing about the meson theory to Pauli in 1938 [32]:

> It was interesting to me, however, that Bhabha as well as Heitler also obtain [in the expression for] the force between neutron and proton a term of the form $\delta(|r_1 - r_2|) \ldots$ [i.e., an infinite contact force]. This result is very agreeable [*sehr sympatisch*] to me, because it shows again that one cannot make further progress without the universal length.

Yoichiro Nambu has characterized Yukawa's theory as the principal paradigm of the theory of elementary particles, just as the cyclotron of E.O. Lawrence is paradigmatic for its experimental side. For Yukawa's prophetically entitled paper, "On the interaction of elementary particles. I" opens the door to a world of high-energy processes involving the creation and annihilation of new and in many cases ephemeral substances (mesons, leptons, strange and charmed particles, quarks, gluons, intermediate vector bosons, etc.), a world of astonishing variety and novelty. Until the present, few of the major problems faced by high-energy physics have *not* been solved by the introduction of new particles. Having arrived at a new Standard Model of elementary particle interactions, it is not impossible that we have by now taken this lesson too much to heart!

Notes to text

[1] Nakayama (1977, p 4). See also Nakayama (1984), Nakayama *et al.* (1974), Bartholomew (1989) and Watanabe (1990).
[2] See also Koizumi (1975).
[3] Riken was a private institution, founded with government support in 1917, devoted to pure and applied research in physical and chemical science, and the development of industries based upon this research (Itakura and Yagi 1974).
[4] See Hirosige (1974) and Takeda and Yamaguchi (1982).
[5] These papers have been organized and catalogued in the Yukawa Hall Archival Library [YHAL].

[6]   We are indebted to the members of the Japan–USA Collaboration for the Study
      of the History of Particle Theory in Japan for much information and other help.
      The authors have spent much time with the Yukawa archive in Kyoto, with the
      assistance of Rokuo Kawabe and Michiji Konuma. For more on Yukawa and
      his school, see Brown *et al.* (1991). For Yukawa, see Brown (1981, 1985, 1986,
      1989, 1990), Kemmer (1983b) and Darrigol (1988), for Tomanaga see Matsui
      and Ezawa (1995).

[7]   Yukawa (1982) mentions Fritz Reiche's *The Quantum Theory* (translated from
      German to English) and Max Planck's course of several volumes *Introduction to
      Theoretical Physics* in the original German.

[8]   Yukawa (1982, p 177).

[9]   Yukawa (1982, p 166).

[10] Emilio Segrè described Fermi's motivation as follows: "There had always been
      a strong spectroscopic tradition [in Italy]. For some time... it had been felt by
      Fermi that physicists would be ready in the near future to attack the problem of
      nuclear structure... The study of hyperfine structure was a very natural bridge
      between the two domains." (Segrè, in Fermi (1962, p 328))

[11] Yukawa (1982, p 173).

[12] The same could be said, of course, for the somewhat earlier quantum treatments
      of the specific heats by Einstein and others.

[13] For the history of quantum theory see, e.g., Jammer (1966) and Mehra and
      Rechenberg (1982, 1987).

[14] Yukawa (1982, p 174).

[15] H. Yukawa, "On the problem of nuclear electrons. I" [YHAL] E05 030 U1.

[16] His first *research* publication was Yukawa (1935).

[17] For a complete translation of Yukawa's note, see Brown (1981, pp 121–2).

[18] [YHAL] E05 060 U01. We have discussed this problematic feature of
      Heisenberg's theory of nuclear forces in chapters 2 and 3. Yukawa's
      manuscript (in Japanese), from which the following discussion is taken, is
      entitled, "The role of the electron for nuclear structure".

[19] See chapters 2 and 3.

[20] The Dirac operator is $(4\pi/\hbar^2)^{-1}(p_0/c - \rho_1\sigma \cdot p - \rho_3 mc)$, using Dirac's notation
      for the matrix vectors σ and ρ.

[21] [YHAL] E05 080 U01; English translation in Brown *et al.* (1991, pp 248–9).

[22] Yukawa (1982, pp 201-2).

[23] Yukawa kept these manuscripts in an envelope, together with the typescript of
      the first meson paper, and they are now labelled as follows: , E01010P01:CL,
      E01020P01L, E01030P01:CL, E01040P01:CL, E01050P01:CL, E01060P01:CL,
      E01070P01:CL, E01080P01:MT, E01092P01:MT, E01100P01:MP and
      E0111P01:MP. CL for calculations, MT for talk manuscript and MP for
      paper manuscript.

[24] Yukawa's notes (mixed Japanese and English), entitled "On the interaction of
      elementary particles" are in [YHAL] E01092P01:MT, 11(+1) pages. Although
      Yukawa claimed in *Tabibito* that he gave the talk at "a regular meeting of the
      Osaka branch [of the Physico-Mathematical Society of Japan]", the branch was
      started only in June 1935 (Kawabe 1991a).

[25] Programme of Regular Meeting of PMSJ, 1:30 pm, 17 November, 1934 (in
      Japanese, except for some titles). [YHAL] E01090P01.

[26] [YHAL] has two manuscripts, E01100P01:MP (14 pp, dated 1 November 1934)
      and E01110P01:MP (18 pp, dated 1 November, but crossed out), plus a
      fragment, E01120P01 (8 pp, in English). It has also a typescript of the published
      article, E01130P01.

[27] See p 48. In this part of the chapter we will cite page numbers from Yukawa
      (1935).

[28] Yukawa (1979, editor's introduction, p vii). See also Takabayasi (1983).

[29] H. Yukawa, *Shizen*, July 1975, pp 28–39, *Special Issue on Forty Years of Meson Theory*. Our translation is by Y. Yoshida (unpublished).

[30] See, e.g., *Exploring the History of Nuclear Physics* (American Institute of Physics, New York, 1972), edited by C. Weiner, pp 170–2.

[31] P.A.M. Dirac, "The prediction of antimatter" The 1st H.R. Crane Lecture, 17 April, 1978 at the University of Michigan, Ann Arbor, Michigan.

[32] Heisenberg to Pauli, 5 April 1938. [WPSC], pp 563–4.

Chapter 6

The Discovery of the Mesotron
(1935–37)

6.1 Introduction

One of the main predictions of Yukawa's paper "On the interactions of elementary particles. I", published in February 1935, was the existence of charged particles having mass about 200 times that of the electron. He pointed out that to produce particles of this large mass would require energy much larger than was achievable in the nuclear laboratory (at that time) and that it would be possible only in cosmic-ray interactions. In that first paper he did not explicitly state that, due to its postulated weak interaction with the electron and neutrino, the U-particle would be unstable, with a very short lifetime, which would also account for its not having been observed earlier [1]. Although Paul Kunze in Germany had already in 1933 identified a cosmic-ray track of a charged particle lighter than a proton and much heavier than an electron (Kunze 1933), it was not heeded until the existence of such particles had been definitively established by American cosmic-ray physicists in 1937.

Until then, practically no attention was paid to Yukawa's 1935 paper, even in Japan among the colleagues of the young Osaka theorist. Yukawa himself continued to work on the theory during the first part of 1935, but then turned to other problems (section 2). However, towards the end of 1936, he began to reformulate his theory of U-quanta, using a new relativistic quantum theory of spinless particles (Pauli and Weisskopf 1934), while working in parallel with a relativistic QFT of massive spin-1 particles that was similar to one proposed in Paris by the Romanian theorist Alexandru Proca (1936). At the same time, Yukawa noticed that some cosmic-ray cloud-chamber tracks observed by Carl Anderson and Seth Neddermeyer could be interpreted in terms of the U-quanta that he had proposed.

The experimental investigations of a possible "heavy electron" were extended in early 1937 and by the middle of the year three cosmic-ray groups agreed that such particles existed (section 3). The particles were referred to by various names, including "mesotron", the name that we shall adopt here provisionally [2]. It was tempting to seek a dynamical role for these new particles, so several Western theorists also called attention to Yukawa's 1935 publication, discussing a possible connection between them and the nuclear forces (section 4). However, there still remained some doubts about whether these new particles really existed (section 5).

On the positive side — inspired by the hope that the experiments would eventually confirm Yukawa's ideas — there began a competition — cooperation to develop the theory further between a group of Western theorists (mainly assembled in England) and Yukawa and his students in Japan, as we shall relate in the chapter 7 [3].

6.2 Yukawa's researches in 1935 and 1936

In his paper on U-quanta, Yukawa said that a fuller account of the theory would be "made in the next paper" [4]. Although he continued to work on it for a few months, the second paper in the series appeared only about two and a half years later (Yukawa and Sakata 1937c). In the interim, he published a number of other papers on electromagnetic and nuclear phenomena, in collaboration with his talented student, Shoichi Sakata. We shall comment on these papers after mentioning Yukawa's *unpublished* work on the U-quanta [5].

A file of "manuscripts of papers, 1935" in the Yukawa Hall Archival Library includes several "memoranda" related to a proposed article "On the interaction of elementary particles. II" as well as a talk with the same title, dated 30 March 1935 and read at the Annual Meeting of the PMSJ (Physico-Mathematical Society of Japan) on 4 April [6]. In a memorandum dated 19 March, Yukawa noted the following "defects" of the published part I (Yukawa 1935) [7]:

(1) Only the exchange force was considered.
(2) The forces between like particles were not considered.
(3) The spin-dependence was not considered.
(4) The range parameter λ and the coupling constant g determined from the collision theory and from cosmic ray bursts become rather large, so the mass of the U-quantum is large.
(5) The interaction of the charged U-quantum with the electromagnetic field was not investigated.

In 1932–33, Yukawa as a lecturer at Kyoto University had given a course of lectures on quantum mechanics. Two of his students, Sakata and Minoru Kobayashi, and Mituo Taketani, who took his course the following year, became his collaborators at Osaka University, to which Yukawa had moved in 1933 while continuing as a part-time lecturer at Kyoto. Meanwhile, Sakata spent the year after his graduation in 1933 at the Institute for Physical and Chemical Research (known by the acronym Riken) in Tokyo, working with Yukawa's former classmate Sin-itiro Tomonaga. He then joined Yukawa at Osaka as a Research Associate. Between summer 1935 and spring 1937 Yukawa published nine papers, seven of them joint publications with Sakata.

The idea for the first of these papers resulted from a discussion with the Austrian physicist Guido Beck, who paid a visit to Osaka in spring 1935 [8]. Sakata later recalled [9]:

> Beck gave a few lectures at Osaka about the scattering of neutrons by nuclei, and we dined with him at the New Osaka Hotel on one occasion. Yukawa told him about his meson hypothesis, but Beck did not at once accept it. But the ideas that Beck told us about β-radioactivity were very suggestive, and I think the later discovery by Yukawa and myself of electron capture by a nucleus owes very much to Beck's remarks.

In their first joint paper, Yukawa and Sakata (1935a) calculated the probability of radiationless electron–positron pair production in radioactive transitions between two nuclear S-levels, from which real γ-emission is forbidden, and compared it with published data on the positron emission from a radium source [10]. Sakata later recalled that [11]

> this process [of pair production] was very similar to the mechanism of β-radioactivity (in Beck's theory it was exactly the same), and when Yukawa and I were discussing this process we discovered the possibility of nuclear transition by means of the capture of an electron by the nucleus.

Sakata and Yukawa presented their paper at a meeting of the Osaka branch of the Physico-Mathematical Society of Japan (PMSJ) on 6 July 1935, where they also read a second paper, in which they considered two isobaric nuclear states with atomic charge numbers Z and $Z - 1$, and calculated the rate of transition by absorbing an atomic electron, a process that had been described previously by Fermi's student Gian Carlo Wick. This paper (Yukawa and Sakata 1935b) had two notable features. First, in applying the weak interaction theory of Fermi, the authors stressed its similarity to Dirac's hole theory, in contrast to Fermi, who had said that there was "no analogy to the production of an electron–positron pair" [12].

Secondly, the paper on β-decay was notable because it contained the first application of the U-quantum theory of nuclear forces after Yukawa's first paper (and the only published reference to it until 1937). While Wick, in estimating the probability of electron capture from an S-state (e.g., an atomic K-shell electron), had approximated the nucleus by a point, Yukawa and Sakata represented it by a field of force, carried both by the scalar and by the vector parts of the U-field. They assumed that the field vanished outside the (finite-sized) nucleus and was constant inside. (The transition occurs, for zero neutrino mass, when the state $Z - 1$ lies higher than the state Z by an amount less than the rest energy of the electron minus its binding energy in the atom.)

In February 1937, the authors submitted a letter to the *Physical Review*, calling attention to their K-capture paper (and also to a later supplement that used a modified Fermi theory due to E.J. Konopinski and G.E. Uhlenbeck), and noted: "It will be possible to test these conclusions by experiment." [13] Regarding the response to this work, Sakata later recalled (Sakata 1965):

Although this problem was a significant test of Fermi's theory of β-radioactivity, and although we announced our calculations about it in November 1935, it was ignored by the world's physicists for more than a year. Experimentalists showed an interest only after [Christian] Møller rediscovered our results in early 1937, and the effect was finally demonstrated by [Luis] Alvarez.

The published record, however, does not entirely support Sakata's recollection. While it is true that Alvarez was the first to demonstrate K-capture experimentally in 1938, the work of Sakata and Yukawa was cited much earlier by the Harvard experimentalists Kenneth T. Bainbridge and Edward B. Jordan, who refer to "a general theory of the stability of adjacent isobars" by the Japanese scientists, and who compare their results with mass spectroscopic data (Bainbridge and Jordan 1936). (This is probably the first reference to a work of Yukawa or Sakata, other than by Yukawa himself.) The Harvard paper was received on 12 June 1936; about a week later at the Seattle meeting of the American Physical Society, Willis E. Lamb Jr discussed some experiments by his colleague at Berkeley, L. Jackson Laslett. The latter had produced the positron emitter ^{22}Na by deuteron bombardment and observed a half-life of about 200 days. Lamb's abstract (Lamb 1936) stated [14]:

> According to the Fermi theory of β-decay, it can also decay by the alternative process of K electron capture and neutrino emission. We have extended the calculations of Yukawa and Sakata to the case of the Uhlenbeck–Konopinski coupling.

Møller repeated the same calculations in autumn 1936, sending a short account of his work to the *Physical Review* while he sent the main paper to be published in the Soviet Union (Møller 1937a,b). Although the Danish author cited neither Lamb nor Yukawa and Sakata, when Alvarez provided in mid-1938 "rigorous experimental proof of the hypothesis for the case of ^{67}Ga", he cited the pioneering theoretical work in the first sentence of his publication (Alvarez 1938):

> The suggestion that positron emitters might decay by the alternative process of electron capture was first advanced by Yukawa [citing Yukawa and Sakata (1935b, 1936)] from considerations based on the Fermi theory of β-decay.

Aside from the paper on K-capture and related phenomena, in which the hypothesis of the U-quantum played at least a modest role, according to a biographical sketch by one of Yukawa's students, Sakata and Yukawa were studying a variety of other topics [15]:

> During the period from 1935 to 1937, Yukawa worked vigorously in collaboration with Sakata on many topics in nuclear physics: the internal pair production, the K-electron capture, the nuclear transformations by a neutron, scattering of a neutron by a deuteron and the efficiency of the γ-ray counter. Some of these subjects were raised by [Seishi] Kikuchi's experimental group and these investigations belong to a special genre in Yukawa's work. Before those days, theoretical and experimental physicists in Japan were completely isolated from each other and there was no mutual stimulation or

Physics Department of Osaka Imperial University. Front row, middle: M. Kikuchi; back row, far left: H. Yukawa; far right: S. Sakata about 1935. Photograph reproduced by permission of Yukawa Hall Archival Library.

collaboration between them. The Western system of study which had been introduced by [Yoshio] Nishina was producing a number of fruitful results at Kikuchi's and Nishina's laboratories, in each of which Yukawa and Tomonaga were theoretical leaders.

In addition to papers on the topics mentioned above by Tanikawa [16], Sakata and Yukawa published a theoretical study of relativistic theories of particles with spin larger than $\frac{1}{2}$, based upon a new work of Dirac (1936) using the general spinor calculus (Yukawa and Sakata 1937b). The Japanese physicists derived some results of the theory that Dirac had not presented explicitly: "Namely the expressions for the velocity, the current density, the spin angular momentum and the electric and magnetic moments". They also noted that in general the equations are not invariant "under the reflection (*Spiegelung*) with respect to the origin", i.e., parity [17].

From these publications, and from the record of talks that Yukawa delivered at seminars and at the monthly meetings of the Osaka branch of the PMSJ, it is clear that he was following much of the important theoretical development between 1935 and 1937 in nuclear physics and in quantum field theory. He composed several unpublished memoranda on the theory of the positron (one draft is dated as early as 27 November 1934). He began writing a letter to the *Physical Review*, dated 21 April 1936, in which he tried to construct a theory symmetrical between positive and negative electrons to replace the hole theory [18]. Using the idea of the Dirac density matrix, he succeeded in deriving results equivalent to Dirac's positron theory.

Yukawa had hoped to cure the divergence difficulties of QED by his method. Indeed, so long as he considered a finite number of particles without external fields, his density matrix was finite. However, when external potentials were present, he encountered infinities just as others had found, and the letter was never sent to the *Physical Review*. Throughout this period (and throughout his scientific career) Yukawa was struggling with the problem of the infinities of QED, which he believed to be closely related to his own QFT of nuclear forces, as well as other QFTs, such as the neutrino theory of light [19].

It is hard to discern any continuing interest in the *U*-quantum theory among the many concerns which occupied Yukawa during this period. Perhaps some of the topics mentioned in a 35-page manuscript hint at some connection, especially a section containing "queries on nuclear physics", in which he refers to the "absence of experiments on proton–proton scattering" [20].

Whatever his own feelings, Yukawa's fundamental theory was ignored by others and perhaps it was this neglect that led him into other areas of research. As Taketani later recalled [21]:

> The work of Yukawa was at that time not recognized by anyone. It was partly because the PMSJ [i.e., their journal] was not well-known abroad. But the main reason why it was neglected should lie in the fact that such a quantum had never been observed. No one was willing to accept a theory which was based on the existence of an unknown particle. In Japan, no people accepted the Yukawa theory, except Nishina and Tomonaga.

But even the two Japanese physicists singled out by Taketani as supporters of Yukawa's *U*-quantum did not refer to it in their publications; they could easily have done so in a paper on neutron–proton interaction in 1936 (Nishina and Tomonaga 1936), in which calculations on neutron–proton scattering by Tomonaga (possibly in 1934, or even earlier) are included. The calculations are based on a potential close to what is now called the "Yukawa potential". Tomonaga sent them by letter to Yukawa, commenting that the result fitted the experiments quite well [22]. In his first paper on *U*-quanta, Yukawa thanked Tomonaga for sending him these calculations [23].

In the autumn of 1936, after a lapse of a year and a half, Yukawa began to work again on a second paper on the *U*-quanta. The programme of the meeting of the Osaka branch of the PMSJ for 28 November 1936 lists the last of 11 five-minute talks as: Yukawa, "On the interaction of elementary particles. II". Two documents in the Yukawa Archive bear this title and carry the notation "read 28 November 1936": one is the manuscript of a talk and the other is the draft of a paper that was not published in that form [24]. (In spite of these notes, we cannot be sure what Yukawa actually said at the

November meeting.) Furthermore, there are two nearly contemporary documents dated 6 and 11 January 1937, dealing with vector and scalar quantum field formalisms, which appeared eventually in part III and part II, respectively [25]. Concerning the question of what stimulated Yukawa to take up the theory of U-quanta again, it was very likely the growing evidence and belief that particles resembling the U-quanta, first hinted at in experimental observations in 1933, indeed formed a significant component of the cosmic rays. We shall discuss these observations in the next section.

6.3 The mesotron discovered (1936)

In 1932 Paul Kunze, at the University of Rostock, carried out studies of highly energetic cosmic rays by means of a cloud chamber operated in a uniform magnetic field of 18000 gauss (produced by a current-carrying iron-free copper coil) and identified positively and negatively charged particles with energies up to about 10^{10} eV (Kunze 1932a, b). In a later paper of May 1933, he analysed individual photographs of tracks; in particular, he drew attention to a double track, which he described in the text as follows [26]:

> The double track shows in the same neighbourhood a thin track of an electron having 37 million volts and another of a positive particle of smaller curvature, which ionizes much more strongly. The nature of the latter particle is not known; it ionizes too little for a proton, too much for a positive electron. This double track is probably a section from a "shower" of particles such as have been observed by Blackett and Occhialini, hence the result of a nuclear explosion.

Kunze was later cited by Patrick Blackett, together with Carl Anderson of the California Institute of Technology, as a pioneer in the cosmic-ray studies of very energetic particles. Blackett also cited the work of the Paris group led by Louis Leprince-Ringuet (Leprince-Ringuet and Crussard 1937) [27].

By the end of 1935, Blackett had constructed his own powerful magnet, making use of an iron yoke, for cloud chamber studies of the cosmic rays (Blackett 1936; Blackett and Brode 1936) [28]. He and his student John G. Wilson submitted a paper in April 1937, which they summarized as follows:

> These results, which are shown to be roughly consistent with the previous results of Anderson and Neddermeyer up to energies of 4×10^8 e-volts and those of Crussard and Leprince-Ringuet, are in striking contrast with the predictions of quantum mechanics which predicts a constant value of the [mean relative energy loss] R about 1.9 for all but the lowest energies. There is no possibility of avoiding this discrepancy by assuming protons in the beam, as all the particles with energy less than 6×10^8 e-volts are recognizable as electrons.

The authors suggested that the radiation energy loss formulae be modified (following a hint by Lothar Nordheim) in such a way "that radiation is not emitted when the field strength exceeds a critical value" [29].

Meanwhile other groups in Europe and America had made related investigations and their results were cited by Blackett and Wilson. For example, Anderson and Seth Neddermeyer used a counter-controlled cloud chamber with a magnetic field of 7900 gauss, in summer 1935, to observe cosmic rays at the summit of Pike's Peak in California and also at Pasadena (Anderson and Neddermeyer 1936). Their paper of 1936 reproduced 12 photographs taken at mountain altitude and two at sea level, with annotations. These showed extensive showers as well as several peculiar tracks, which the authors carefully analysed. A 0.35 cm plate placed in the middle of the chamber allowed a study of the energy loss as a function of the energy of the shower particles, leading to the following conclusion: "Within the accuracy of measurements the experimental energy losses and those calculated from the theory are in agreement up to energies $\sim$300 MeV." [30] Thus they agreed with the Bethe–Heitler formulae derived from standard QED to an energy much higher than they had been supposed to hold (Bethe and Heitler 1934).

While verifying QED for the shower electrons, Anderson and Neddermeyer found some puzzles among the more penetrating particles, i.e., the charged particles, apparently primaries, which were able to traverse absorber equivalent to three atmospheres. It seemed most obvious to consider these particles as protons, but the Caltech physicists said that their observations had "revealed no tracks which could be ascribed to primary protons near the end of their ranges" [31]. (The masses of such slow particles could have been determined.) They also found a number of other puzzles in discussing individual photographs. For example, a secondary particle with the range of a 1.5 MeV proton had a radius of curvature in the magnetic field that was only a third as large as a proton of that energy should have had. They said about this particle:

> If the observed curvature were produced entirely by magnetic deflection it would be necessary to conclude that this track represents a massive particle with an e/m much greater than that of the proton or any other known nucleus. As there are no experimental data available on the multiple scattering by low energy protons in argon [that being the chamber gas] it is difficult to estimate to what extent scattering may have modified the curvature in this case. The particle is therefore tentatively identified as a proton.

On 12 December 1936, Carl David Anderson was awarded the Nobel Prize in Physics for the discovery of the positron (sharing the prize with Victor Franz Hess, for the discovery of cosmic rays). At the end of his Nobel address, he mentioned the penetrating particles which proved so puzzling; his concluding prophetic sentence reads: "These highly penetrating particles, although not free positive and negative electrons, will provide interesting material for future study." [32]

Anderson later remarked that he "received no reaction at all" to what he regarded as an announcement of the new particles. However, exactly one month earlier he had given "evidence for the existence of new particles of intermediate mass" in a colloquium at Caltech. This announcement of the Nobel Laureate elect had been duly reported by the press. Watson Davis of *Science Service* started his account with this [33]:

Discovery of an unknown particle that may prove to be as important as the positron was made known by Dr. Carl Anderson to his colleagues at the California Institute of Technology just a short time after he was notified of his sharing in the Nobel physics prize for his discovery of the positron.

Furthermore, an article in *Science* on Anderson's colloquium emphasized its importance [34]:

This latest report by Dr. Anderson was received with great and serious interest by his colleagues. He has often produced conclusive proofs of startling results, and he has the reputation for careful interpretation of results. The experiments which led Dr. Anderson gradually to his conclusion were carried out in collaboration with Dr. Seth Nedermayer [*sic!*] and Dr. R.A. Millikan, head of the California Institute of Technology, himself a Nobel Laureate.

We note, in this connection, that a paper by J. Robert Oppenheimer and his student J. Franklin Carlson (received by the *Physical Review* on 8 December 1936) also mentions a "radical alternative" to explain the puzzling penetrating particles: "If they are not electrons, they are particles not previously known to physics." [35]

6.4 The discovery is confirmed (1937)

We recall (from section 2) that in the autumn of 1936, Yukawa began again to work on the theory of U-quanta and paid close attention both to theoretical and to experimental work on the cosmic rays. Thus, he gave several talks on cosmic-ray theory in Osaka [36] and met regularly with the experimental group led by Kikuchi [37]. Feeling that the puzzling penetrating tracks of Anderson and Neddermeyer's paper of 1936 could best be explained by his hypothesis of heavy quanta, on 18 January 1937 Yukawa sent a letter to *Nature* to make this bold conjecture and to call attention to his theory of U-quanta [38]. The editor of *Nature*, however, rejected it.

Yukawa's letter was entitled "A consistent theory of the nuclear force and the β-disintegration" and it began with a paragraph attacking the prevailing Fermi-field theory of nuclear forces [38]:

In spite of many attempts to develop the so-called "β-hypothesis of the nuclear force" [here he cites Bethe and Bacher, von Weizsäcker, and Iwanenko and Sokolov], there still remains in the current theory the well known inconsistency between the small probability of the β-decay and the large interaction of the neutron and the proton. Hence it will not be useless to give on this occasion a brief account of one possible way of solving this difficulty which was proposed by the present writer about two years ago.

He concluded his letter with a reference to Anderson and Neddermeyer's 1936 article [39]:

Now it is not altogether impossible that the anomalous tracks discovered by Anderson and Neddermeyer, which are likely to belong to unknown rays with e/m larger than that of the proton, are really due to such quanta [i.e., the U-

quanta], as the range–curvature relations of these tracks are not in contradiction to this hypothesis. At present, much reserve is, of course indispensable owing to the scantiness of the experimental information.

Pike's Peak 1935. Top, from left to right: C.D. Anderson (1905–1991), R.A. Millikan (1868–1953) and S. Neddermeyer (1907–1988). Bottom: C.D. Anderson at control panel for his cloud chamber. Photographs reproduced by permission of AIP Emilio Segrè Visual Archives.

Still, the year 1937 saw the definite confirmation of the presence of particles of intermediate mass in the cosmic rays, which finally led physicists in the West to take note of Yukawa's theory. The first of the confirmations, and in some ways the most comprehensive, was a new paper (Neddermeyer and Anderson 1937), published in the *Physical Review* of 15 May 1937. To analyse the penetrating particles further, the authors replaced the earlier 0.25 cm lead plate in their chamber by a 2 cm thickness of platinum. They found that the measured particles (55 in all) fell into two distinct groups: shower particles of high absorbability, on the one hand, and on the other hand single particles, most of which lost a small fraction of their energy (although four particles lost more than 60%).

As a result of their measurements, they drew the following strong conclusion [40]:

> The present data appear to constitute the first experimental evidence for the existence of both penetrating and nonpenetrating character in the energy range extending below 500 MeV. Moreover, the penetrating particles in this range do not ionize perceptibly more than the nonpenetrating ones, and cannot therefore be assumed to be of protonic mass.

In the penetrating (and weakly ionizing) group, the particle showing the largest curvature in the magnetic field, and hence having lowest momentum, would have ionized 25 times as strongly, had it been a proton.

Although the existence of the "anomalous" penetrating tracks had for a time caused the Caltech experiments to speak of "red" and "green" electrons, Anderson later asserted, that [41]:

> In the summer of 1936, Neddermeyer and I were quite firmly convinced that all the data on cosmic rays as known at that time nearly forced on us the conclusion that the penetrating sea-level particles could be neither electrons nor protons and must therefore consist of particles of a new type.

Indeed, that was the import of their paper of May 1937, although they could not determine the masses of the particles or even establish that there was a unique value.

Jabez Street and E.C. Stevenson of Harvard University had also been trying to get a handle on the penetrating particles, and for this purpose they had set up an elaborate apparatus that they described briefly in an abstract as follows (Street and Stevenson 1937a):

> To investigate the penetrating rays a vertical column was set up as follows: a counter, 10 cm Pb, a second counter, a cloud chamber in a magnetic field, a third and fourth counter, 3 cm Pb, and finally a cloud chamber containing three separated lead plates 1 cm thick. The counter telescope selected particles directed toward the visible region of the lower chamber where their absorption and shower production was observed.

Their conclusions, based on 90 tracks with large penetration, agreed with that of Anderson and Neddermeyer, namely, "that the penetrating particles cannot be described as electrons obeying the Heitler theory [*sic!*] nor can an appreciable fraction be protons" [42].

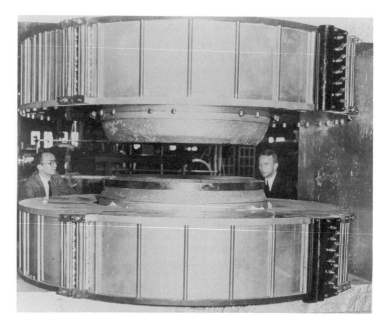

J.C. Street (born 1906) and K.T. Bainbridge (born 1904) at the Harvard cyclotron, 1930. Photograph from the Smithsonian Institute and reproduced by permission of AIP Emilio Segrè Visual Archives.

Clearly, we have been witnessing a *Gestalt* switch or, perhaps, a paradigm shift, as well as the priority race that often accompanies one [43]. The anomalous penetrating tracks had been ever-present in cosmic-ray observation since 1933, but only at the end of 1936 had it become acceptable that they were the tracks of "particles not previously known to physics", which now became thought of throughout the physics community as "heavy electrons".

The Caltech physicists once again used the press to communicate their results early and to a wide public. On 26 April 1937, the *Pasadena Post* reported the "discovery of a new atomic particle" at Caltech and said that "the sensational discovery may at last unlock the mystery of the true nature of the cosmic rays". The *Los Angeles Times* also gave full coverage, quoting the opinion of Caltech physicist R.A. Langer, who said, "In my opinion this is a greater and far more difficult discovery than that of the positron." [44]

At about the same time, and independently, Walter Heitler in Bristol turned his attention to the penetrating cosmic-ray component (Heitler 1937). He considered as two possibilities, that the "hard" particles were (a) not electrons or (b) electrons of very high energy and commented [45]:

Assumption (a) has one great advantage: according to the theory large showers can only be produced by electrons of very high energy. Thus, if (b) be right, we should have to fall back on a completely new theory for large showers, which is not the case if (a) is right. On the other hand, (a) necessitates the assumption of a new sort of particles for which so far no direct evidence exists.

After thoroughly reviewing the cosmic-ray evidence, however, Heitler found that assumption (a) carried the day [46].

In 1935, the Nishina group at Riken began the construction of a large (40 cm) cloud chamber with strong magnetic field, comparable to that used by Blackett and Wilson to measure the momentum spectrum of the cosmic rays. To operate the magnet required a large direct current power source, for which purpose, two members of the Nishina group, Masa Takeuchi and Tarao Ichimiya, transported it to Yokusuka Naval Arsenal, where they could use a generator normally used to charge submarine batteries. As Takeuchi later reported [47]:

> In 1937 Blackett published his results on the energy spectrum obtained by his cloud chamber... prior to our publication. We found that our result agreed completely with his result, and presented our result only orally at the meeting of I.P.C.R. [Riken], but did not publish it.

In the spring of 1937, they focused their attention on the penetrating component, putting a lead plate (of either 3.5 or 5 cm) in their cloud chamber.

Their research may have been influenced by discussions with Niels Bohr, who was on a world tour which included America and Japan. On 15 February 1937, Nishina wrote to Dirac about this trip [48]:

> Bohr is to arrive here from America on 15 April together with Mrs. Bohr and Hans [their son]. They will stay in Japan for about a month and then travel to China, where they intend to spend about a week. They will travel back to Copenhagen via Canada.

According to Taketani, Bohr's visit resulted in the following exchange [49]:

> In Kyoto, Yukawa and Nishina met with N. Bohr. Yukawa talked to him about the meson theory, on which Bohr was not attracted. He asked Yukawa, "Why do you want to create such a new particle?". At this question we all were not happy. However, before Bohr returned to his native country from Japan, we were informed from the United States that a new charged particle having mass $\approx 200 m_e$ (m_e being the mass of the electron) was found in the cosmic rays by Anderson and Neddermeyer and also by Street and Stevenson.

In contrast to this, Takeuchi recalled that Bohr told Nishina *everything* that he had learned about the cosmic rays in America, including the new cosmic-ray particle, and stated that this information influenced their work (Takeuchi 1975):

> Since we too were photographing with an absorber of 3.5 cm of lead, it was decided to re-analyse all of the traces photographed. They were mostly of energy above 10^9 eV and were not suitable for determining the mass. Later, luckily, a trace with a particle of about 10^8 eV was found from which the mass could be calculated; and it was in August that we reported the mass to be 1/7 to 1/10 the mass of the proton.

Y. Nishina (1890–1951) with N. Bohr (1885–1961) and, top, S. Kikuchi (1902–1974). Photograph from the Nishina Memorial Foundation and reproduced by permission of AIP Emilio Segrè Visual Archives.

The Japanese cosmic-ray workers thus provided the first quantitative estimate of the mass of the new particle, which was published in the *Physical Review*, but only after an unfortunate delay. Meanwhile, another mass determination, submitted nearly two months later than theirs, had been published (Street and Stevenson 1937b). As Takeuchi later complained (Takeuchi 1975):

> Although the *Physical Review* received the paper on 28 August, it was sent back because of its length and it was finally published in the issue of 1 December. But Street and Stevenson's letter of 10 October [actually it was dated 6 October], reporting a mass of 160 times that of the electron, was published a month earlier and we were very disappointed.

Although some of the Japanese experience in the 1930s with the Western scientific establishment was thus less than satisfying, it also should be pointed out that some American physicists were very helpful to their Japanese colleagues. Ryokichi Sagane, of the Nishina Laboratory at Riken, learned much of the necessary cloud chamber and counter techniques from a stay at the University of California at Berkeley. There was extensive correspondence regarding cosmic-ray matters between Nishina and Millikan, and the latter sent cosmic-ray counters of his own invention to be used by the Nishina group [50]. This assistance, both personally and by correspondence, contributed significantly to the establishment of nuclear and cosmic-ray research in Japan in the years before the Second World War.

6.5 The mesotron and the Yukawa theory – hopes and doubts (1937)

As we stated at the end of section 2, Yukawa resumed work on his theory of U-quanta in the autumn of 1936, encouraged by the anomalous cloud chamber tracks reported (Anderson and Neddermeyer 1936). The manuscript on which Yukawa based his talk in Osaka on 28 November 1936 has two parts, the first part in English and the second in Japanese; it was summarized this way by Satio Hayakawa [51]:

> In section 1 he gave a programme to improve on the first meson paper, as follows: (1) relativistically invariant formalism, (2) introduction of the Majorana force and (3) modification of the interaction of the meson [i.e., the U-quantum] with the electron and the neutrino so as to obtain the β-decay interaction of George Gamow and Edward Teller [1936]. For the first programme he introduced a four-vector and a relativistic six-vector [52]. In section 2 he tried to formulate scalar meson theory, following Pauli and Weisskopf [53]. Without completing the formulation, however, he turned to discussing observable features of this hypothetical particle... He made numerical estimates to determine whether cloud chamber tracks obtained by Anderson and Neddermeyer might be those of heavy quanta.

The three points that Hayakawa mentions as being discussed in Yukawa's section 1 (entitled: "Recapitulation and Revision of the Previous Paper") are closely linked to each other. In order to obtain an exchange force of the Majorana type (interchange of charge and spin coordinates), the source term, to be introduced on the right-hand side of (5.10), must transform like the nucleon spin vector (i.e., as an axial vector). Thus the complex U-field is complemented by a complex vector field B. However, U and B do not form a relativistic four-vector. The relativistic generalization is for U to become a four-vector U_μ and for B to become a second-rank Minkowski antisymmetric tensor B_{ij}, i.e., a six-vector. For ordinary nuclear forces, only the non-relativistic limits U and B would be effective. As applied to the β-decay interaction, the presence of B in the exchange-current source of the electron and neutrino fields produced the Gamow–Teller (or spin-flip) contribution to that interaction. There is no doubt that Yukawa was pleased that all this seemed to be fitting together in such a natural way.

Further insight into Yukawa's thinking at this time is provided by a draft of a paper entitled "On the interaction of elementary particles. III" [*sic*] and dated 6 January 1937. The contents of its 22 pages are as follows [54].

(1) Linear Generalized Maxwellian Equations for the New Field. *Vertauschungsrelationen* [8th page] Supplementary Conditions [14th page]

(2) Interaction of Heavy Quanta with the Electromagnetic Field [18th and 19th pages]
(3) Linear Field Equations of Proca Type [20th page]
(4) Interaction of the U-Field with the Heavy Particle [55].

Completing this survey of the earliest documents showing Yukawa's revived interest in his nuclear force theory, we note that there is also a draft headed "On the interaction of elementary particles. II". Seven pages in length and dated 11 January 1937, it deals with the quantization of the relativistic scalar field [56].

Earlier in this section we have discussed Yukawa's unpublished letter to *Nature* of 18 January 1937, in which he suggested that the new cosmic-ray particles might well be the massive charged quanta envisaged by his theory [57]. Later, Yukawa recalled: "Soon the manuscript was sent back with the reply that it could not be printed in that journal because there was no experimental evidence to support my idea." [58] However, in the first months of 1937, as we have seen, the experimental evidence for the new particles in the cosmic rays became much stronger. On 1 June, J. Robert Oppenheimer and Robert Serber wrote a letter about them to the *Physical Review*, which was followed by a letter from Ernst Carl Gerlach Stueckelberg of Geneva, dated 6 June. Both letters called particular attention to Yukawa's 1935 paper on U-quanta. Finally, in a letter, dated 2 July, to the *Proceedings of PMSJ*, with contents similar to those of his unpublished January letter to *Nature*, Yukawa recalled his own theoretical prediction of the existence of such particles. We shall consider these three letters, which marked the start of a world-wide interest in the meson theory of nuclear forces, beginning with Yukawa's letter, designated by the journal as a "Short Note" (Yukawa 1937).

After referring to the failure of the "β-hypothesis of the nuclear force", Yukawa's letter continued:

> More than two years ago, the present author introduced a new field of force, which is responsible for the short-range interaction of the neutron with the proton, but it is something different from the so-called "neutrino–electron field". This field was considered to interact with the light particle also, inducing the transition of the latter between the neutrino and the electron states. Thus we could get rid of the above difficulty by assuming the interaction between the field and the heavy particle to be large compared to that between the former and the light particle.

He went on to say that the "mathematical formulation of the theory was preliminary and incomplete in many respects" and that other types of forces than those deduced previously are possible (e.g., exchange forces of the Majorana as well of the Heisenberg type and β-decay forces *à la* Konopinski–Uhlenbeck as well as of the Fermi type). However, he stressed that:

> The most important and at the same time inevitable consequence of the theory was that the field was to be accompanied by new sorts of quanta obeying Bose

statistics and each having the elementary charge $+e$ or $-e$ and the proper mass m_U about 200 times as large as the electron mass. It was shown further that such quanta could never be produced by the ordinary nuclear reaction, by which the energy liberated was certainly smaller than $m_U c^2$, whereas they might be present in the cosmic ray.

Yukawa then cited the experimental evidence, analysed it briefly and concluded that "if the above theory [i.e., Yukawa's U-field theory] is correct, we can expect that at least a part of the penetrating component [of cosmic rays] consists of the heavy quanta considered above".

The letter of Oppenheimer and Serber was published in the issue of 15 June of the *Physical Review* (Oppenheimer and Serber 1937), and constituted the first response to Yukawa's theory in the West. Although the authors correctly claimed this distinction on various future occasions, their letter was in fact largely negative in tone, claiming that the Yukawa theory neither gave an adequate explanation of nuclear forces nor predicted the new particles.

At a historical symposium in 1980, Robert Serber discussed the letter of the Berkeley theorists, as follows [59]:

In May 1937, Anderson and Neddermeyer and, simultaneously, Jabez Street and E.C. Stevenson, announced the identification of a meson in the cosmic rays. A month later, in June 1937, Oppenheimer and I made this promised [*sic!*] suggestion: The particles discovered by Anderson and Neddermeyer and Street and Stevenson *are those postulated by Hideki Yukawa to explain nuclear forces.* (Anderson and Neddermeyer were wiser: they suggested "higher mass states of ordinary electrons".) Yukawa's paper came out in 1935, but we know of no reference to it before our 1937 paper, and a very conscious purpose of our paper was to call attention to Yukawa's idea.

Oppenheimer, who was teaching at Caltech as well as at Berkeley, was certainly aware that the cosmic-ray physicists at Caltech had observed intermediate-mass particles [60]. The letter began by mentioning the particles, commenting that the nuclear force range suggested a mass in the range 50–100 MeV/c^2. However, it cautioned: "These observations themselves, however, could be equally well interpreted if the particles had a quite wide variation in mass; nor do they exclude values considerably less than 50 MeV." Then came the reference to Yukawa:

In fact, it has been suggested by Yukawa that the possibility of exchanging such particles of intermediate mass would offer a more natural explanation of the range and magnitude of the exchange forces between proton and neutron than the Fermi theory of the electron–neutrino field. Yet in trying to account in detail along these lines for the characteristics of nuclear forces, one meets with difficulties hardly less troublesome than in the various forms of electron–neutrino theory which have been proposed.

Oppenheimer and Serber mentioned some of these problems — achieving "saturation" of the forces, equal like- and unlike-particle forces, and the magnetic moments of the neutron and proton — saying that a "reconciliation" of these problems could be achieved in Yukawa's theory "only by an extreme artificiality". Their conclusion followed [61]:

> These considerations [Yukawa's] therefore cannot be regarded as the elements of a correct theory, nor serve as any argument whatsoever for the existence of the particles.

Stueckelberg's letter appeared in the next issue of the *Physical Review*; referring to the new particles as "heavy electrons". More positive in tone than the Oppenheimer–Serber letter, it began (Stueckelberg 1937a):

> Different observers believe that they have found evidence for the existence of charged particles whose mass amounts probably to about fifty times the electron mass. The writer wishes to call attention to an explanation of the nuclear forces, given as early as 1934, by Yukawa, which predicts particles of this sort.

Stueckelberg quickly pointed out that he "independently of Yukawa arrived at the same conclusion". During 1936, the Swiss theorist had constructed a "unitary field theory" in which he had described the known fermions (i.e., electron, neutrino, proton and neutrino) by a single 16-component spinor Ψ, and had supplemented the electromagnetic four-potential A_μ by a field B_μ, "created by the heavy particles" (thus making what in today's language would be called a *grand unified theory*). However, it is hard to see from the published papers how this would necessarily lead him to "the same conclusion" as Yukawa (Stueckelberg 1936a, b, c).

In his letter of 6 June, Stueckelberg replaced his A-field and B-field by a single tensor field of five components (calling the new field A), which satisfied the generalized d'Alembert equation:

$$(\Box - \Sigma \cdot 1/\lambda^2)A = -e\Psi^\dagger \Lambda \Psi,$$

where Σ and Λ are appropriate 5×5 matrices. Four components of A were to be associated with electrodynamics and the photon, while the fifth component gave a nuclear interaction, which in the static limit was the Yukawa potential. The author then claimed:

> A suitable choice of the generalized Dirac matrices Λ gives the electrostatic interaction between charged matter particles plus the Heisenberg, Majorana, Wigner and Bartlett interactions between heavy and light matter particles (β-decay, etc.) discussed by the author. The Heisenberg interaction seems to demand a second order tensor field.

Finally, he concluded [62]:

> It seems highly probable that Street and Stevenson, and Neddermeyer and Anderson, have actually discovered a new elementary particle, which has been predicted by theory. This particle is unstable and can only be of secondary origin, its mass being greater than the sum of the masses of electron plus neutrino.

We end this section by discussing a second unpublished letter of Yukawa (and his collaborators Sakata and Taketani), which he composed and sent to the *Physical Review* on 22 October 1937 [63]. By the time this letter was written, the three attention-getting notes discussed above had appeared and Yukawa's letter begins by citing these works:

From left to right: H.A. Kramers (1894–1952), P. Morse (1903–1985), Wang, E.C.G. Stueckelberg (1905–1984) and H. Barton in the 1930s. Photograph reproduced by permission of AIP Emilio Segrè Visual Archives (Uhlenbeck Collection).

As already suggested by several authors [footnote 1], the existence of the new particle in cosmic rays, if confirmed, will be a strong support to the theory which had been proposed by one of the present writers and recently by Stueckelberg. Thus it will not be useless to give here a brief account of further consequences of the theory and their bearings on cosmic ray and nuclear phenomena.

Footnote 1 is a revealing response to the negative tone of the first Western recognition. After citing the three letters discussed above, it added:

It should be noticed that the criticism of Oppenheimer and Serber is not well founded, since many of the difficulties in the current theory [i.e., the Fermi-field theory] do not appear in our theory, as will be shown in the following paragraphs.

In the next paragraph, Yukawa and his collaborators gave the equations of motion satisfied by the U-field and its complex conjugate, showing explicitly the Heisenberg exchange "current" as a source, and then pointed out how these elements can be modified [64]. For example, by letting the source depend upon the spins of the heavy particles, they could obtain both Heisenberg and Majorana nuclear forces. They could also derive like-particle forces, but these were (as stated in their footnote 2) "only about 1/10 of the unlike particle force". They suggested, therefore, that it may be necessary to "consider neutral heavy quanta" in order to obtain like- and unlike-particle forces approximately equal in strength.

By the addition of another source term, with a much smaller coupling constant (the analogue of the electric charge), they "obtain a theory of β-disintegration essentially equivalent to that of Fermi". However:

> The well known modification due to Konopinski and Uhlenbeck can also be adopted in this theory, if we assume the source terms to contain derivatives of the wave functions for the light particle.

Next the Japanese authors addressed the "important and inevitable consequence of the above theory", namely the existence of massive charged quanta with mass estimated at 200 times that of the electron. (A footnote states: "According to the preliminary result of the experiment of Nishina, Takeuchi, and Ichimiya, the mass of the new particle in the cosmic ray is of the order of 1/10 of the protonic mass in fair accord with the theory.")

Assuming "the self energy of the heavy particle due to the U-field to be responsible for the whole mass M of the heavy particle", they estimated its radius to be $g^2/(Mc^2)$ and its additional magnetic moment (of the proton or neutron) to be ± 2.5 times the nuclear magneton, "which is in good agreement with the observed value of $\pm 2e\hbar/(2Mc)$".

The authors concluded on an optimistic note:

> In this way, the above theory seems to be very promising for consistent interpretation of the nuclear phenomena as well as of the cosmic ray, although it cannot be avoided for the time being that the mathematical scheme becomes more and more complicated, as we want to fit the theory better to the experiment. Detailed account of the whole subject will be given in later issues of the *Proc. Phys.-Math.Soc.Japan.*

Two months later, however, they received a disappointing response. In a letter dated 2 December 1937 and signed "J.W. Buchta, Assistant Editor" they were given the following report of an (unnamed) associate editor:

> The Letter suggests that the theory proposed is far more adequate to account for the facts of nuclear physics than it actually is. The theory as presented gives 1) like-particle forces too small by a factor 10^{-20}, 2) wrong spin-dependence, and 3) non-saturating like-particle forces. It also gives no account of the anomalous magnetic moments of the proton and neutron. None of the suggested modifications are acceptable in detail. A factor of 4π is omitted on page 2.

That is all, except for the close by Buchta: "In view of these criticisms we thought it best to return the paper to you for consideration." [65]

One wonders what the result might have been, had the same critical standards been applied to the letter of Stueckelberg in which he claimed to have "independently of Yukawa arrived at the same conclusion" as Yukawa. Buchta's letter does not reveal which associate editor gave this referee report, but it certainly is in agreement with the Oppenheimer–Serber criticisms. At the Stanford meeting of the American Physical Society, 17–18 December 1937, Serber made the same four points given in the rejection letter and added another criticism, namely: "The force between neutron and proton in a state symmetric in their coordinates turns out to be repulsive rather than attractive." (Serber 1938). The last difficulty and the relative weakness of the forces that resulted between like particles he considered to be "intrinsic", i.e. irremediable, whereas the other objections, he conceded, might be overcome by "further *ad hoc* assumptions".

6.6 Conclusions

As 1937 drew to a close, the nuclear force theory of the U-field had a number of difficulties still to overcome before it could supplant the Fermi-field theory. The most important theoretical problem was probably to find a natural way to make the forces charge-independent; this appeared to call for a new *neutral* particle of intermediate mass, for which there was no experimental evidence whatever [66]. Many physicists were even doubtful about the identification of the anomalous cloud chamber tracks by the Caltech physicists as new particles, which they had decided to christen "mesotrons" [67].

For example, Heisenberg wrote to Niels Bohr in July 1937 [68]:

An interesting news item, which I do not yet know whether I should believe, is the unstable "heavy electron" that Anderson claims to have found. *A priori* I am not against such a possibility; however, I do not know how well it is substantiated experimentally. Blackett at least has raised strong arguments *against* the assumption that the penetrating rays are "heavy" electrons. He claims that most of the particles with energies higher than 6×10^8 eV are "penetrating" in lead but create cascades in copper and aluminium; he therefore believes the Bethe–Heitler formula to fail for high energies, the limiting energies being different for different elements. I do not have any definite opinion about this question, which will be decided only by experiment.

Blackett retained his doubts at least until the end of 1937, because in November he raised the issue with another theoretician friend, Paul Dirac [69]:

Dear Paul,
I want to ask your views about these cosmic-ray particles. I do not know whether you have been following the rather obscure literature on the subject but I think the position is roughly as follows: all the particles over two hundred

million volts, only about 5% or less, are radiating electrons. The rest are much more penetrating, but when they slow down below two hundred million volts, become radiating electrons. It looks, therefore, as if the radiation formula does hold for all elements up to high energies, but that only a fraction of the cosmic rays are normal electrons. The rest, that is 80 or 90% of all the rays at sea level, lose much less energy until they become slow, when they become identical with radiating electrons. The question is what physical property characterizes the difference between the normal and abnormal particles. Anderson's suggestion of a greater mass leads to a mass which changes gradually or abruptly with energy. Are there any other properties which might affect the radiating power of electrons? [Arthur] Bramley [Bramley 1937] has suggested that electrons with a spin of more than $\frac{1}{2}$, as discussed in your recent paper in the *PRS* [Dirac 1936] would have much less radiation loss. Then Swann has discussed the possibility of the Lorentz force of an electron in a magnetic field going wrong, as a possible cause of the apparent reduced energy loss [Swann 1936]. What do you think about that? I would much value any ideas you have as to what is plausible and what is not plausible. Since I believe that radiation loss probably changes gradually with the energy of the particle, I do not feel attracted to the heavy electron theory. The electron is in a different state, e.g., characterized by a different spin or some difference in some new property seems a rather more likely possibility. (?)

Dirac answered Blackett as follows [70]:

After Shankland [whose experiments had recently raised a false alarm about the validity of the photon concept, causing Dirac to have doubts about the conservation of energy] I feel very skeptical of all unexpected experimental results. I think one should wait a year or so to see that further experiments do not contradict the previous results, before getting worried about them. If electrons pass through lead plates more easily than the electron theory predicts, I do not think it will help at all to assume a different state for the electron. The difficulty is to get the electric charge through the lead plate, and I would prefer to try to explain the result by supposing the electron to change to a photon, or some other penetrating particle, soon after entering the lead, and to change back again before leaving. One could check this by seeing how the energy loss varies with the thickness of the plate. I do not have any other ideas on the subject at present.

This exchange of letters again confirms the desire of Western physicists to keep the number of fundamental constituents of matter as small as possible. The Japanese physicists, marginally located, were less influenced by this European tradition and were more liberal on this point. Mituo Taketani, one of Yukawa's collaborators, developed a philosophy of three stages — we shall outline it shortly in the next chapter — which, combining Eastern and Western ideas, seemed to provide some guidelines for meson theory.

Whatever the difficulties and doubts, and whatever the philosophy or lack thereof, the mesotron was here to stay and so was Yukawa's nuclear theory, but whether the two new parts of physics had anything really to do with each other was but one of the still outstanding questions to be decided in the decade that followed.

Notes to text

[1] The first person to point out the instability of the heavy quantum was Ernst Stueckelberg (1937a); then Homi Bhabha gave a detailed discussion of the decay, including the relativistic time-dilation effect (Bhabha 1938a); finally Yukawa and Shoichi Sakata calculated the mean life (Yukawa and Sakata 1939a).

[2] For other names, see the above introduction to part B.

[3] In writing the present chapter 6, we have benefited from some related historical treatments. These include a broad pioneering study by Visvapriya Mukherji (1974), an account of the Japanese contributions by Satio Hayakawa (1983) and a work by Peter Galison on aspects of the discovery of the mesotron (Galison 1983a). Furthermore, we have used unpublished reminiscences of Nicholas Kemmer, of Yukawa and of others in Japan. We are especially grateful to Professors Rokuo Kawabe and Michiji Konuma for assistance at YHAL in Kyoto and for many useful discussions.

[4] Yukawa (1935, p 49).

[5] See Brown *et al.* (1991).

[6] This document is [YHAL] F03 090 P12. We use the catalogue descriptions such as "memorandum" and "draft".

[7] [YHAL] F03 104 P12.

[8] In 1933 Beck left Prague, where he had held a faculty position at the Karl-Ferdinand University, in response to anti-Semitic student actions. After a few months in Copenhagen, he spent a year in Odessa, then at the University of Kansas at Lawrence. In 1935 he visited Japan on the way back to Odessa.

[9] Sakata (1956). The term "discovery" used here is a bit exaggerated because the K-capture process had been noticed and its rate roughly estimated by G.C. Wick (Wick 1934), work to which Yukawa and Sakata refer in their paper.

[10] The source was Ra(C+B). Naturally occurring heavy elements do not emit positrons in β-decay, so the process involves an internal γ-ray, as in the process of electron emission by internal conversion.

[11] See note 9.

[12] Fermi (1934b, p 162). The Japanese authors viewed the usual β-process as the promotion of a negative energy electron, leaving a neutrino "hole" or antineutrino. Similarly, in positron β-decay a negative energy electron is promoted to a positive-energy neutrino, leaving an electron "hole" or positron. The same picture had been advocated earlier by Wick (1934).

[13] Yukawa and Sakata (1937a, p 467).

[14] Apparently Lamb was unaware of Yukawa and Sakata's "Supplement" on the Konopinski–Uhlenbeck theory result (Yukawa and Sakata 1936).

[15] Yasutaka Tanikawa, editor, in Yukawa (1979, p xvi).

[16] These are papers numbered 2–10 in Yukawa (1979) except for number 7. All but two of these articles are co-authored by Sakata (and in one of the latter Sakata is thanked for his assistance).

[17] Yukawa and Sakata (1937b, p 91). Note, however, that in 1938 Markus Fierz disputed the validity of these results (Fierz 1939, p 4).

[18] H. Yukawa, "The density matrix in the theory of the positron" [YHAL] E08 010 U04.

[19] See chapter 4. Yukawa discussed his ideas on the neutrino theory of light in a memorandum dated 24 February 1936: [YHAL] E08 130 U04.

[20] The manuscript is that of a projected scientific paper. It is not dated, but it was filed with papers of the year 1936. That was the year of the "American experiments" (as Pauli referred to them), which showed that like-particle forces among neutrons and protons were as strong as the unlike-particle force and led

to the concept of charge-independence of the nuclear force. (However, this topic was not among Yukawa's seminar subjects listed for 1936.)

[21] Taketani (1971, p 18).

[22] Tomonaga to Yukawa, [YHAL] F02 080 C01. The letter is not dated, but it is contained among manuscripts of 1934.

[23] Yukawa (1935, p 52, note 6). Incidentally, Tomonaga's old calculations used a potential very close to what was called the Yukawa potential. (See Tomonaga to Yukawa, undated but contained among manuscripts of 1934, YHAL F02 080 C01.)

[24] The contents of the "manuscript of the talk" [YHAL] E02 060 P12 are discussed in Hayakawa (1983, pp 88–9).

[25] These documents are, respectively, [YHAL] E03 060 P13 and [YHAL] E02 080 P12. They have been mentioned in a very thorough study (Kawabe 1991b).

[26] Kunze (1933, p 10).

[27] According to Leprince-Ringuet (1983, p 178): "First, we found the predominance of positive particles in the totality of the penetrating radiation. Second, with J. Crussard, we found that there were particles of both charges and of relatively low energy (less than 300 MeV) that were nevertheless able to traverse a large thickness of lead (14 cm). These particles could not be protons, and it was extremely improbable that they were electrons. We therefore made the hypothesis that these could be new particles, but we were unable to confirm their existence, because at that time the properties of very energetic electrons were not well known. Nevertheless, these results were evidence in favor of the existence of the meson, which was identified by Anderson two years later".

[28] The new magnet at Birkbeck College, London, allowed the determination of particle energies up to 2×10^{10} eV (Wilson 1985).

[29] For doubts concerning the validity of QED, see Galison (1983a) and Brown and Hoddeson (1983, editors' introduction and chapters 3 and 13).

[30] Anderson and Neddermeyer (1936, p 267).

[31] Anderson and Neddermeyer (1936, p 269).

[32] This sentence from his Nobel address and Anderson's comments on it are in Anderson and Anderson (1983, p 147). The latter article reviews the history of the discovery of the mesotron, going back as far as the London Conference of 1934.

[33] W. Davis in *Science Service*, 13 November 1936.

[34] Unsigned report: "Particles in cosmic rays similar to but different from the electron", in *Science*, November 20 1936, p 9 of the *Supplement*.

[35] Carlson and Oppenheimer (1937, p 220). On the strength of this remark, Galison comments (1983a, p 303): "Of course, there is a sense in which Carlson and Oppenheimer discovered the muon, since they first suggested the existence of a particle of intermediate mass in 1936". There is an unconscious irony here, for Neddermeyer told one of us (LMB) in Seattle in 1978 that Oppenheimer had argued so strongly against particles of intermediate mass that Anderson and he were delayed for nearly two years; they published an announcement on the existence of the particles only after Oppenheimer, on a visit to Cambridge, Massachusetts, told J.C. Street of MIT about their results and on returning to California, reported that Street was about to publish similar results.

[36] E.g., [YHAL] F05 030 T02.

[37] Seishi Kikuchi moved from the laboratory of Masaharu Nishikawa at Riken, where he had worked especially on electron diffraction, to become a professor at Osaka University in 1933 and to start a group in nuclear physics (Kaneseki 1974, especially p 239). T. Watase of Kikuchi's Osaka group began to do cosmic-ray research in collaboration with J. Itoh in 1935, using counters in coincidence. (See Hayakawa (1983, p 88) and the "Discussion" in Brown *et al.*

(1991, pp 55–69); see also Sakata (1965).) Another cosmic-ray group in Japan was that of Nishina at Riken. (For its work, see Takeuchi (1985).)

[38] [YHAL] E06 U02. This letter and another one, [YHAL] E16 U05 (sent to the *Physical Review* on 4 October 1937, and also rejected by the editor), are reproduced and discussed in Kawabe (1991b). According to Kawabe's article, the news item about Anderson's Caltech colloquium in November 1936 (Note 34) was reviewed in the Japanese science magazine *Kagaku* (January 1937 issue, p 45), but Kawabe wrote: "We have not been able to determine whether Yukawa had this information when he sent his letter to *Nature*."

[39] Brown *et al.* (1991, pp 262–5). A "Short Note" having similar content, dated 2 July 1937, was published later (Yukawa 1937).

[40] Neddermeyer and Anderson (1937, p 886).

[41] Anderson and Anderson (1983, pp 146–7).

[42] For more details on Street and Stevenson's work see Galison (1983a).

[43] There is an extensive literature based on this notion, to which we shall *not* refer here. A single recent example must suffice: the observation of neutral currents in the weak nuclear interaction, see Pickering (1984) and Galison (1983b).

[44] *Los Angeles Times*, 25 April 1937: "Cosmic Ray's Secret Key Believed to Be Found. Caltech Scientists Report Discovery of New Form of Matter Vital to Physics".

[45] Heitler (1937, p 263).

[46] Heitler (1937, p 281). In a note added in proof on 3 June 1937, Heitler referred to the new paper by Blackett and also mentioned Neddermeyer and Anderson (1937) as having "pointed out a few arguments in favor of a new sort of particle with mass between an electron and a proton".

[47] Takeuchi (1985, p 140).

[48] Dirac Archives, Churchill College, Cambridge. According to Moore (1970, p 208), by whom this trip has been described, the Bohrs returned home via the USSR.

[49] Taketani (1971, p 20). A similar story has been told by Sakata (1965).

[50] Millikan Archive, Caltech.

[51] Note 24, pp 88–9.

[52] In the electrodynamic analogy, the four-vector would be the four-potential, while the six-vector would be the antisymmetric 4×4 electromagnetic field tensor, which has as its six independent components the electric and magnetic field intensities. Evidently, Yukawa was pursuing simultaneously two distinct generalizations of the first U-quantum paper. In the first instance, he used a generalized form of the four-vector field (of which the U of part 1 was the "spinless" fourth component) in order to describe massive quanta of spin 1. Since the quanta were massive, no gauge principle applied; therefore (unlike the case of QED) the "potential" and the "field" could be considered independent.

[53] This was Yukawa's second generalization making use of Pauli and Weisskopf (1934). Yukawa did not know this work when he formulated his first theory of "scalar" U-quanta (1935).

[54] [YHAL] E03 060 P13, on p 266 of Brown *et al.* 1991.

[55] Item (4) is a separate draft of three pages, designated as [YHAL] E03 060 P13.

[56] [YHAL] E02 080 P12. See Kawabe (1991b). We are greatly indebted to Professor Kawabe for supplying us with copies of these and many other [YHAL] documents, together with his excellent annotations.

[57] See Note 38. Yukawa remembered posting the letter to *Nature*, and indeed, its envelope is so marked, but no letter of acknowledgment has been found.

[58] Yukawa (1971, p 243). The English translation of this passage is given in Kawabe (1991a, p 176). Fermi had received a similar rejection from *Nature* (F. Rasetti in Fermi (1962, p 540)): "Fermi intended to announce the results of his beta-decay theory in a letter to *Nature* but the manuscript was rejected by the

Editor of that journal as containing abstract speculations too remote from physical reality to be of interest to the readers."

[59] Serber (1983, especially p 12) (our emphasis).

[60] See note 33.

[61] Oppenheimer and Serber (1937). It is difficult now to understand the defensive tone of this letter. Possibly it has to do with the attitude of the Caltech cosmic ray experimenters. Serber said (when this point was raised to him by LMB) that Oppenheimer changed the letter after discussing it with Millikan. Both Anderson and Neddermeyer have stressed that their particle discoveries, i.e., both the positron and the mesotron, were made independently (and without any prior knowledge) of any theoretical predictions (e.g., see Anderson and Anderson (1983, p 149)).

[62] Stueckelberg (1937, p 42).

[63] It is entitled "On the theory of the new particle in cosmic rays". The other letter was that to *Nature*, dated 18 January 1937. See Kawabe (1991b), in which both letters are reproduced.

[64] The *U*-field, since it describes charged particles, must be complex, as opposed to the electromagnetic field, which is real. The term "current" here refers to the analogue of the electromagnetic current.

[65] This letter is also reproduced in Kawabe (1991b).

[66] Neutral π-mesons were first observed in 1950.

[67] Anderson recalled in 1980: "At first the new particles were known by various names, such as baryon, Yukawa particle, x-particle, heavy electron, etc. One day Neddermeyer and I sent off a note to *Nature* [Anderson 1938] suggesting the name *mesoton* (*meso* for intermediate). At the time, Millikan was away, and after his return we showed him a copy of our note to *Nature*. He immediately reacted unfavorably and said the name should be *mesotron*. He said to consider the terms *electron* and *neutron*. I said to consider the term *proton*. Neddermeyer and I sent off the *r* in a cable to *Nature*." (Anderson and Anderson 1983, p 148, see also Millikan 1939 and Anderson 1961).

[68] Heisenberg to Bohr, 5 July 1937 (Niels Bohr Archive, Copenhagen).

[69] Blackett to Dirac, 16 November 1937 (Dirac Archive, Churchill College, Cambridge.)

[70] Dirac to Blackett, 2 December 1937 (Dirac Archive, Churchill College, Cambridge).

Chapter 7

The Development of the Vector Meson Theory in Britain and Japan (1937–38)

7.1 Introduction

After launching the meson theory of nuclear forces with his first article, read in November 1934 and published in February 1935 [1], Yukawa wrote nine papers on other subjects (as discussed in chapter 6) before taking up the U-field again in the autumn of 1936. His primary motivation for returning to it may have been that increased knowledge and understanding of the composition of the cosmic rays made the existence of heavy quanta appear more likely. By the middle of 1937, indeed, physicists had started to believe that the so-called penetrating component of the cosmic rays consisted of charged particles, positive and negative in approximately equal numbers, whose mass (or perhaps, masses) appeared to be intermediate between those of the proton and the electron (in particular, estimated at about 100–200 times the mass of the electron) [2]. These particles were for a time called "mesotrons" and they soon became recognized as candidates for Yukawa's heavy quanta. Another likely motive for Yukawa's return to his theory of heavy quanta was that there were advances in quantum field theory (QFT) going beyond quantum electrodynamics (QED) [3], advances to which Yukawa and Sakata had given careful attention and had also contributed.

This new phase of meson theory began publicly on 28 November 1936, when Yukawa, at a meeting of the Osaka branch of the Physico-Mathematical Society of Japan (PMSJ), gave a talk entitled "The interaction of elementary particles. II", reminding the audience of his U-quantum theory and giving a revised alternative version of it [4]. Only two days earlier, he had given a colloquium at Osaka Imperial University on "The present status of the theory of cosmic rays", exhibiting a detailed knowledge of that subject which showed evidence of exhaustive study [5].

The lecture began with a historical review, starting from the discovery of the extraterrestrial origin of cosmic rays by V. Hess in 1912, and discussing the work of R.A. Millikan, J. Clay, W. Bothe and W. Kolhörster, A.H. Compton, etc., as well as the latitude, altitude and directional effects, including the analysis of the effect of the earth's magnetic field by T.F. Störmer. Yukawa went on to discuss the work of B. Rossi, C.D. Anderson, P.M.S. Blackett and G.P.S. Occhialini, the hard and soft components, ionization and radiation, absorption, scattering, showers and bursts.

Yukawa concluded by referring to recently published cloud chamber pictures taken by Carl Anderson and Seth Neddermeyer, which contained anomalous tracks (Anderson and Neddermeyer 1936) [6]:

> Recent pictures of Anderson show heavy tracks of positive charge with e/m larger than the proton's. Accordingly, in cosmic ray showers many unknown processes are concealed and even unknown elementary particles might be present. In short, the cosmic rays are the vanguard problem for theoretical physics and the success of explanation will extend the applicability limit of the theory. In this sense, cosmic ray phenomena are gaining more and more importance.

Further signs of Yukawa's renewed interest in U-quanta were two incomplete drafts of sequels to the original paper (Yukawa 1935). Labeled II (11 January 1937) and III (6 January 1937), they reformulate the meson theory in, respectively, relativistic spin-0 and spin-1 versions [7]. During 1937, Yukawa and his students, Sakata, Mituo Taketani and Minoru Kobayashi, elaborated these two new theories, predicted nuclear forces in better agreement with experiment and answered some objections raised against Yukawa's original paper [8].

Based upon his own analysis of the anomalous tracks in Anderson and Neddermeyer's paper of 1936, Yukawa wrote a (rejected) letter in January 1937 to the editor of *Nature* calling attention to his theory of U-quanta and suggesting near the end: "Now it is not altogether impossible that the anomalous tracks discovered by Anderson and Neddermeyer, which are likely to belong to unknown rays with e/m larger than that of the proton, are really due to such quanta, as the range–curvature relations of those tracks are not in contradiction to this hypothesis." [9]

By 1937 Yukawa and his associates, who formed a small but vigorous school in Osaka, had developed two different forms of the meson theory, scalar (section 2) and vector meson field theories (section 3). By the end of the year some scholars in Europe also entered into competition with the Japanese, although there was also considerable cooperation through the exchange of letters and manuscripts. In sections 3–6 we examine the output both of the Japanese and of the European groups and study their relations with each other and with the experimental community [10].

7.2 The formation of the Yukawa school and the scalar field theory

Yukawa's first collaborator, Shoichi Sakata, graduated from Kyoto Imperial University in 1933, four years later than Yukawa and Sin-itiro Tomonaga. Sakata was related by marriage to Yoshio Nishina, who was then the dominant figure in Japanese nuclear and cosmic-ray physics. Nishina had

S. Sakata (1911–1970) (left) and H. Yukawa (1907–1981) in 1942. Photograph reproduced by permission of Yukawa Hall Archival Library.

spent nearly a decade studying and working in Europe, especially at Bohr's institute in Copenhagen, and he was competent both in theoretical and in experimental physics [11]. Through Nishina, Sakata became personally acquainted with Yukawa and Tomonaga; he also attended the quantum mechanics course taught by Yukawa at Kyoto in 1933. After graduation, Sakata spent a year in Nishina's laboratory at Riken in Tokyo. There he worked on problems in nuclear theory with Tomonaga and Nishina.

In 1934, Sakata became Yukawa's assistant at Osaka Imperial University; he also became Yukawa's first and most important collaborator [12]. Although they worked together on a variety of subjects, it was not until 1937 that Sakata began to help to develop the theory of Yukawa's heavy quanta [13]. Yukawa's notes indicate that two separate papers were originally intended, although eventually Yukawa and Sakata published their work as co-authors. Two of Sakata's memoranda on heavy quanta, dating from that time, have been found among Yukawa's papers [14].

While Sakata was working in Tokyo at Riken, he met a student from Kyoto who had taken Yukawa's course the year after him; he also became a collaborator of Yukawa on the meson theories. As Sakata later recalled (Sakata 1965) [15]:

> When I was in Riken, Mr Yamanashi introduced Taketani to me, and after that
> I gradually became intimate with him. He had a fellowship at Kyoto University

after his graduation and, when I moved to Osaka University to accept my assistantship, he visited me in my office about once a month. He had excellent ideas about nuclear theory and we used to have discussions until late at night. Sometimes he talked about Hegel's logic and Yukawa would listen to him with great interest.

Mituo Taketani, at a time when guilt-by-association witch-hunting was the order of the day, collaborated with scholars from Kyoto University's Department of Literature, in editing and in writing for a leftist magazine called *Sekai Bunka* (*World Culture*) [16]. He wrote semi-popular articles on quantum mechanics, which he analysed from a Marxist dialectical viewpoint: he also wrote about his own "three-stage theory" of scientific methodology, according to which (Sakata 1965):

> Nature develops in a spiral, passing successively through the following three stages. The first is the *phenomenological* stage, describing phenomena as they appear in nature. The second is the *substantialistic* stage, analysing the material structure of the phenomena. The third is the *essentialistic* stage, giving the mathematical laws and interactions governing the matter and producing the phenomena... Beginning research on theoretical nuclear physics in our group, Taketani applied his three-stage theory; he decided that we were at the substantialistic stage and that we were seeking to enter the essentialistic stage.

The next phase of the development of the meson theory of nuclear forces was the following. In early summer 1937, Yukawa and Sakata had read the letters in the *Physical Review* referring to Yukawa's *U*-quantum paper (Oppenheimer and Serber 1937; Stueckelberg 1937a) and, according to Sakata, this encouraged them to make a thorough review of the theory. Sakata further noted that (Sakata 1965) [17]:

> Just two years earlier, Pauli and Weisskopf had studied a "scalar electron theory". The electron, in Dirac's ingenious theory, was described by a spinor wave function, but Pauli and Weisskopf considered the case where a particle is described by a scalar wave function. Therefore this research was of purely formal interest at that time; however, we could literally take over their results in quantizing the *U*-field. Using that formalism, we performed a derivation of the nuclear forces and also calculated some processes induced by mesons.

Since, as Sakata indicated, the quantization of the scalar *U*-field by the method of Pauli and Weisskopf was largely taken over from those two authors (and is now standard), we shall discuss only a few minor, but not insignificant, aspects of their work as contained in their published paper, Yukawa and Sakata (1937c) (which we shall now refer to as *Interaction II*). They began by recalling Yukawa's first paper on the *U*-field [18]:

> This field turned out to be accompanied by quanta each with the elementary charge either $+e$ or $-e$, the mass about 1/10 of that of the proton and zero or integer spin, obeying the Bose statistics. It was shown further that such quanta, if they ever existed, could not be produced by ordinary nuclear reactions, but might be present in the cosmic rays as the primary or the secondary.

The only notable point here is that they seem to be unaware of the instability of the heavy quantum, for that would certainly rule out the possibility of the *U*-quantum being present in the *primary* cosmic rays [19].

The authors then drew attention to the growing evidence for new particles of about the expected mass in the cosmic rays and mentioned the physicists who had accepted the existence of these [20]. Reacting to earlier criticism of Yukawa's theory by the theorists, they argued that, if the new particles were really the predicted heavy quanta, then [21]:

> If this is true, the only serious drawback in our theory disappears, although there are still many points to be completed or improved in the course of development of the theory, which will be made in this and subsequent papers. We hope that the whole phenomena of the nuclear transformation, β-disintegration and cosmic rays will be explained in a unified way.

The young Japanese physicists thus projected a series of "improvements" (and a series of papers), in the course of which they expected to find a "unified" explanation of a host of puzzling phenomena at high and low energies. It is remarkable that this bold programme was to a large extent actually carried through by the "Yukawa school" (as it may now be appropriate to call them).

The next step taken by Yukawa and Sakata was the canonical quantization of the complex relativistic scalar U-field, *à la* Pauli and Weisskopf, expressing the charge-exchange interaction between the "heavy particles" (i.e., the proton and neutron) in terms of the creation and annihilation operators of the charged quanta of the U-field. Using this interaction in the second order of perturbation theory, they obtained the static interaction potential as

$$V_{pn} = J(r)P_{12}^{H}\beta_1\beta_2, \tag{7.1}$$

where the exchange integral $J(r)$ is

$$J(r) = g^2 e^{-\mu r}/r \tag{7.2}$$

and P_{12}^{H} is "Heisenberg's exchange operator".

However, the expression (7.1) led to a difficulty [22]:

> In non-relativistic approximation, β_1 and β_2 reduce to 1, so that (1) becomes the same [as] the result in [*Interaction I*] except the sign. In order to obtain a result exactly the same as that of I [necessary to get the right spin for the deuteron ground state], we have to change the sign of [the Hamiltonian for the free U-field], which will obviously lead to serious difficulties of negative energy for the U-field. Whether or not this defect can be removed by introducing non-scalar field will be discussed in [*Interaction*] III etc.

A nuclear force of the Majorana type (i.e., pure space-exchange) could also be obtained as a second order perturbation result by "introducing terms involving the spin of the heavy particle" into the interaction part of the Hamiltonian. On the contrary, ordinary (non-exchange) forces between unlike particles, as well as the forces between like particles, would appear only in fourth, or higher, order. Calculation then showed that in the static limit, these fourth-order forces would be of shorter range and smaller by about a factor of 10 than the second-order forces between unlike particles (contrary to experiment, which showed the like and unlike forces to be nearly equal).

To solve the last difficulty, Yukawa and Sakata noted [23]:

> ... it is not certain whether or not we have to introduce neutral heavy quanta also, in order to account for the approximate equality of two forces assumed in the current theory ... Such speculations are, of course, premature at present.

The authors concluded their paper by considering the scattering of the U-quantum by a heavy particle: "The calculation can be performed in a manner similar to that for Compton scattering of the light quantum by a free electron." [24] They also considered the absorption of a negative U-quantum by a nucleus, leading to the emission of a neutron (a process analogous to the photoelectric effect in atomic physics). Finally, they mentioned other interesting problems, such as β-disintegration and anomalous magnetic moments, and promised that they would be discussed in subsequent papers.

7.3 The vector meson theory in Japan

Partly because of the shortcomings of the scalar meson theory (for example, the incorrect sign of the forces in the deuteron) and partly in the spirit of exploration, Yukawa and Sakata, soon joined by Taketani, took up what appeared to be the chief other candidate for a theory of heavy quanta [25]. As Sakata later described their first attempts (Sakata 1965):

> Yukawa tried to generalize Maxwell's equations of the electromagnetic field, while I examined the wave equations proposed [for various spins] by Dirac a year earlier. Taketani joined our group about this time and we included him in our studies. We three developed a theory, today called "vector meson theory", and showed that it agreed with the experimental results.

Sakata's recollections agree with the documentary evidence. As early as January 1937, Yukawa considered generalized Maxwell equations for the U-field [26]. Before him, the Romanian physicist Alexandru Proca had constructed a theory of massive particles described by a vector field, under the misapprehension that it might describe the electron (although it would have endowed that spin-$\frac{1}{2}$ particle with a spin of 1). His theory (Proca 1936), based upon the Pauli–Weisskopf quantization and apparently taking Pauli's "anti-Dirac" injunction to heart, was probably known to Yukawa when he began to generalize QED to the case of massive quanta. On the other hand, the approach that Sakata adopted made use of Dirac's generalized wave equations, based on the spinor calculus [27].

The preparation of the third paper on U-quanta went on throughout the year 1937. Yukawa himself wrote several drafts and memoranda, and in November he began to compose the manuscript of a joint paper to carry also the names of Sakata and Taketani [28]. This material formed the basis for

Top: V.F. Weisskopf (born 1908) in 1934 (photograph from Paul Ehrenfest, Jr); bottom: P.A.M. Dirac (1902–1984). Photographs reproduced by permission of AIP Emilio Segrè Visual Archives.

colloquia and talks given at meetings of the PMSJ. One problem with which Taketani was especially concerned was that of the magnetic moments of the neutron and the proton [29].

Finally, we come to two drafts of *Interaction III*, bearing the names of the three authors and written in Yukawa's hand. The first, four pages long, is marked "Read 25 September 1937" (but was written after that date); the second, 33 pages in length, is marked "Read 25 September 1937 and 22 January 1938" (but was written after the later date). The published paper (*Interaction III*) carries the same dates as the second draft [30].

We turn now to the content of *Interaction III*: The authors refer to the two earlier papers "introducing a new field of force" in four-vector and in relativistic scalar versions, but assert that neither of them was "ample enough for the derivation of complete expressions for the interaction of the heavy particles and their anomalous magnetic moments". Describing the formalism which they would employ, the authors said [31]:

> In this paper, we begin with the construction of the linear equations for the new field, which can be considered as a generalization of Maxwell's equations for the electromagnetic field. The field is thus described by two four-vectors and two six-vectors, which are complex conjugates to each other respectively. It is interesting that this system of equations written in spinor form reduces to a special case of Dirac's wave equations for the particle with the spin larger than $\frac{1}{2}$. Meanwhile, it came to our notice that our formulation was equivalent to a method of linearization of wave equations for the electron, which had been developed by Proca as an extension of the scalar theory of Pauli and Weisskopf.

Regarding the discussion of the anomalous magnetic moments of the heavy particles, the introduction refers to the Japanese paper of Taketani and a paper by Herbert Fröhlich and Walter Heitler (Fröhlich and Heitler 1938b). References to the British paper were obviously added in the later preparation of the Japanese paper.

As an aside, we mention that, just before posting *Interaction III* to the journal, Yukawa received a letter from Fröhlich and Heitler in Bristol (written in German), which began [32]:

> Many thanks for the interesting manuscripts that you have been sending us, especially the last one. We quite believe that your theory is correct in principle. We ourselves have considered a great deal about the heavy electron and have formulated a theory of its interaction with the nucleus (together with Kemmer). From the discussion of the *spin-dependence* of the proton–neutron force we have arrived at the conviction that the field [of the heavy electron] must be a *vector field*, as you have also assumed in your last Japanese note. The wave equation is the same as yours and has already been proposed by Proca (*Journal de Physique* 1936).

We can see that the East–West competition–cooperation on the meson theory was well under way by the beginning of 1938. The level of sophistication of the Japanese authors had risen considerably since 1935, as regards both field quantization and nuclear physics. For example, that the heavy quanta have spin 1 is inferred from noting that all ten of the field quantities depend on only three independent components. Similarly, the unit charges are inferred from the gauge transformations of the field. Again: "It is likely that the anomalous magnetic moments of the neutron and the proton can be attributed to the virtual presence of the heavy quanta in the intermediate states." Finally, although the paper does not consider "the possible existence of the neutral heavy quanta, which seems to be important in connection with the problem of the forces between two neutrons and between two protons", it treats quantitatively the decay of a heavy quantum in free space, "the possibility of which was pointed out by Bhabha" and its "bearing on the problem of the hard component of the cosmic ray is discussed" [33].

After the general introduction, the authors began by introducing vector fields **F** and **G**, "in analogy with the electric and magnetic vectors in electrodynamics". Each component of the free field satisfies the generalized d'Alembert equation (just as the scalar field U of *Interaction II* did:

$$(\Box - \lambda^2)\begin{pmatrix} \boldsymbol{F} \\ \boldsymbol{G} \end{pmatrix} = 0, \tag{7.3}$$

with $\lambda = m_U c/\hbar$, m_U being the mass of the U-quantum. To obtain linear equations of the Maxwell type, whose iteration yields the above second-order equations, it is necessary to introduce another four-vector of fields U_0, U, analogous to the four-potential in electrodynamics). One can then write ten linear Maxwell equations, each one generalized by the inclusion of a term proportional to λ. There are in addition five supplementary conditions, an example of which is: div $\boldsymbol{G} = 0$. The fields are complex and an analogous set of equations holds for their complex conjugates.

The canonical quantization procedure is then carried out, and the Hamiltonian and the field equations can all be expressed in terms of the four-vector potential function U_0, U, as is the case in QED. The Hamiltonian that describes the free U-quanta after quantization is

$$\overline{H}_U = \sum_k \sum_j k_0 \hbar c (N_{jk} + M_{jk} + 1), \tag{7.4}$$

where N_{jk} denotes the number of heavy quanta with positive charge, energy $\hbar c k_0$ and momentum $\hbar c k$, and M_{jk} denotes the number with negative charge, energy $\hbar c k_0$ and momentum $-\hbar c k$. The indices j and k can take on three values, and thus the fields describe quanta of spin 1. On introducing the electromagnetic field in the usual gauge-invariant way, the resulting equations for the four-vector U_μ "show that the heavy quantum has the magnetic moment $e\hbar/(2m_U c)\boldsymbol{\sigma}$, where $\boldsymbol{\sigma}$ is the 3×3 matrix three-vector for spin 1".

This deduction of the magnetic moment of the heavy quantum now permitted a qualitative understanding of the anomalous moments of the heavy particle, which was assumed to spend a certain fraction of its time "dissociated" into a heavy quantum and the other kind of heavy particle. The rough estimation (which followed earlier discussions by Taketani and by others [34]) was given as follows [35]:

> The fraction of time during which the neutron is splitting up into a proton and a heavy quantum with the negative charge virtually, is roughly given by $g^2/(\hbar c)$,

where g is the constant characterizing the strength of the interaction of the heavy particle with the heavy quantum. Thus the contribution of the heavy quanta to the magnetic moment of the neutron has the negative sign and the magnitude of the order

$$[g^2/(\hbar c)] \cdot [e\hbar/(2m_U c)] = [g^2/(\hbar c)] \cdot [M/(mc)] \cdot \mu_n \,,$$

where μ_n is the nuclear magneton and M is the mass of the heavy particle... It is clear from the symmetry considerations that the extra magnetic moment of the proton due to the virtual presence of the heavy quantum with the positive charge has approximately the same magnitude as that of the neutron, but the sign is positive. These results agree with the experiment both in sign and in the order of magnitude.

The next two sections dealt respectively with "Interaction of the U-field with the heavy particle" and "Deduction of exchange forces between the neutron and the proton". These are fairly elaborate calculations, one result of which is the value of the U-field at a point r_1 due to the presence of a heavy particle at the point r_2:

$$(r_1) = -g_2 \mathrm{curl}(\boldsymbol{\sigma}^{(2)} Q_2^* \cdot \mathrm{e}^{-\lambda r}/r) \tag{7.5}$$

and an expression for the canonically conjugate field:

$$\bar{U}^+(r_1) = [g_1/(4\pi\lambda c)]Q_2^* \mathrm{grad}(\mathrm{e}^{-\lambda r}/r). \tag{7.6}$$

The quantity r is the magnitude of $r_2 - r_1$; Q_2^* represents the operator that changes a proton into a neutron. Note that two different strong coupling constants are involved, because the interaction of the quantum with the heavy particle can have a part which is spin-independent (with coupling g_1) and another part which is spin-dependent (with coupling g_2).

Making use of the U-field of a static nucleon, as given in equations (7.5) and (7.6), the authors then obtained an expression for the Hamiltonian of interaction between two heavy particles, which is similar to one given previously by Kemmer. (*Interaction III*'s result contains only two constants, but Kemmer's interaction has three — see below!) The Japanese authors commented on their result as follows [36]:

> As is well known, this is a combination of exchange force of Majorana and Heisenberg types between the neutron and the proton. It should be remarked, however, that the force thus obtained is not strictly central, so that we can separate S state, P state etc. only in the first approximation.

Although the force is satisfactory for unlike particles, it is necessary to perform a much more complicated fourth-order calculation to obtain like-particle forces, or ordinary (i.e., non-exchange) forces between unlike particles. However, the authors noted [37]:

> It was shown, however, in [*Interaction II*], that these forces was [*sic*] smaller by a factor 10 than the exchange force of the second order, so that the introduction of the neutral heavy quanta was felt necessary in order to reproduce the approximate equality of the like particle and the unlike particle forces assumed in the current theory. These conclusions appear to be true in the present case

and indeed, it is not difficult to consider the field accompanied by the neutral, heavy quanta and described by the linear equations similar to those considered above. The detailed discussions of these subjects will be made in the next paper.

In the final section of the three-man paper, they considered the creation and the annihilation of the heavy quanta. Creation occurs by collision of the primary cosmic rays with matter, if energy exceeding $m_U c^2$ is available, whereas annihilation can occur when the quantum collides with matter. However:

> It should be noticed, further, that a heavy quantum disappears even in the free space by emitting a positive or negative electron and a neutrino or an antineutrino simultaneously according [to whether] the charge of the heavy quantum is positive or negative, as already pointed out by Bhabha [1938a].

Because the lifetime is small compared with the time of passage of the heavy quantum through the atmosphere, "those which have been observed on sea level should have been created, for the most part, in the atmosphere" [38].

The authors performed the actual calculation of the lifetime under the rather arbitrary assumption that the interaction with the light particles has the same form as that with the heavy particles, with smaller values substituted for the coupling constants g_1 and g_2. Using the value of 100 electron masses for the mass of the heavy quantum, they estimated the lifetime at rest to be 5×10^{-7} s. Because of the relativistic time-dilation effect (as Bhabha had pointed out), the lifetime of a fast meson can be much longer; but "even the heavy quantum with the energy 10^{12} eV can travel only a distance smaller than the radius of the earth, before it changes into an electron and an antineutrino" [39].

There followed a brief discussion of the "bearing of this conclusion on the interpretation of the hard component of the cosmic ray" and of the question whether the observed frequency of heavy quanta can be reconciled with the cosmic primaries consisting "exclusively of the positive and the negative electrons". The number of heavy quanta estimated on the latter basis "seems to be a little too small" [40].

7.4 Three refugees from Hitler take up Yukawa's theory

On 13 January 1938, Wolfgang Pauli wrote from Zürich to his former assistant Victor Weisskopf in America a letter full of news, which read in part [41]:

> Last week Bhabha and Kemmer were here, and we had (with the essential participation of [Gregor] Wentzel) a kind of theoretical conference on cosmic radiation. These two gentlemen and also Heitler intend soon to flood *Nature* and the *Proceedings of the Royal Society* with their intellectual products, which deal with the so-called "Yukawa theory" of nuclear forces.

There followed a two-sentence account of the charge exchange process involving Yukawa's heavy quantum (which Pauli called the "ϵ-particle"), leading in second approximation to the Yukawa potential. Pauli continued:

> At the same time, nothing at all is assumed about the interaction of the ϵ-particle with the light particles, so that the theory of nuclear forces is made independent of that of β-decay. In fourth approximation one obtains also (finite and not small) proton–proton forces. If that doesn't suit one, one can also invent further electrically neutral "γ-particles" with Bose statistics. By applying enough learnedness — say, by describing the ϵ-particle by a vector field instead of a scalar field — one can further derive arbitrary spin-dependence and the correct sign of the nuclear forces. In contrast to the second approximation of the Fermi theory of β-decay, here all collision cross-sections and the binding energy of the deuteron remain finite, on account of the only very weak and harmless singularity of the potential energy for $r = 0$, going as $1/r$.

The reader may discern that Pauli had not quite grasped all the subtleties of the Yukawa theory (e.g., that it did make assumptions about the interaction of the heavy quanta with the light particles); still, his announcement to Weisskopf was big news, signalling a potentially very important advance in the theory of nuclear forces and the understanding of cosmic-ray phenomena.

Among the points mentioned by Pauli, one that may well have surprised Weisskopf was the "flood" of papers that would soon appear in British journals not noted for being very open to theoretical works. For example, during the 1930s not more than 20% of the papers in the *Proceedings of the Royal Society*, the leading British journal for physics (it having taken over that position from the *Philosophical Magazine* after the First World War), were theoretical. Paul Hoch wrote about this period [42]:

> The predominant attitude to physics in Britain at that time was that it was — and should be almost as a matter of morality — an experimental science (at Oxford it was still known as "experimental philosophy"). Cambridge had a considerable mathematical aspect to its physics at least since Maxwell, which embraced in part the work of Kelvin, Rayleigh and J.J. Thomson among others. However, by the 1930s it was assumed even within this tradition that Cavendish physicists did their own experiment and that those not doing experiments were not physicists and belonged in the mathematics faculty.

In 1938, however, a single volume of the *Proceedings*, namely **A166**, was to carry twice as many theoretical papers as usual, including the "flood" of four papers on Yukawa's theory. Moreover, this departure from tradition was mainly due to the work of non-British authors, three refugees from Hitler's Germany and an Indian physicist.

After the takeover of the German government by the National Socialist German Workers' Party (NSDAP) in 1933 and the issuing of anti-Semitic laws which soon followed (the Civil Service Law of April 1933 and the Nuremberg Laws of September 1935), more than 100 physicists, mainly Jewish or part-Jewish, emigrated, preferably to English-speaking countries [43]. Three of the authors of the Yukawa-theory papers mentioned by Pauli belonged to this group of émigrés; they had been educated at the leading centres of theoretical physics in Germany. We give here a brief biographical sketch of each of them.

H. Fröhlich (1907–1989) in 1961. Photograph reproduced by permission of AIP Emilio Segrè Visual Archives (Physics Today Collection).

Walter Heitler, born in 1904 in Karlsruhe, studied in Berlin and in Munich, where he obtained his doctorate in 1926 in Sommerfeld's institute, supervised by Karl Herzfeld. (The latter also emigrated to America, but before the Nazis came to power.) Heitler then went to work with Erwin Schrödinger in Zürich, where he collaborated on the theory of chemical binding with Fritz London (who also emigrated to England after 1933). Having obtained his *Habilitation* in 1927–28 in Göttingen with Max Born, Heitler got a Research Fellowship in the autumn of 1933 from the Academic Assistance Council, which had been organized to help academic refugees from Germany, and he moved to the University of Bristol. There he quickly achieved a reputation as a theoretician, especially in the applications of QED at high energy, on which he wrote a path-breaking book (Heitler 1936) [44].

Nicholas Kemmer, one of the visitors mentioned by Pauli to Weisskopf in his letter of January 1938, was born in Petrograd (St Petersburg) in 1911, but in 1933 he was a German citizen. He had moved with his family to England in 1916, where his father's business took him, and where Kemmer received his early education. After the war, the family moved to Germany; from 1931 onwards, Kemmer studied with Wentzel in Zürich and with Born in Göttingen. In the summer of 1933, he returned to Zürich to do a doctoral thesis with Wentzel, which he completed in 1933; then, for a time, he was Pauli's assistant. In the autumn of 1936, he assumed a Beit Scientific Research Fellowship at Imperial College of Science and Technology, London [45].

Herbert Fröhlich was a collaborator of Heitler at Bristol and eventually also of Kemmer. Born in 1905 in Rexingen (in the Schwarzwald), Fröhlich was of Jewish origin. He studied in Munich with Sommerfeld and did his doctoral thesis on the theory of metals. Although he obtained his *Habilitation* in Freiburg im Breisgau, ordinarily qualifying one for a university position, this was of no value to a Jew in Nazi Germany. Despite having gone to Leningrad to continue work on the theory of the solid state with Jacov Frenkel, the deteriorating political situation in the Soviet Union persuaded him not to return to Leningrad from a trip abroad in the summer of 1935 [46].

Instead, Fröhlich joined Heitler in Bristol, where they began collaborating on various topics of theoretical physics. Thus, in early 1936 they submitted a paper on the process of cooling by adiabatic demagnetization of paramagnetic matter (Fröhlich and Heitler 1936). This was part of a theoretical study in which the Bristol *experimental* physicists took great interest. Even greater interest was shown by the Oxford experimenters, Nicholas Kurti and Francis Simon (both of them Jewish refugees from Germany), who were steadily pushing closer to the absolute zero of temperature.

It was characteristic of the work of Heitler and Fröhlich (also of Homi Bhabha, see below, and to some extent of Kemmer, although his interest was more mathematical), that they chose topics of research that were closely related to the leading edge of experimental advance. That was not typical of British theoreticians in general, as Hoch argued (Hoch 1990). They thus filled an empty niche, so their British colleagues did not necessarily feel that they were in competition with them [47].

Curiously, Fröhlich and Heitler's first collaboration was to have a bearing on the determination of the anomalous magnetic moment of the proton, the same topic which three years later became their entry point to the theory of heavy quanta. An earlier treatment of the problem of magnetic cooling by Heitler and Edward Teller (another refugee from Germany, of Hungarian origin), had shed doubt on a method employed by two Russian physicists, who had obtained the magnetic moment of the proton by measuring the magnetization of solid hydrogen as a function of temperature below 4.22 K (Lasarew and Schubnikow 1936; Heitler and Teller 1938a). One of the results of Heitler and Teller appeared to imply that the Russian result was without significance; but a re-examination by Heitler, together with Fröhlich, showed that the Heitler–Teller result held for ionic or atomic lattices, but not for the molecular lattice of solid hydrogen (Fröhlich and Heitler 1936). An improved result by the Russian experimentalists (Lasarew and Schubnikow 1937), namely 2.7 ± 0.2 nuclear magnetons, turned out to be a very good determination.

When Otto Stern and collaborators had shown in 1933 that the proton's magnetic moment was larger than the nuclear magneton assigned to it by Dirac's relativistic equation (which had given a good result for the electron), this proved to be one of the more disturbing and puzzling aspects of the elementary particles [48]. By 1937, Stern and collaborators at the Carnegie–Mellon Institute in Washington and Isidore Rabi and collaborators at Columbia University in New York agreed that the proton had a magnetic moment that was close to two nuclear magnetons, whereas the neutron's moment was about minus three nuclear magnetons [49]. This was generally thought to mean that the heavy nuclear particles were not really describable by the Dirac theory. (This pleased some, who found the hole theory distasteful.)

Gian Carlo Wick had tried to explain the anomalous magnetic moment of the proton on the basis of the Fermi-field theory of nuclear forces and had estimated that the virtual dissociation of the proton into a neutron plus a positron plus a neutrino might give the right value (Wick 1935). However, in a more thorough treatment of Wick's suggestion, Hans Bethe and Robert F. Bacher concluded in 1936 [50]:

> If we insert the ordinary interaction derived from the probability of β-disintegration itself, we are faced with the same difficulties as... when trying to account for the nuclear forces: if we accept the β-interaction as it stands, the expression for the magnetic moment of the neutron will diverge. If we avoid the divergence by "cutting-off" the β-interaction for high energies of electron and neutrino, we shall obtain much too small a value for the magnetic moment.

That was the status of the magnetic moment problem when Fröhlich and Heitler, in late 1937, thought of the possible involvement of the new cosmic-ray particle. The cosmic rays were at the centre of Heitler's attention at that time. In December of 1936, he had submitted a paper with Bhabha on electromagnetic shower production, which had illuminated the issue of the soft component of cosmic rays (Bhabha and Heitler 1937). He then had begun to analyse the penetrating component (Heitler 1937). At that time, he noted that, although there was "no direct evidence" for associating the hard component with the new particles, assuming their existence led to "at least qualitative agreement with the observations" [51].

By the end of 1937, however, even those most skeptical about the new particle (e.g., Niels Bohr) were convinced, together with the others [52]. At the annual conference in Copenhagen in September 1937 it was certainly discussed, since, in addition to Heisenberg and Pauli, it is probable that Heitler attended [53]. Bohr had returned from visiting America, where he had learned personally about the experimental evidence for the mesotron, and from Japan, where he had discussed Yukawa's theory with him.

Some participants heard Yukawa's name for the first time in Copenhagen, at least that is what Heitler stated later: "I do not remember who drew the attention of people to this paper [of Yukawa], I only remember that it was... at a Copenhagen Conference." [54] On another occasion, Heitler recalled that, after he had learned of Yukawa's "short paper published in an

UNIVERSITY OF BRISTOL.

H. H. WILLS PHYSICAL LABORATORY,
ROYAL FORT.
BRISTOL 8.

Tel. No: Bristol 23749.

[Eine Übereinstimmung mit dem Experiment zu erzielen,]

[daraus folgt, daß] $g \simeq f \simeq 5e.$

[Das magnetische Moment wird ?]

$$\frac{\mu}{\mu_0} = \frac{f^2}{tc} \cdot \frac{M}{u} \cdot \frac{1}{\lambda} \int_0^{k_0} dk \qquad \qquad \mu_0 = \frac{et}{2Mc}$$

[(u = Masse des schweren Elektrons). Die richtige Größenordnung erhält man für] $k_0 \simeq \lambda = \frac{uc}{t}.$

[Wir glauben nicht, daß man die gesamte Ruhenergie des Protons Mc^2 als Selbstenergie durch Wechselwirkung mit dem schweren Elektron deuten soll. Mit dem obigen Wert für k_0 wird die Selbstenergie von der Größenordnung]

$$\frac{g^2}{4u} \cdot \frac{1}{\lambda^2} \int_0^{k_0} k^2 dk \ll Mc^2.$$

[Unsere Arbeit wird in den Proc. Roy. Soc. erscheinen.]

[Mit freundlichen Grüßen]   W. Heitler

H. Fröhlich.

From a letter of Heitler and Fröhlich to Yukawa on the vector meson theory, dated 5 March 1938. Reproduced by permission of Yukawa Hall Archival Library.

obscure place", he and Fröhlich proceeded to "extend Yukawa's theory" and: "The first [extension] in which we succeeded was a theory of the magnetic moments of neutron and proton, which are well-known to deviate strongly from the values predicted by the Dirac equation." [55]

Fröhlich's recollections of the same period agree generally with those of Heitler, but raise some questions about their knowledge of Yukawa's paper when they wrote on the anomalous moment problem. Thus, Fröhlich recalled [56]:

> In 1937, Heitler and I were interested in understanding the deviation of the magnetic moment of the proton and the neutron from the Dirac moment in terms of virtual absorption and emission of a "heavy" particle satisfying Bose statistics. One of the main problems arising here concerned the symmetry of this particle. Only later did we realize that Yukawa's general idea would involve this possibility.

There is thus a difficulty in deciding whether Fröhlich and Heitler had Yukawa's theory in mind in November 1937, when they treated the anomalous moments. In any case, in their letter to *Nature* (Fröhlich and Heitler 1938b) they did not cite Yukawa at all. Instead, they referred to the previous unsuccessful treatments performed on the basis of the Fermi-field theory. Using a similar approach, they assumed that the virtual emission of a "heavy electron" will cause a neutron to transform into a proton (or vice versa): "hence it would follow that heavy electrons have no spin and satisfy Bose statistics" [57]. Provided that the (unspecified) interaction could induce a spin flip as well as a change of charge, there would be a contribution to the magnetic moment arising from the orbital angular momentum of the heavy electron.

The authors then pointed out that their idea permitted an estimate of the mass of the heavy electron through a fit of their *Ansatz* to the measured moments. Thus, if α is the fraction of time that the heavy nuclear particle spends dissociated, then (in units of the Bohr nuclear magneton, with M the mass of proton and m that of the heavy electron)

$$\mu_P = 1 - \alpha + \alpha M/m, \qquad (7.7a)$$

$$\mu_N = -\alpha - \alpha M/m. \qquad (7.7b)$$

"Inserting the observed values $\mu_P = 2.6$, $\mu_N = -1.75$, we obtain $M/m = 22$ or $m = 80$ electron masses" [58].

7.5 The first vector meson theories of Kemmer and Bhabha

The letter of Fröhlich and Heitler on the "heavy electron" contribution to the nucleon magnetic moments was sent from Bristol on 24 November 1937. On 8 December Kemmer also submitted a letter from London to *Nature*, which made use of what *he* called the "Yukawa particle" to obtain the nuclear forces (Kemmer 1938a). On 13 December, Homi Bhabha sent to *Nature* a letter from Edinburgh (which, like Kemmer's letter, cited Yukawa's paper of 1935) which contained Bhabha's own version of the theory (Bhabha 1938a).

The Indian physicist Homi Jehangir Bhabha, the fourth of the meson theorists working in Britain, was not a refugee, neither was he a member of the British establishment. Born in 1909 in Bombay, he was related by marriage to the famous Tata family of industrialists. He entered Gonville and Caius College, Cambridge, obtaining a first class degree in the Mechanical Sciences Tripos, and then transferred to physics (Dirac was one of his tutors). As the holder of an Isaac Newton Studentship in 1934, he travelled in Europe, and studied with Bohr, Pauli and Fermi. He obtained his Ph.D. in 1935 [59]. In his most important work prior to that on the meson theory, Bhabha had collaborated with Heitler to explain the development of cosmic-ray cascade showers (see chapter 4). We shall now discuss his and Kemmer's letters to *Nature*, both of them proposing vector-meson theory independently of the Yukawa school.

Kemmer's attention had been drawn to Yukawa's paper by a letter from Wentzel, with whom Kemmer had remained in contact after leaving Zürich. Both men had been working on the proper formulation of a charge-independent theory of nuclear forces [60]. Wentzel had written two papers, in which he had proposed the existence of virtual heavy nucleons of integer spin, after which he had tried to add "neutral" supplements to the Fermi-field, namely electron–positron and neutrino–antineutrino pairs (Wentzel 1937a, b, c). He had not achieved his aim, a charge-independent theory, thinking it impossible except in a *purely* neutral theory, but Kemmer showed that it was possible by constructing a correct theory (Kemmer 1937); see chapter 3.

In an informal talk in 1970 at the Scottish Universities Summer School in Physics, Kemmer, who was then a professor at the University of Edinburgh, told how he had become interested in Yukawa's theory via his work on extending what he called (in 1970) "this electron-neutrino field nonsense" to obtain a charge-independent theory (Kemmer 1971). Regarding Yukawa's 1935 paper, he said:

> In fact I think I have heard someone say that his paper was read, but as far as I was concerned I knew nothing about it until 1937 when of course there came the Anderson mesotron – a charged particle of just about the right mass for Yukawa's theory. As soon as that was known, we were off and running on Yukawa's ideas.

Kemmer felt that he was particularly well equipped by his background and training to handle these theoretical notions [61]:

> I was in the fortunate position of having dealt with the whole business of charge independence with a different kind of field and so I could set off without delay on doing the same thing for the meson. But I also had another bit of good fortune, for reasons which have nothing to do with this subject (they were an exercise in trying to work with various extended versions of General Relativity, including Mie's theory), it happened that I knew all about spin 1 equations! I knew that Proca had used them in a paper where he was applying them to the electron, which of course wouldn't work, and I also knew how to handle the appropriate Lagrangian–Hamiltonian formulation, and how to quantize them. So there it was – there was the Yukawa idea and there were two bits of technique that I had in my hands. I knew the Proca equations and I knew about charge independence.

N. Kemmer (born 1911) in the K.B. Library. Photograph reproduced by permission of N. Kemmer.

In his letter to *Nature*, which appeared in January 1938, Kemmer pointed out that "the scalar relativistic Schrödinger equation" gave the wrong spin to the deuteron (spin 0, rather than the observed spin 1), as Sakata and Yukawa already knew (*Interaction II*). However, "a more satisfactory relation can be obtained if one admits a *vector* wave function for the new particle, such as was used by Proca in a different connection". Then:

Using the most general combination of possible interactions of the Yukawa and Proca type, the potential is found to be

$$V(r) = [A + B(\boldsymbol{\sigma}_N \cdot \boldsymbol{\sigma}_P) + Ck^{-2}(\boldsymbol{\sigma}_N \cdot \mathrm{grad})(\boldsymbol{\sigma}_P \cdot \mathrm{grad})]\mathrm{e}^{-kr}/r, \qquad [7.8]$$

where k is $2\pi c/h$ times the rest mass of the new particle, $\boldsymbol{\sigma}_N$ and $\boldsymbol{\sigma}_P$ the spin operators of neutron and proton respectively, and r the distance between these particles. A, B and C are constants.

A satisfactory spin-dependence was obtained with the choice $C = 0$ and $A : B = 3 : 5$. However, one achieved a charge-independent theory only "by assuming that the new particle also has a charged and an uncharged state" (Kemmer 1938a).

By the time Kemmer's letter to *Nature* appeared, he had discovered that Fröhlich and Heitler had also considered the possible role of the "heavy electron" in nuclear physics (Fröhlich and Heitler 1938b). Perhaps more unexpectedly, Kemmer soon found that he had a friendly rival in Bhabha. The latter, an expert in cosmic-ray theories, who with Heitler had done much to explain the nature of the soft component (Bhabha and Heitler 1937), had also tried (as had Heitler, earlier) to analyse the penetrating component (Bhabha 1938b; Heitler 1937).

P.M.S. Blackett, the dominant experimental cosmic-ray physicist in Britain, preferred to believe (even late in 1937) that the penetrating cosmic-ray particles were electrons of very high energy, whose electric field would be strongly concentrated relativistically and whose radiation energy loss was modified so "that radiation is not emitted when the field strength exceeds a critical value" [62]. Nevertheless, Bhabha made bold to suggest that "there are sufficient grounds to justify us in considering the presence of new particles of electronic charge and mass between those of the electron and proton at least as a possibility" [63]. His detailed analysis of the experiments then showed that [64]:

> a "breakdown" theory for radiation loss of electrons [i.e., the view held by Blackett] cannot explain (1) the latitude effect at sea level from latitudes of 35–50°, (2) the large number of particles found at sea level ... of 10^{10} e-volts, (3) the shape of the transition curve for large bursts. All these facts can be explained by assuming that the penetrating component consists of new particles with masses between those of the electron and proton.

Now in the issue of *Nature* of 15 January 1938, immediately following Kemmer's letter, there appeared a letter of Bhabha (dated 13 December 1937) which began, "We have generalized a theory put forward by Yukawa showing that nuclear forces can be explained by assuming the existence of new particles of mass about two hundred times that of the electron." (Bhabha 1938a). A month later (on 28 February), he submitted a paper with detailed calculations to the *Proceedings of the Royal Society*, in which a footnote stated [65]:

> I am very much indebted to Dr. W. Heitler for a discussion on cosmic radiation in which he drew my attention to the theory of Yukawa. This discussion formed the starting point of the considerations of this paper.

Two other footnotes give acknowledgment to Kemmer for the private communication of important technical information, showing that the "meson theorists" working in Britain formed a collaborative–cooperative group by themselves (in addition, as we shall see, with the Yukawa school).

One aim of Bhabha's letter and his paper that followed was to point out the apparent advantages of the relativistic four-vector (spin-1) version of the theory over Yukawa's spin-0 version. However, in the brief space of the letter, Bhabha emphasized the important physical consequences, some of which had been "mentioned by Dr. Heitler in a conversation with me on the original Yukawa theory". One of the most important of these concerned the instability of the heavy quantum [66]:

> A positive *U*-particle at rest may disintegrate spontaneously into a positive electron and a neutrino. This disintegration being spontaneous, the *U*-particle

may be described as a "clock", and hence it follows merely from considerations of relativity that the time of disintegration is longer when the particle is in motion. We believe that this may have to do with the fact observed by Blackett and others that below 2×10^8 eV most cosmic ray particles are electrons, above this energy heavy electrons. In a previous paper [67] we have shown that the experimental evidence requires that heavy electrons can apparently turn into ordinary electrons. Our U-particles are then to be identified with the heavy electrons, and it follows that most of the heavy electrons have been created either in the earth's atmosphere or not very far from it.

The other consequences of the theory noted by Bhabha are essentially the same as those given by the Yukawa group, by Kemmer, and by Fröhlich and Heitler. Bhabha also points out that the charged U-particles "cannot explain the close-range proton-proton interaction". For this, one needs a neutral partner of the charged U-particles.

When Bhabha and Kemmer had announced to Pauli the "flooding" of the British journals with papers on the meson theory, they had scarcely exaggerated. Volume A166 of the *Proceedings of the Royal Society* contained four substantial papers by the four "Britons" espousing meson theory. Besides Bhabha (1938c), to which we have already referred, one was Kemmer (1938b), one was Heitler (1938a), and one by the last two with Fröhlich (Fröhlich *et al.* 1938).

There is considerable overlap of the contents of these papers and internal evidence of communication. We shall discuss them in the next section. Kemmer gave his recollection of how this came about [68]:

> Now how did Fröhlich and Heitler get into this? Well they were in Bristol and I was in London. Pauli wrote to Heitler saying "Here is a young chap you had better meet some time and work with him". Indeed we did meet in London. There used to be a thing called the Physics Club which met in the Royal Society rooms and there was a talk by somebody or other and I met Heitler and reminded him that I had known him before — I had sat at his feet as a student in Göttingen for a semester or so. I met him and he said "What are you working on?" and I said, "I'm working on this Yukawa stuff". "Oh, so are we!". I said "Oh well, I've got a way of introducing the spin dependence in the forces...". "So have we — that's very interesting — well, what have you done?" and I told him. "But we have done something different!". When I saw what they had done I felt confident enough to shout Heitler down and to tell him it was rubbish — and do you know the reason I shouted them down? Because their version of the theory violated parity! What they had done — they were working with spin zero — was simply to add a straight ordinary Yukawa term and another term which we would now call the pseudovector coupling of the pseudoscalar meson... They just put those two together.

After a considerable argument, Kemmer convinced Heitler that he was correct in his objections. Then the three men published a joint paper using Kemmer's method of achieving a spin-dependence of the nuclear forces.

On the other hand, Fröhlich had only this to say about the collaboration [69]:

> At this time we met Kemmer who had developed the quantum field theory of the sixteen relevant Bose fields. Jointly with him we then decided that Yukawa's

idea of nuclear forces, together with an interpretation of anomalous magnetic moments could be best expressed through a vector field and we developed a formal theory.

7.6 British papers on the vector field theory of nuclear forces

We turn now to the four papers dealing with the meson theory that appeared in the volume A166 of the *Proceedings of the Royal Society of London*. Kemmer was the sole author of the first of these, and co-author, with Fröhlich and Heitler, of the second, which appeared immediately following. The two articles could have been combined — to some extent they are redundant — but evidently there was a difference in viewpoint that led to two papers rather than one. Kemmer's paper was more mathematically complete, more formal and less speculative than the three-man paper.

Both articles began with references to Heisenberg's charge-exchange hypothesis for the neutron–proton force, and both rejected the Fermi-field as the exchanged entity, on account of the small value of the Fermi constant. Characteristically, Kemmer stated at the outset: "As an alternative and simpler description of the nuclear field, Yukawa put forward the idea that the interaction is transmitted by charged particles obeying Einstein–Bose statistics".

In contrast, the three-man paper began with experimental observations: "A new hope for such an 'exchange theory' of the properties of nuclei is offered by the probable existence of a hitherto unknown type of particle constituting the hard component of cosmic radiation." Only after some discussion of the cosmic-ray evidence do the *three* authors refer to the theory of Yukawa and his deduction of the mass of the "heavy electron" from the range of nuclear forces. They considered as "strong support" that the mass determined from the nucleon magnetic moments by two of the authors (Fröhlich and Heitler 1938b) was in agreement with Yukawa's estimate.

In Kemmer's solo paper he set as his goal to consider the most general (linear) forms of Yukawa's "treatment", that is, all possible Einstein–Bose or integral spin fields and all possible generalizations of their nuclear interaction. He found that there were two independent possibilities for each of spin 0 and spin 1, differing in their behaviour under space reflection [70], remarking: "On the other hand, a consistent theory for higher values of [spin] does not appear possible." For the nuclear forces, he stated [71]:

> The expression for the interaction of the neutrons and protons with the Bose particles will also be taken in a very general form, a generalization being possible even in the case treated by Yukawa [i.e., spinless U-quanta]. It will be shown that the similarity of all the cases studied is so great that *a priori* it seems impossible to discriminate in favour of any of them. A decision can, however, be reached if one evaluates the neutron–proton force in all the different cases. Comparing with experimental data, only one possibility proves tenable, and it is important to note this is a case with spin and not the scalar case hitherto usually considered.

Kemmer next proceeded to the quantization of Proca's equation, obtaining the same results as the Japanese authors had done independently [72]. Then, in order to obtain the most general set of equations to describe particles of integral spin, he turned to the general spinor theory of Dirac (1936), just as the Japanese authors (somewhat earlier) had done. Noting that each of Dirac's spinor equations had two independent representations (of opposite reflection character), which correspond to four different types of wavefunction when expressed in tensor notation, he wrote four Lagrangians from which their equations can be derived and determined their energy-momentum tensors, observing [73]:

> In each case the energy density is essentially positive; this is a special characteristic of these simple cases. If we had proceeded from any more complicated examples of Dirac's spinor equations, we would not in general have obtained this result... A theory for spin values greater than unity therefore immediately leads to serious difficulties. On the other hand... it would not appear justified to exclude any of [the four simple cases] *a priori*.

In constructing the interaction Lagrangian, it is necessary to form a relativistic scalar by combining one of the fields with one of the five covariant quantities of the form $\Phi_N^* \beta \Gamma \Phi_P$, where Φ is either a proton or a neutron spinor and Γ is one of the four relativistic covariants that can be formed from products of the Dirac matrices ($\boldsymbol{\alpha}$ and β). It is assumed furthermore that no explicit derivatives shall be included. Each of the two spin-0 and the two spin-1 cases can then have interactions with the nucleon covariants which have two independent terms and thus depend upon two assigned coupling constants f and g [74].

Next, Kemmer calculated the neutron–proton exchange potentials resulting from his four interactions in the static limit (neglecting nucleon recoil) and gave the results as

$$V^a(r) = -[c\lambda/(4\pi)]g_a^2 Y(r),$$

$$V^b(r) = +[c\lambda/(4\pi)]\{g_b^2 + f_b^2[(\boldsymbol{\sigma}_N \cdot \boldsymbol{\sigma}_P) - (\boldsymbol{\sigma}_N \cdot \mathrm{grad})(\boldsymbol{\sigma}_P \cdot \mathrm{grad})]\} Y(r),$$

$$\begin{aligned} V^c(r) = -[c\lambda/(4\pi)]\{g_c^2(\boldsymbol{\sigma}_N \cdot \mathrm{grad})(\boldsymbol{\sigma}_P \cdot \mathrm{grad}) + f_c^2[(\boldsymbol{\sigma}_N \cdot \boldsymbol{\sigma}_P) \\ -(\boldsymbol{\sigma}_N \cdot \mathrm{grad})(\boldsymbol{\sigma}_P \cdot \mathrm{grad})]\} Y(r), \quad (7.9) \end{aligned}$$

$$V^d(r) = +[c\lambda/(4\pi)]g_d^2(\boldsymbol{\sigma}_N \cdot \mathrm{grad})(\boldsymbol{\sigma}_P \cdot \mathrm{grad}) Y(r),$$

with $Y(r) = \mathrm{e}^{-\lambda r}/r$. By comparing the signs of the various contributions for the singlet-S and triplet-S states of the neutron-proton system with those required by the known properties of the ^{3}S ground state of the deuteron and the low-energy scattering data, Kemmer concluded that "no other case but (b) agrees with experience". He added: "The detailed discussion of this fact is the subject of the paper with Fröhlich and others." [75]

In the last section of his paper, Kemmer considered the self-energy of the proton and neutron, finding that the results from all four cases gave a quadratic divergence (i.e., worse than the electromagnetic divergence, which is only logarithmic). He also gave a "more rigorous" derivation of the magnetic moments than "had been studied for case (b) in the paper by Fröhlich and others" by considering the third-order perturbation result involving the emission and absorption of a "heavy electron" by a neutron or proton in the presence of a magnetic field. Again, he found that the result was quadratically divergent in three of the cases, but only linearly divergent in case (b). Without doubt, Kemmer's paper was the most complete and general investigation of the quantum field theory of Bose–Einstein particles up to that time. He had attempted to exhaust all reasonable possibilities and arrived at a unique spin-parity (polar vector, i.e., spin 1, parity odd) and a unique form of interaction (although it would eventually be superseded in a more complete theory). He did all this on his own.

It is clear that Kemmer had much to contribute to the paper with Fröhlich and Heitler. The latter was perhaps the leading cosmic ray theorist in Britain, rivalled only by Bhabha, who was his junior. In 1973 Heitler had occasion to compare his two collaborators. Speaking of Fröhlich, he said [76]:

> Herbert's strength consisted in his wealth of *anschaulich* ideas, by which he grasped the physics without much reference to the underlying mathematics. Formalism was not his strength... Such a gift is today unfortunately rare. Of formalists there are enough.

Of his association with Kemmer, he wrote [77]:

> [When Kemmer joined us] he was working in London, but we met often. That gave a very happy balance to our efforts, for Kemmer had mastered the formalism of quantum field theory much better than we. The three of us could then establish a meson theory (as we call it today). It was a vector meson theory, in analogy to the Maxwell theory. Only much later was it recognized that the meson field was a pseudoscalar. We could also predict the existence of a neutral pi-meson (from the proton–proton interaction) which was later discovered. From that Kemmer then developed the charge-symmetric theory, which was the basis of the concept of isospin. The vector meson theory was developed in three different places at the same time and in almost identical ways: by Yukawa, Sakata and Taketani and by Bhabha [and by ourselves].

Since we agree with Heitler on the last point, we shall only call attention to the distinctive differences in the three treatments. We have already mentioned the difference of emphasis between the introduction to Kemmer's solo paper and that with his collaborators. The latter paper pointed out: "The heavy electrons are certainly not stable." [78] However, it is clear from the paragraph of which this is the leading sentence that the authors refer to nuclear absorption as the means of disappearance of the heavy electrons. Nowhere in the paper is there any reference to the radioactive decay of the free heavy electron or to its role as an intermediate virtual particle in β-decay (neither is there in Kemmer's solo paper) [79]. Thus these authors deliberately ignored a major feature of Yukawa's theory, in contrast to the

Japanese authors of *Interaction III* and Bhabha. After citing the agreement of the mass of the heavy electron deduced by Yukawa from the range of nuclear forces and by Fröhlich and Heitler from the nucleon magnetic moments, Fröhlich, Heitler and Kemmer wrote: "We think, therefore, that it might be a reasonable policy to try to link up the nuclear properties (forces and magnetic moments) with the cosmic-ray phenomena of the hard component rather than with the β-decay... It will be seen that such a scheme is possible and that it leads to a consistent theory of the nuclear forces and of the magnetic moments." [80]

In setting up the formalism for calculating the neutron–proton interaction, the authors cited the work of Yukawa and Sakata on scalar particles (*Interaction II*), stating, "It turns out, however, that such a theory cannot account for the magnetic moment of the proton and the neutron and leads to the wrong spin dependence of the nuclear forces". They thus assumed "a vector wave function for the heavy electron satisfying the equations given by Proca". Other possibilities were disregarded on the strength of Kemmer's analysis "in an accompanying paper" [81]. Because they had reached a decision on the form of the theory (and thus not requiring the most general expressions), the quantization was carried out in a simpler manner than that in Kemmer's accompanying paper. Noting that the theory was a generalization (for massive charged quanta) of Maxwell's equations, they stated that "the quantization may be performed in close correspondence to the well-known procedure in radiation theory". In particular, they adopted much of the notation in Heitler's book (1936) on that subject.

They obtained the same free-field Hamiltonian as Kemmer had (and also Yukawa, Sakata and Taketani in *Interaction III*) and then considered the interaction of the meson with the nuclear particles. Here again, they adopted Kemmer's conditions and adopted his choice, namely the vector interaction, Kemmer's type (b) [82]. They then stated: "For our purpose we have simply to take the nonrelativistic parts of this interaction." These are the parts for the transverse and longitudinal waves, with coupling constants f and g, respectively. Their results were obtained for the static case, that is with the neutron or proton taken to be at rest.

For the neutron–proton potential from the longitudinal waves they found the simple result

$$V_{II} = g^2 P Y(r), \qquad (7.10)$$

$Y(r)$ being the Yukawa potential $e^{-\lambda r}/r$ and P the charge-exchange operator. The transverse wave potential was given as

$$V_I = f^2 P Y(r)\{(\boldsymbol{\sigma}_N \cdot \boldsymbol{\sigma}_P)[1 + 1/(\lambda r) + 1/(\lambda^2 r^2)]$$
$$-[(\boldsymbol{\sigma}_N \cdot \boldsymbol{r})(\boldsymbol{\sigma}_P \cdot \boldsymbol{r})/r][1 + 3/(\lambda r) + 3/(\lambda^2 r^2)]\}. \qquad (7.11)$$

The authors then proceeded to find the mean values of the neutron–proton potential energy in the two lowest states of this system: the spin-triplet ground state of the deuteron and the spin-singlet "virtual" scattering state.

(They assumed that the D-wave contribution to the ground state arising from the tensor force was negligible.)

These potentials were shown to be

$$^3\text{S}: \qquad V_{NP} = -Y(r)(g^2 + 2f^2/3),$$

$$^1\text{S}: \qquad V_{NP} = -Y(r)(2f^2 - g^2). \qquad (7.12)$$

Choosing $f \approx g$, they concluded [83]:

> We have found that our scheme makes it possible to account in a reasonable way for the nuclear forces, including their right spin dependence...Only the vector scheme leads to a qualitative agreement with the experiments.

When they calculated the proton–proton force, a fourth-order interaction, they found it to be always repulsive. Because of the large value of the coupling constant inferred from the earlier neutron–proton force calculation, the proton–proton force can be large enough, but the experiments indicated that the force was an attractive one. They wrote [84]:

> If this [experimental] result is established it would inevitably lead to the conclusion, that we have also to introduce *neutral particles* ("Neutrettos" [*sic!*]) of the same mass as and similar properties to those of the heavy electrons.

The paper concluded with an elaborate but inconclusive treatment of the anomalous magnetic moments of the neutron and proton.

7.7 Bhabha's paper and the application of meson theory to cosmic-ray phenomena

Like Heitler, Bhabha's interest in the heavy electrons originated in his analysis of the penetrating component of the cosmic rays. It was Heitler who drew his attention to Yukawa's theory. Bhabha learned from the article of Yukawa and Sakata on the relativistic spin-0 particle that that theory gave the wrong neutron–proton force, whereas Kemmer taught him that only the vector field could be correct [85]. Nevertheless, in his contribution to volume A **166** of the *Proceedings*, submitted from Cambridge a few weeks after his three colleagues in London and Bristol had submitted theirs, he produced some independent and original results.

Unlike Fröhlich, Heitler and Kemmer, for whom the decay of the heavy electron was not important enough for even passing mention, Bhabha emphasized that the heavy mesons must somehow be removed as they passed through the atmosphere, referring to another recent paper of his (Bhabha 1938b). In the opening paragraph of the new paper he stated that "a breakdown of the theory of electrons of very high energy" cannot account for known cosmic-ray effects and that [86]:

> It has further been shown in the same paper that it does not seem sufficient just to postulate another particle behaving exactly like an electron of larger mass,

but that the experimental evidence demands further that under certain circumstances a single heavy electron must be able to change its rest mass in the absence or presence of particles constituting ordinary matter. Indeed the energy loss measurements of Blackett and Wilson indicate that most particles below about 2×10^6 e-volts are electrons, whereas most particles above this energy are heavy electrons, so that at about this energy there must be a large probability of a heavy electron changing or losing its identity.

This work, as did Bhabha's letter to *Nature* that preceded it, favoured the decay over the absorption of the heavy meson [87]. Thus Bhabha welcomed the proposal of Yukawa to make use of the U-quantum as an intermediate particle in the process of β-decay, even though the Fermi theory, to which Yukawa had shown his theory to be equivalent, was being questioned at the time.

In a lengthy introduction, Bhabha also discussed the very different issues of the proton–proton force and of Heisenberg's postulated explosive showers. With regard to the first item, he wrote [88]:

> It is obvious that the U-particles, being charged, can never lead to a proton–proton force except as a fourth-order process, corresponding to the emission of U-particles by each of two protons and their absorption by the other... In order to have a proton–proton interaction of the same order as the proton–neutron force, it appears necessary to introduce a neutral particle N obeying Bose statistics which may be emitted or absorbed when a proton jumps from one energy state to another.

Although, on grounds of analogy with the other known particles (such as the trio e^+, e^- and neutrino), Bhabha considered such a neutral particle "not improbable", he did not consider it *"necessary"* (his emphasis), since the coupling constant was not small, so that in a perturbation expansion the fourth-order interaction might well be as large as the second.

Regarding explosive showers, the introduction concluded [89]:

> Finally, it may be mentioned that this theory also leads to showers of Heisenberg's type, with this difference that these showers would consist mostly of heavy electrons and some proton–neutrons, but few electrons. The existence of these showers still further invalidates the usual methods of calculation, since it shows that for high energies higher order processes are just as important as the first-order process.

The formal developments in Bhabha's paper differed in notation and in other minor respects from those of the other authors, but after some manipulations he arrived at a neutron–proton interaction equivalent to Kemmer's case (b). As in Kemmer's treatment, there are two coupling constants, denoted as g_1 and g_2' by Bhabha. Thus [90]:

> The interaction is therefore just of the required form consisting of Heisenberg and Majorana forces of the right sign so as to allow one to make the triplet state of the deuteron the lowest stable state. We should emphasize the fact that since only the squares of g_1 and g_2' enter into this expression, the sign of the Majorana force is beyond our control, and it is to be looked upon as a strong argument in favour of this theory that it allows only that sign of the force which actually occurs in nature.

Bhabha's paper concluded with some approximate relativistic calculations of the neutron–proton scattering and of the scattering of U-particles by "proton–neutrons" (i.e., nucleons), the latter calculation being an analogue of the Compton scattering of light quanta by electrons. The result for U-particle scattering depended on the relative sign of g_1 and g_2, whereas the neutron–proton scattering did not. Bhabha promised to give the applications of these results to cosmic radiation in a forthcoming paper.

However, it was not Bhabha, but Heitler who first applied the vector theory of the heavy quantum to cosmic ray problems and it is with a brief discussion of his paper (Heitler 1938a) that we conclude this chapter. Heitler claimed no more than a "qualitative explanation of a number of cosmic-ray facts connected with the penetrating radiation". That was because: "For higher energies the theory leads to serious mathematical difficulties (diverging self-energy, diverging nuclear forces of higher orders, etc.)." A footnote stated that only *charged* heavy electrons and not massive neutral particles are considered, because: "The existence of the latter is, although very probable, not yet proved by direct experiments." [91]

The motivation of Heitler's paper was this [92]:

> Even for comparatively small energies of only μc^2 [i.e., the rest energy of the heavy electron] the theory leads to a number of processes giving rise to the creation of various types of showers which all seem to have been observed in connection with the penetrating component of cosmic rays.

These processes are the following.

(1) A neutron captures a positive heavy electron, becoming a proton and emitting an energetic light quantum, which can in turn initiate an electromagnetic cascade shower. The analogous process occurs also when a proton captures a negative heavy electron.

(2) Multiple production of heavy electrons resulting from the collision of a heavy electron with a nucleon: "They lead to penetrating showers produced by penetrating particles." [93]

(3) In a heavy nucleus, the capture of a heavy electron: the nucleus "subsequently evaporates emitting a few protons and neutrons (proton shower) accompanied possibly by electrons and also heavy electrons" [94].

(4) "The inverse process to (1) leads to the creation of heavy electrons by light quanta. It is shown that the order of magnitude of the cross-section is sufficient to explain all heavy electrons at sea-level as secondaries produced in the high atmosphere." [95]

(Process (4) is now referred to as *photoproduction*, whereas process (1) is called *radiative capture*.)

Heitler's paper clearly marked the beginning of a new phase in the discussion of theories of Yukawa type. Although the previous papers of the Japanese and the British schools had tried to provide rigorous mathematical foundations for the quantum field theory of the U-quantum (or heavy electron), its application to the cosmic rays made manifest the presence of the characteristic field-theoretic difficulties already noted in QED, notably the divergence problems, which were even worse in the *U*-quantum case.

In the late 1930s, theorists were divided about how to deal with these difficulties. Pragmatists like Heitler and Bhabha went ahead with explicit calculations, arguing that qualitative insights might be gleaned even from divergent results. An archetype for the cut-off type of argumentation was the anomalous magnetic moment estimate of Fröhlich and Heitler, which had marked their entry into the heavy-electron field. (Fröhlich and Heitler 1938b). More rigorous thinkers like Pauli, however, found cut-off physics distasteful and were highly critical about drawing any conclusion whatsoever from those calculations.

The next part of the story would involve a set of puzzling discrepancies between the properties expected for the heavy electron as the quantum of the nuclear force field, on the one hand, and the behaviour of the cosmic-ray *mesotron* (as it came to be called), on the other. A larger circle of theorists and cosmic-ray phenomenologists were drawn into this controversy, while experimenters set out to measure such properties as the nuclear capture probability and the decay lifetime of the mesotron. These problems were not resolved until after the Second World War.

Notes to text

[1]   Yukawa (1935). We refer to this paper as *Interaction. I.* For Yukawa and his introduction of the *U*-field, see chapter 5.
[2]   See chapter 6 for details.
[3]   Pauli and Weisskopf (1934), Dirac (1936), Proca (1936) and Yukawa and Sakata (1937b).
[4]   The manuscript is [YHAL] E020 60 P12.
[5]   [YHAL] FO50 30 T02. The material in this document of about 18 pages is extensive; it appears suitable for a good deal more than one lecture.
[6]   Note 5. For the translation of the contents of this (and many other) documents, we are very grateful to Professor Kawabe.
[7]   These are: 6 January, [YHAL] E03 060 P13 and 11 January, [YHAL] E02 080 P12. See chapter 6. For a summary of the talk at the PMSJ, Note 4, see Hayakawa (1983, especially pp 88–9).
[8]   Yukawa and Sakata (1937c) and Yukawa *et al.* (1938a). Hereafter we shall refer to these papers as *Interaction II* and *III.*
[9]   [YHAL] E06 U02. This letter and a second one, sent to the editor of the *Physical Review* on 4 October 1937 and also rejected, were reproduced and discussed by Kawabe (1991b) and treated in chapter 6.
[10] Previous studies of this stage of meson theory have been made by Visvapriya Mukherji (1974) and by Satio Hayakawa (1983); we have profited from both. We have also made use of published recollections as well as personal interviews. The publications include Kemmer (1965), Sakata (1965), Sekido and Elliot (1985), Brown *et al.* (1991) and Colloque International (1982). Personal interviews were held by LMB with the following physicists: G.C. Wick in Geneva on 29 October 1979; N. Kemmer in Edinburgh on 25 and 26 July 1984; W. Heitler in Zürich on 29 September 1979; M. Taketani in Tokyo on 13

October 1978 and 27 September 1984; K. Husimi in Tokyo on 13 October 1978; M. Fröhlich in Stuttgart on 26 July 1989. MR and M. Konuma interviewed M. Taketani in Tokyo on 12 October 1990.

[11]   See the article on Nishina in *Dictionary of Scientific Biography* [DSB 1976]. Also, see articles on Sakata, Tomonaga and Yukawa in *DSB*, Suppl. II. [DSB 1990].

[12]   We have discussed some of the early collaborative work of Yukawa and Sakata in chapter 6.

[13]   There is a manuscript [YHAL] E02 130 P12, in Yukawa's hand and in English, entitled "On the Interaction of Elementary Particles. II. Generalization of the Mathematical Scheme". It is marked "Read Sept \(blank)\ 1937" but that date is overwritten as November 1936. In it Sakata is cited as having in press a "detailed discussion" of the magnetic moments of the proton and the neutron, based on the theory of heavy quanta. This manuscript of Yukawa could have been written no earlier than the summer of 1937, because it cites two letters to the *Physical Review* which were published in June and July, respectively: Oppenheimer and Serber (1937) and Stueckelberg (1937a).

[14]   These are an 8 page memorandum in Japanese on the scalar meson theory [YHAL] E02 110 P12 and a second, entitled "Quantization of the Yukawa Field" [YHAL] E02 122 P12.

[15]   See also Taketani's account of their conversation at Osaka in Taketani (1971, especially p 16).

[16]   "The militarists...sanctioned and encouraged a veritable witchhunt, for all persons whose slightest word or deed could be considered to be *lèse majesté*...Even the two great Imperial Universities at Tokyo and Kyoto, which had always enjoyed great prestige and considerable academic freedom were condemned for harbouring 'red' professors and were subjected to purges." (Reischauer 1964, especially, pp 174–5).

[17]   Pauli and Weisskopf did not, of course, expect their spin-0 theory to describe the electron, but Pauli was delighted to find a self-consistent theory that did not satisfy the postulates that Dirac had assumed to derive his electron theory (and which had led many physicists to believe that all elementary particles *must* have spin $\frac{1}{2}$). Pauli liked to call his new theory the "anti-Dirac" theory. (See Weisskopf (1983, especially p 70).)

[18]   *Interaction II*, p 1084.

[19]   Stueckelberg (1937a) stated explicitly that Yukawa's quanta were unstable, in view of their weak interaction with the electron and neutrino.

[20]   These were Anderson and Neddermeyer (1936, 1937) and Street and Stevenson (1937a). Yukawa and Sakata also quote "preliminary results of Nishina and others", giving the mass of the new particle as about 1/10 of the proton mass. Nishina's letter had been submitted to the *Physical Review*, but was delayed in publication by a referee report. It appeared finally as Nishina *et al.* (1937). The theorists quoted were Stueckelberg (1937a) and Oppenheimer and Serber (1937). The latter authors had stated that "the reconciliation of the approximate saturation character of nuclear forces and with the magnetic moments of neutron and proton could be achieved [in Yukawa's theory] only by an extreme artificiality".

[21]   *Interaction II*, pp 1084–5.

[22]   *Interaction II*, p 1088.

[23]   *Interaction II*, p 1090.

[24]   It is perhaps worth noting here that Nishina was an author of the first correct calculation of the Compton scattering cross-section using Dirac's relativistic theory of the electron: Klein and Nishina (1929).

[25]   What turned out to be the preferred theory after the discovery of the pion in 1947, namely the spin zero *pseudoscalar* theory, was not considered at that time.

[26] H. Yukawa, "On the Interaction of Elementary Particles. III. (January 6, 1937)", [YHAL] E03 060 P13, 22 pages, and an undated addendum, [YHAL] E03 070 P13, 3 pages. The contents are given in chapter 6.

[27] As we mentioned above, Sakata and Yukawa had already written a paper deducing some of the properties of particles with spin greater than $\frac{1}{2}$ and satisfying Dirac's theory (Dirac 1936; Yukawa and Sakata 1937b).

[28] Note 26; also H. Yukawa, "Interaction of the Heavy Quanta with the Electromagnetic Field", [YHAL] E0 80 P13, and "Interaction of the Neutron and the Proton", [YHAL] E0 90 P13; H. Yukawa, S. Sakata and M. Taketani, "On the Interaction of Elementary Particles. III" (Read 25 September 1937), [YHAL] E03 100 P13, manuscript of 5 pages, dated November 1937.

[29] For example, a colloquium talk by Yukawa at Osaka University on "The magnetic moment of the neutron", 10 June 1937, [YHAL] F05 140 T11; also, a talk by Taketani at the Kyoto branch of the PMSJ on 25 September 1937, mentioned in a postcard of Taketani to Sakata, 26 September 1937, [YHAL] E02 120 P13. Taketani had published separately in Japanese: "Quantization of proton and neutron fields and their magnetic moments", *Kagaku* 7 (1937), pp 532–3.

[30] [YHAL] E03 100 P13 and E03 110 P13. The published paper is *Interaction III*.

[31] *Interaction III*, p 319. They also refer to other relevant work that uses "Proca's scheme", namely Kemmer (1938a) and Bhabha (1938a). In fact, Bhabha does not refer to Proca, apparently considering his vector field equations to be an obvious generalization of the spin-0 theory of Pauli and Weisskopf. In a letter to one of us (Kemmer to LMB, 24 May 1989), Professor Kemmer has written that he had "lived through the birth of the Pauli-Weisskopf paper", and so Proca's equation was already known to him when he saw the paper cited. Although Kemmer realized that it could never be applied to the electron (as Proca attempted), he felt that he should nevertheless cite "Proca's otherwise quite misguided paper". Kemmer thinks that he may be responsible for that now-common attribution. Although the Yukawa group said in the quoted paragraph that it was only later that Proca's work "came to our notice", it should be noted that the earliest partial draft of *Interaction III*, dated 6 January 1937, has a section headed "Linear Field Equations of Proca Type" (see note 26). *Interaction III* also quotes a Russian spin-1 paper (E. Durandin and A. Erschow 1937).

[32] Heitler and Fröhlich to Yukawa, 5 March 1938 [YHAL].

[33] All quotations are from *Interaction III*, pp 319–21. The Bhabha reference is to Bhabha (1938a).

[34] See Fröhlich and Heitler (1938b); also see Bhabha (1938a, pp 335–6).

[35] *Interaction III*, pp 329–30.

[36] *Interaction III*, pp 335–6. Actually, parity conservation prevents the mixing of states of even and odd orbital angular momentum, but mixing of S- and D-states does occur as a result of non-central forces. It is responsible, e.g., for the quadrupole moment of the deuteron.

[37] *Interaction III*, pp 336–7.

[38] *Interaction III*, pp 337.

[39] *Interaction III*, pp 339.

[40] *Interaction III*, pp 339–40.

[41] Pauli to Weisskopf, 13 January 1938; published in [WPSC], pp 547–50, especially p 548.

[42] Hoch (1990, especially pp 24–5). Hoch's remarks about Cambridge physics are borne out by Dirac, Britain's leading theorist, who occupied a chair in applied mathematics. However, Hoch does not take note of the strong ties between Ralph Fowler, Stokes Lecturer in Mathematical Physics and later occupant of the Plummer chair, and Britain's foremost experimental physicist, his father-in-

law Ernest Rutherford. See also Mott (1984); on p 126, Mott wrote: "At that time theory had no recognized place in the Cavendish. No space was assigned to theorists, who were supposed to work in their college rooms or in their lodgings. Theorists were members of the faculty of Mathematics".

[43]   In addition to Hoch (1990), see Rider (1984) and Stuewer (1984), which deals with the refugees' rather different reception in the USA.

[44]   Biographical information on Heitler can be found in Jost (1983) and Rasche (1980).

[45]   Biographical information on Kemmer was obtained by one of us (LMB) in interviews with him in Edinburgh during July 1984.

[46]   Haken (1975). Information also from an interview with Fröhlich by L.M. Brown in Stuttgart, July 1989.

[47]   It should be noted that German refugees in the USA met stronger competition in phenomenological theoretical physics. American theorists were oriented towards treating problems arising from the latest research in nuclear and cosmic ray physics (in which American experimental physics was outstanding). For the competitiveness issue, see, e.g., Stuewer (1984) and Wheeler (1979).

[48]   Frisch and Stern (1933) and Estermann and Stern (1933a,b). The last paper was received by *Z. Phys.* on 19 August 1933. At that time Stern and Estermann had to leave Germany (Frisch had already left for Denmark) and they continued their investigations in the USA, reporting them to the American Physical Society meeting in Cambridge, Massachusetts, on 17 March 1934.

[49]   Kellogg *et al.* (1936). These authors gave the results: $\mu_p = (2.85 \pm 0.15)\mu_0$ and $\mu_d = (0.85 \pm 0.03)\mu_0$, where μ_0 is the nuclear magneton of the proton. Hence $\mu_n = (-2.00 \pm 0.18)\mu_0$ (Estermann *et al.* 1937). Frisch also returned to the problem of the neutron's magnetic moment in 1937.

[50]   Bethe and Bacher (1936, p 205).

[51]   Heitler (1937, p 82).

[52]   See [WPSC], pp 522–6, for Pauli–Heisenberg letters on the subject.

[53]   Pauli to Weisskopf, 3 August 1937 in [WPSC], p 533: "By the way, will Heitler come? This would be desirable because of the discussion on cosmic radiation."

[54]   Heitler (1985, p 10).

[55]   Heitler (1973, p 422).

[56]   Fröhlich (1985, pp 9–10).

[57]   Fröhlich and Heitler (1938b). Their argument does not really rule out integer spins other than zero.

[58]   Note 57, p 38. One can also obtain the fractional dissociation time α from this argument (not given by the authors) as about 8%.

[59]   Biographical details on Bhabha have been obtained from Cockcroft (1967) and Blanpied (1986, especially p 42).

[60]   LMB interview with Kemmer, 25 July 1984.

[61]   Kemmer (1971), quotations from pp 10–12.

[62]   Blackett and Wilson (1937, especially p 322).

[63]   Bhabha (1938b, p 259).

[64]   Bhabha (1938b, p 293). One should not overemphasize, however, the apparent rejection of Blackett's hypothesis by Bhabha, who still had trouble with the data of Blackett and Wilson (1937). He speculated (on p 290) that "a later and more complete theory may allow particles to exist whose rest mass may take on one of an infinite number of possible values of which only a few may turn out to be stable".

[65]   Bhabha (1938c, footnote on p 504).

[66]   Bhabha (1938a, p 118).

[67]   Indicated as "*Proc. Roy. Soc.*, in the press" this is probably Bhabha (1938b).

[68]   Kemmer (1971, pp 12–13).

[69]   Fröhlich (1985, p 9).

[70] These are respectively (in modern terminology) scalar and pseudoscalar for spin-0, and vector and pseudovector (or axial vector) for spin-1. In terms of relativistic tensors, the spin-0 fields are of rank zero and completely antisymmetric fourth-rank tensors, respectively, whereas the spin-1 fields are first-rank and antisymmetric third-rank tensors, an antisymmetric second-rank tensor and its dual. These are not all independent, so that there are four cases of free fields. In interaction, however, there are six cases to consider. For example, in relativistic electrodynamics we can consider interaction with the six-vector (antisymmetric tensor) field strength or with the four-vector potential.

[71] Kemmer (1938b, p 128).

[72] In his interview, note 45, Kemmer mentioned that he had decided to call the generalized d'Alembert equation describing the four-vector field, $\Box\phi_\alpha = k^2\phi_\alpha$, the "Proca equation" even though he was already familiar with the theory of the equation and Proca had "misguidedly" written that it could be applied to the electron.

[73] Kemmer (1938b, pp 136–7).

[74] In working out the resulting interaction, Kemmer found that there were terms linear and quadratic in the f and g terms. The quadratic terms, however, corresponded to direct point-like interaction of the nucleons, which he had shown could be eliminated. In a footnote on p 142, Kemmer thanks Pauli for permission to state that Pauli had also proved that the point-like interaction played no role in binding or in scattering.

[75] Kemmer (1938b, p 147). "Fröhlich and others" refers to Fröhlich et al. (1938). Kemmer's conclusion that only case (b), the vector meson case, fitted the empirical neutron–proton interaction, was no longer valid in the context of a charge-independent (also called isospin invariant or symmetric) meson theory. In that case, the correct choice would have been the pseudoscalar meson theory. This became the standard theory in the 1950s, after the discovery of the neutral pion, and the demonstration of the pseudoscalar nature and charge-symmetric interaction of the charged and neutral pions. (See Kemmer (1971, pp 16–7), for wartime correspondence between Kemmer in Britain and Pauli in the USA in which Pauli pointed out that pseudoscalar mesons were required by a charge-independent theory and stressed the importance of including the tensor force in the problem of the deuteron ground state.)

[76] Heitler (1973, pp 422–3).

[77] Heitler, note 76.

[78] Fröhlich et al. (1938, p 155). The paragraph continues: "They seem to be absorbed strongly as soon as they reach an energy of less than 200×10^6 e-volts. It is probable that this absorption is due to some sort of *nuclear* processes [original emphasis]... There is no need for the introduction of any neutrino."

[79] Curiously, however, Kemmer believed that the strong nuclear interaction would give rise to explosive showers of the type that Heisenberg inferred from the Fermi theory. On pp 147–8 of Kemmer (1938b), he pointed out that the perturbation expansion arising from the longitudinal part of the vector field is "an expansion in powers of a fundamental length", and that, according to Heisenberg, "multiple processes such as cosmic ray showers should occur... when sufficiently high energies are available".

[80] Fröhlich et al. (1938, pp 155–6). It is clear that at least two of the authors took the magnetic moment results a bit too seriously (when we note that the divergent results required an arbitrary cut-off and that, even in 1990, no quantitative explanation of the magnetic moments has been given). At the same time, all three ignored the radioactivity of the mesotron, which played a significant role in cosmic-ray phenomena. (Hindsight *does* work better than prescience!)

[81] The quotations are from Fröhlich et al. (1938, p 156).

[82] At this point they noted that Bhabha had also obtained the same field equations and the relativistic interaction of type (b).

[83] Fröhlich *et al.* (1938, p 166).

[84] Fröhlich *et al.* (1938, p 170).

[85] This information is contained in the three footnotes on p 504 of his article (Bhabha 1938c).

[86] Bhabha (1938c, p 501).

[87] In this he differed from Bhabha (1938b), in which he said on p 290: "This change in the rest mass may be spontaneous, or caused by an external agency. The former possibility is not very interesting as far as cosmic radiation is concerned, for if the probability is large, the change will take place before the particle reaches earth, if small, then the chance of its taking place in the very short time taken by the particle in penetrating the earth's surface [atmosphere?] is also negligible." Evidently, Bhabha was assuming that the heavy electrons belonged to the primary cosmic rays.

[88] Bhabha (1938c, p 504–5).

[89] Bhabha (1938c, p 505).

[90] Bhabha (1938c, p 524).

[91] Heitler (1938a, p 529).

[92] Heitler (1938a, p 530).

[93] Heitler (1938a, p 542).

[94] *Ibid.*

[95] *Ibid.*

Part C

The Meson Takes its Place Among the Elementary Particles

The title of Yukawa's first research paper, "On the interaction of elementary particles. I", clearly established the highest of claims. His theory was an attempt to provide the basis for all forces existing between the elementary constituents of matter, other than electromagnetism and gravitation. Incorporating the already known proton and neutron, negative and positive electrons, and the conjectured neutrinos, he introduced the new U-particles, soon to be called mesons (as well as other names). The period following the identification of new charged particles of intermediate mass in the cosmic rays, roughly 1937 to 1941, saw efforts to prove the principal assertion of Yukawa's theory, that one new elementary particle, the meson, would suffice to account for all the unexplained phenomena of nuclear and cosmic-ray physics.

To prove a theory in physics implies setting up predictions for the properties and the interactions of the system of objects described, and submitting those predictions to experimental test. In the case of the free Yukawa particle, an unstable object, its mean lifetime was, perhaps, the most obvious property to submit to the theory–experiment confrontation. The difficulty soon arose that the mean decay lifetime of the cosmic-ray particle, associated with the hard component, was a factor of 10–100 larger (perhaps even more) than the prediction from the Yukawa theory. This will be discussed further in chapter 8.

The definiteness of the lifetime discrepancy was not grasped immediately, because the theory at first left room to vary its formulation; this made possible a range of prediction that was capable of accommodating differing experimental outcomes (and the experiments themselves were not very accurate). However, between 1938 and 1941, the experiments improved, and theorists began to gain confidence in their knowledge of what the theory could or could not accommodate without extreme artificiality. Also doubts arose about whether one could maintain a unified treatment of all nuclear phenomena, or even a unified theory of meson decay and β-decay.

We deal with the further development of the theory in two chapters. Topics related to the then standard vector-meson theory are discussed in chapter 9, which is mainly devoted to its response in Europe (England, and especially Germany and Switzerland) and, with a year's delay, in the United States (Schweber 1986). This chapter concludes with the global recognition of Yukawa's achievement and his first journey to the West in 1939. Invited to present talks at several international conferences in Europe, the outbreak of war led to their cancellation and Yukawa returned to Japan, visiting the United States *en route*.

In the last year before the war, considerable effort was invested to explore the general properties of relativistic quantum fields associated with different spin values, as we shall describe in chapter 10. Notably, Pauli and Fierz established the spin-statistics relation for free particles. Other theoretical work concerned the problem of interaction: quantum-theoretical perturbation theory broke down for most meson processes (it was worse than the similar divergence problem of QED); Heisenberg, among others, suggested that a classical field approach be tried. The Heisenberg–Pauli report prepared for the (cancelled) Eighth Solvay Conference summarized the pre-war situation. After September 1939, Pauli improved his presentation of the free-field theories, while other physicists, mainly in the United States (also in the Soviet Union) tried to establish the spin of the cosmic-ray meson from its observed electromagnetic interactions.

Chapter 8

Decay of the Meson — Experiment Versus Theory (1937–41)

8.1 Introduction

Yukawa had attacked the difficulties of the Fermi-field theory of nuclear forces by boldly replacing the Fermi-field by a new field of force, the U-field with its massive quanta. He also introduced two coupling constants, instead of the one Fermi constant. The U-field coupling to nucleons and the mass of the U-quantum (the meson) were adjusted to fit the observed strength and range of the nuclear force, whereas its coupling to the electron–neutrino pair was fitted to the observed β-decay lifetimes. At the energy scale of β-decay, Yukawa's theory gave the same results as Fermi's (although it gave different predictions for higher energy). Yukawa postulated the interaction of the meson with the electron and neutrino because he wished to maintain a relationship between the strong and weak nuclear forces, as in the Fermi-field theory. This, in turn, implied an instability of the free meson, giving it a short lifetime (of the order of a microsecond or less; it also explained why mesons were not found in ordinary matter).

The assumption that the cosmic-ray mesotrons undergo radioactive decay led to the explanation of a number of otherwise puzzling cosmic-ray phenomena. An approximate mean decay time was deduced from cosmic-ray phenomenology, and confirmed and improved later by direct observation. However, its value disagreed with that given by the then standard vector-meson theory of β-decay, the discrepancy being a factor of the order of 100. This theory versus experiment confrontation led, by the year 1941, to the conjecture that there might be two separate, though linked, charged cosmic-ray particles of intermediate mass. One may, therefore, consider the discovery of the mesotron to be a forerunner of the "particle explosion" of the 1950s and 1960s.

In this chapter we investigate the history of the decay of the mesotron, beginning in 1937, after the discovery of the intermediate-mass cosmic-ray particle and its subsequent association with Yukawa's U-quantum. (See chapter 6 above.) Although meson decay could qualitatively explain many phenomena associated with the hard component of cosmic rays and an experimental value for the decay time could be inferred from them, the value calculated from the standard vector-meson theory deviated appreciably from it (sections 2 and 3). A direct measurement of the decay time (first results of which became known in autumn 1939) confirmed the value that had been found indirectly (section 4). Further theoretical analysis of meson decay, on the one hand (section 5), and a discussion of meson decay versus meson capture, on the other hand (section 6), led to serious doubts about whether the cosmic-ray mesotron and Yukawa's meson were the *same* particle. This question was resolved only several years later with the discovery of the pion.

Meson decay (together with the comparison of the mass and the probabilities of different types of interaction) led to the perception of a discrepancy between the properties postulated for the Yukawa meson and those of the cosmic-ray mesotron, a typical theory–experiment confrontation, about which many interpretations have been made in social and philosophical studies of science. We observe that, in our case, the standard "scientist's account" of what was going on during the decade in question is actually not far off the mark.

In the account given below, we try to combine here a narrative of several theoretical and experimental developments [1]. It is based on a historico-critical analysis of the scientific articles published on these topics in European, Japanese and American journals, supplemented by selected correspondence and documents, especially from the Yukawa and Heisenberg Archives. The published recollections of Bruno Rossi throw light on some aspects of the experimental determination of the cosmic-ray particle's decay time (Rossi 1990).

8.2 Estimating the lifetime of meson decay (1937–38)

When the earliest conclusive evidence was presented for particles of intermediate mass (Neddermeyer and Anderson 1937; Street and Stevenson 1937a; Nishina *et al.* 1937), they were at once considered as candidates for Yukawa's U-particles (Yukawa 1937). On this account, it was natural to regard them as subject to weak decay, though in 1937 only Stueckelberg stated [2]:

> It seems highly probable that Street and Stevenson and Neddermeyer and Anderson have actually discovered a new elementary particle which has been predicted by theory. This particle is unstable and only can be of secondary origin, its mass being greater than the sum of the masses of electron plus neutrino.

The possibility of new elementary particles that might be unstable—which engaged the attention of the few physicists who were thinking in terms of a unified theory of elementary particles, such as Stueckelberg and Heisenberg [3]—was not deemed important in the minds of others who were mainly concerned with seeking a theory of nuclear forces. For example, the note of Oppenheimer and Serber preceding that of Stueckelberg does not mention meson decay (Oppenheimer and Serber 1937). Even Yukawa in his *U*-quantum paper of 1935 did not explicitly mention the instability of the new particles, although it could have explained their absence in ordinary matter, which, instead, he attributed to the absorption of mesons in the atmosphere [4].

The first person to make a detailed study of the meson lifetime and its consequences for cosmic-ray phenomena was Bhabha. In his letter to *Nature*, dated 13 December 1937, he stressed that the lifetime for decay of the cosmic-ray meson (or *mesotron*, as Anderson called it) would depend on its speed in the atmosphere, for it would be subject to relativistic time dilation, and thus a higher energy would imply a longer range (Bhabha 1938a). Yukawa mentioned the problem of meson decay neither in his first nor in his second paper (Yukawa 1935, Yukawa and Sakata 1937c). Only in the third paper, based on the vector-meson theory, submitted in spring 1938, did Yukawa, Sakata and Taketani note [5]

> that a heavy quantum disappears even in the free space by emitting a positive or negative electron and a neutrino or an antineutrino simultaneously..., as already pointed out by Bhabha...Owing to the small interaction of the heavy quantum with the light particle, the possibility of occurrence of the above process is so small that the mean free path of the high speed heavy quantum in free space is large compared with the dimension of the measuring apparatus, but is not small enough, in many cases, to make the mean free path larger than the height of the whole atmosphere.

Yukawa and his collaborators thus concluded that any heavy quanta observed at sea level must be produced in the atmosphere. They also implied that observing the decay would be difficult, but not impossible. Their evaluation of the probability per second of a meson decaying at rest gave (with neglect of electron and neutrino masses)

$$w_0 = (2g_1'^2 + g_2'^2)[m_0c^2/(6\hbar^2c)], \tag{8.1}$$

where g_1' and g_2' are the coupling constants of the meson to the electron–neutrino field, and m_0 denotes the rest mass of the mesotron. The decay probability per second of a meson moving with velocity v is then

$$w = w_0(1 - v^2/c^2)^{1/2}. \tag{8.2}$$

Estimating $g_1' = g_2' = 4 \times 10^{-17}$ (cgs units) on the basis of known β-decay lifetimes and taking $m_0 = 100m_e$, Yukawa *et al.* obtained the mean lifetime at rest to be $\tau_0 = 10^{-7}$ s. The mean free path for decay of a meson of 10^{12} eV energy came out to be 3000 km.

Without having seen the results of the Yukawa group, Heisenberg estimated (in a letter to Pauli of 5 April 1938) [6], that there would be an appreciable probability for "a heavy electron to decay while traversing a Wilson chamber, a thing that has never been observed so far". Pauli replied that Gregor Wentzel had used Yukawa's theory to obtain a mean lifetime of 10^{-6} s, "so that it would practically *not* decay in a Wilson chamber" (Pauli to

Heisenberg, 11 April 1938) [7]. A few days later, Heisenberg wrote to agree with Wentzel's reported estimate and said that: "For rays of about 200 million volts, therefore, follow ranges of only a few hundred metres, which would be very important for discussion of the cosmic radiation." [8]

At the beginning of May 1938, Heisenberg had in his hands a copy of the vector-meson paper of the Yukawa group (Yukawa *et al.* 1938a); and on 4 May, in a letter to Pauli, he wrote his opinion that "the Yukawa Hamiltonian is more reasonable than that of Kemmer or Bhabha" [9]. Via a letter from Sin-itiro Tomonaga, who was then visiting Leipzig, Yukawa learned "that Heisenberg appreciated the third paper and that Euler obtained the lifetime of two microseconds to explain cosmic-ray phenomena" [10].

Heisenberg's student, Hans Euler, had been studying cosmic-ray data for his *Habilitation* thesis, stimulated by his teacher's interest and by the arrival in Leipzig of the cosmic-ray expert Gerhard Hoffmann, to assume the experimental physics chair previously held by Peter Debye. In September 1937, at the Bad Kreuznach meeting of the German Physical Society, Euler gave one of the plenary lectures: he dealt with the theory of cosmic rays and the various elementary processes responsible for the loss of energy of charged particles as they proceed through the atmosphere (Euler 1937). Although Euler mentioned the "heavy electrons of Anderson" in passing, his analysis did not really take them into account in discussing the intensity of cosmic rays as a function of height.

However, after the existence of the mesotrons had been established, and especially after Bhabha had called attention to their decay, Euler reconsidered cosmic-ray absorption and included the decay of the mesotrons. By spring 1938, he regarded mesotrons as constituting the hard component, and, through their decay, also the source of the soft component of the cosmic rays. Thus the ratio of the soft to the hard component at sea level gave a measure of the mean lifetime of the mesotron. He wrote this up in a short note and provided more details in the second part of his *Habilitation* thesis (Euler 1938a, b) [11].

In the abstract of the longer paper, Euler set forth his assumptions and conclusions:

It will be assumed that the penetrating component of cosmic rays consists of heavy electrons (Yukawa particles) and that the soft component at sea level contains predominantly electrons that were produced by the decay of heavy electrons. With these assumptions, one obtains according to Yukawa's theory the number of soft particles at sea level having momentum greater than p as

$$F(p) = F_s(p)(3/8)[\lambda_0\mu/(\tau p)],$$

where μ is the mass of the Yukawa particles, τ their mean lifetime, $F_s(p)$ the number of Yukawa particles above the momentum p and $\lambda_0 = 275$ m. [λ_0 is a characteristic length for radiation in air, incorrectly given as "275 sec Luft" in the paper.] From the empirical intensity ratio of soft and hard components at sea level follows a mean lifetime for the heavy electrons of

$$\tau = (2 \pm 1) \times 10^{-6} \text{ s},$$

which agrees in magnitude with the value calculated by Yukawa from the theory of β-decay, $\tau = 0.5 \times 10^{-6}$ s.

Meanwhile, the Leipzig theoreticians learned another idea which allowed them to sharpen Euler's determination of the decay time. Helmuth Kulenkampff of Jena was an x-ray specialist, who had for some time shown an interest in the cosmic rays [12]. In a talk at the Breslau meeting of the German Physical Society in June 1938, he gave an explanation (Kulenkampff 1938) of a surprising observation by Alfred Ehmert, made at Lake Constance (Ehmert 1937a, b). Namely, Ehmert had found that the absorption of the cosmic-ray intensity in water of a given thickness was less than that in air of the same "effective thickness" (i.e., thickness measured in g cm^{-2}).

Euler, who attended the Breslau meeting, reported to Heisenberg the findings of Kulenkampff [13]. They incorporated them into their masterful review article on cosmic-ray physics (Euler and Heisenberg 1938). After deriving the equations for the momentum spectrum of the penetrating component as a function of depth — taking into account ionization energy loss and meson decay — they argued: "The formulae provide, according to Kulenkampff, a simple explanation of the striking observation of Ehmert that the absorption in water at first occurs more slowly than in air." [14] The reason is that the *distance* in air corresponding to 1 g cm^{-2} is far greater than the corresponding distance in water, which gives a much larger "absorption" due to meson decay. By comparing the slopes of the two absorption curves, Euler and Heisenberg obtained an improved meson decay time of $\tau = 2.7 \times 10^{-6}$ s.

Although that mean lifetime was about five times larger than Yukawa had estimated from his theory, they concluded [15]:

> Considering the uncertainty of many details in Yukawa's theory, especially of the mass of the heavy electron, the agreement is quite satisfactory. It should be further emphasized that what the experiments determine directly is the ratio of the decay time to the mass of the Yukawa particle; hence a change of the mass causes a corresponding change of the experimental decay time.

On obtaining these satisfactory results, Heisenberg wrote to Yukawa on 16 June (in German) [16]:

> Since you have been so kind as to send me your paper on the heavy quantum, I am glad to thank you and tell you how pleased I am about the great progress that you have made by your work. We have often spoken here in the seminar about your papers and have found that your hypothesis clarifies many aspects of the cosmic rays that could not be understood earlier. In particular, it appears to us now that the hypothesis of the radioactive decay fits the cosmic-ray observations well. Euler has discussed the measurements of Ehmert and similar measurements on the basis of your hypothesis, and has obtained for the decay time of the heavy electrons 2×10^{-6} s. This value does differ by about a factor 4 from your theoretical value, but it is certainly questionable how quantitatively certain one can consider experimental numbers in the field of cosmic rays.

H. Euler (1909–1941) in the mid-thirties. Photograph reproduced by permission of Werner-Heisenberg-Archiv.

Heisenberg went on to ask whether Fermi's theory of β-decay could be improved on the basis of Yukawa's work [17]. If one tried to fit the shape of the spectrum by introducing derivatives in the β-decay interaction, one then calculated too small a lifetime for the meson. Heisenberg inquired whether Yukawa knew a way out of this difficulty.

Yukawa replied (in English) on 15 July that he was honoured by Heisenberg's praise of his work ("which, I fear, is far higher than it deserved") and that he was also pleased by Euler's lifetime estimate. However, he wrote: "It is a pity that there was an error of factor 2 in our calculation, so that the mean lifetime for the heavy electron with mass $m_U = 100m$ [m is the electron mass] becomes 0.25×10^{-6} s, which makes the agreement with Euler's value a little worse." If one dropped the Fermi-type theory in favour of that with derivatives (Konopinski and Uhlenbeck 1935), he reported further, then his group obtained a mean lifetime four times as large as that in the Fermi case, provided that one adopted at the same time the mass $m_U = 200m$ [18].

The new result of 10^{-6} s would have been again in reasonable accord with Euler's estimate, except, as Yukawa soon reported in another letter, that the calculation was wrong [19]. As Heisenberg had stated before, the Konopinski–Uhlenbeck (K.–U.) theory lifetime was not larger than that of the Fermi theory, but shorter, and, as Yukawa's second letter showed in some detail, by a factor 10^4, "so that it is impossible to reconcile the cosmic-ray phenomena with the theory of the K.–U. type". Since there had appeared (from the cosmic-ray data analysed by Euler, Heisenberg and others) some possible evidence for the decay of the heavy quantum, Yukawa wrote:

> Under these circumstances, it seems to be reasonable to retain the main features of our theory of β-decay and to search for the origin of the asymmetry of the β-ray not in the above way of interpreting the β-disintegration by the introduction of the derivatives of the neutrino wave functions, although no good idea occurs to me for the time being.

Heisenberg replied, after the second letter, that he found Yukawa's arguments to be very reasonable (*"sehr vernünftig und plausibel"*) [20].

8.3 Mesotron decay and the resolution of some cosmic-ray puzzles (1938)

The Euler–Heisenberg review created a breakthrough for Yukawa's meson theory, at least among the European cosmic-ray physicists, because it explained some puzzling observations. One of the outstanding experts in cosmic rays was Patrick Blackett in Manchester, to whom Heisenberg had sent a copy of the manuscript of the review article. Blackett used its results in an address to the British Association, Section A, on 20 August 1938 and wrote to Heisenberg on 10 September from a ship taking him to America for a two-month tour. After apologizing for not having written earlier, he wrote [21]:

> I am entirely delighted with your beautiful explanation of the mass absorption anomaly as due to the decay of the barytrons (what are they to be called?). It is so simple and obvious, that I can't understand why we didn't all see it as soon as the decay theory was put forward! I gave special emphasis to your explanation (in a very simplified form) in my address to the British Association.

In the letter, as in the British Association address (Blackett 1938a), Blackett pointed out that Ehmert's research on cosmic-ray absorption (Ehmert 1937a, b) had been anticipated by English work done in a London underground station (Follet and Cranshaw 1936). Blackett's report in *Nature* gave a very clear account of the relationship between the meson decay time and the "mass absorption anomaly", namely [22]:

> By observations at ground level and in the "tube" station at Holborn, Follet and Cranshaw found that the intensity of the rays under a large thickness of air was about one half that found under the same mass of clay. Ehmert found a similar discrepancy between the absorption by air and by water. The effect was studied in considerable detail by Auger and his coworkers (Auger *et al.* 1937), who obtained results which can be interpreted as showing the greater absorption of air at low pressure compared with air at normal pressure. Heisenberg and Euler show how such observations can be explained

quantitatively by the spontaneous decay of the new particle. This follows from the fact that in air the distance travelled by the particle during its mean time of life is rather less than its range as defined by its ionization loss, whereas in dense materials the latter range is very much smaller. Thus in air the decay increases the apparent absorption, while in dense materials it has little or no effect.

After roughly estimating the lifetime on this basis, and comparing it with Yukawa's value, Blackett continued:

There seems, therefore, to exist definite evidence for the spontaneous decay of the new particle. The accurate determination of this time of decay and of the mass of the particle is now one of the outstanding problems of cosmic-ray research.

Blackett concluded his address by pointing out that little or nothing was known about the production processes of the new particles, or about their disappearance by absorption, rather than by decay. In any case, Blackett's talk showed no remaining doubt about the existence of a new particle, which he had been "slow in accepting" [23].

Two letters, by Blackett and by Bruno Rossi, respectively, appeared in *Nature* of 3 December, 1938 and provided "further evidence for the radioactive decay of the mesotron" (Blackett 1938c; Rossi 1938b). Blackett's letter referred to Rossi's letter that followed his own, and to the experiments of the Auger group (Auger *et al.* 1937), "who measured the zenith angle distribution of the penetrating rays at different altitudes above sea level and found that the absorption of the inclined rays was greater than that of the vertical rays under the same thickness of absorber" [24]. Of course, the effect was again related to the difference in the mean density of the air traversed during the mesotron's flight time. So, to the "mass absorption anomaly" was added the "inclination anomaly", as well as the "temperature effect". Blackett averaged the three inferred lifetimes using $m_U = 150m_e$, and again got $\tau_0 = 1.7 \times 10^{-6}$ s.

Rossi's letter drew attention to some data that he and Sergio De Benedetti had obtained some years earlier in Asmara, Eritrea at 2370 m above sea level (Rossi 1934; De Benedetti 1934). They had measured the intensity of cosmic rays coming either from the vertical or from an inclined direction, both with and without lead absorber. (At that time, they were seeking, and found, an East–West difference, which showed that the primary cosmic rays were mainly positively charged particles.) Now, years later, Rossi said (Rossi 1938b):

The difference between the air and the lead absorption, for which no satisfactory explanation had been found at that time [1933], can easily be accounted for by the disintegration hypothesis. The inclined rays have, of course, to travel a greater distance than the vertical ones before they reach the counters.

Assuming all the mesotrons to have roughly the same energy, Rossi thus estimated the lifetime as $\tau_0 \approx 2 \times 10^{-6}$ s (within a factor of two either way).

B. Rossi (1905–1933) (left) with O. Stern (1888–1969) (middle) in 1931. Photograph reproduced by permission of AIP Emilio Segrè Visual Archives (Goudsmit Collection).

Paul Ehrenfest, Jr, a son of the famous theoretician, was an experimental cosmic-ray physicist in the group of Pierre Auger. The Auger group had also been studying the absorption anomaly, which they dramatized as the *"paradoxe d'absorption"*. Being equipped with a counter telescope at the Jungfraujoch (at 3400 m above sea level), they investigated the effect with "different local absorbers". On 20 November 1938 Ehrenfest wrote to Heisenberg and Euler, having heard about their work from Hendrik Casimir [25]. (The latter had attended a lecture in Cambridge by Blackett, referring to the Leipzig work.)

The result given in Ehrenfest's letter to Leipzig, and also in the article that followed (Ehrenfest and Fréon 1938) was $\tau = (4 \pm 2) \times 10^{-6}$ s for a mean energy of 10^9 eV, about 10–20 times larger than the estimate of the Yukawa group (for the same assumed mass of $200m_e$) [26]. In the same letter, Ehrenfest included a photograph of a mesotron "in which one also sees the disintegration electron" and which yielded a mass of about 240 electron masses. He further mentioned that Paul Kunze ("who was of course, the first to have found a heavy electron") had published a picture (Kunze 1933, figure 5), which Ehrenfest now interpreted as containing a backwards-emitted electron of 37 MeV from a "mesotron nearly at rest", from which he estimated $m_U > 150m_e$.

On 15 November 1938, Alexandru Proca conveyed the French results in a letter to Yukawa, sent to thank him for a reprint of the fourth part of Yukawa's meson paper (Yukawa *et al.* 1938b) which, like the third part

(Yukawa *et al.* 1938a), used the Proca field. Proca wrote [27]:

> I was most impressed by your treatment of the β-disintegration problem and the queer impossibility to satisfy at the same time the asymmetry in the β-ray distribution and the cosmic-ray results—as well as by your suggestion pertaining to the creation of heavy quanta (or, as we now say in Europe, of "mesotrons"). Yesterday just when I received your paper I happened to be with Ehrenfest and I think you will be interested in some brand new results about meson-disintegration, which he had just obtained as well in Paris as in the Alps, at the Jungfraujoch.

Proca then gave the lifetime estimate of Ehrenfest and Fréon quoted above and asked "how much [do] you stress the results and what was the permissible approximation?".

Yukawa replied that the Ehrenfest result was in striking agreement with the value 2×10^{-6} s obtained by Euler and Heisenberg. He continued [28]:

> Thus the spontaneous disintegration of the mesotron seems to be an established fact, although the numerical value differs from the theoretical mean lifetime by a factor ten... The matter will be settled, however, when the cloud chamber photograph showing the ejection of the fast electron with the predicted energy from the end (in the gas) of the track of the mesotron will happen to be obtained.

The situation was summarized in a lecture presented on 1 December 1938 in Hamburg by Heisenberg, who cited especially the discovery of the mesotron and its finite lifetime (Heisenberg 1939a):

> Actually we can then understand the entire experimental material, for the time being, rather qualitatively. We may thus rightly expect, from the quantitative development of these ideas... to have a confirmation of the key to a complete understanding of the complex nature of the cosmic rays.

At about the same time Robert Millikan, having seen Blackett's *Physical Review* letter (Blackett 1938b), was reminded that Robert Oppenheimer had discussed something similar in March 1937 and wrote to the latter on 20 December. To this Oppenheimer replied [29]:

> You were right about the radioactivity of the mesotron; I have been thinking of it for two years now, and gave a seminar on it here [in Berkeley] while Bohr was with us. The only evidence we had at that time came from Rossi's work in Eritrea, which he has just recently interpreted in this sense in a letter to *Nature*. We felt that the extension of your own earlier work on air and water absorption would provide a much clearer and less ambiguous test of the idea (which as you can guess rests on no very sure theoretical basis) so that, perhaps mistakenly, we did not publish it, but just urged Bowen to get the air–water experiment done. I do hope that next summer that will be possible. It would put this whole subject on a much firmer foundation.

This exchange of views between Oppenheimer and Millikan appears a bit strange, if one reminds oneself of the rather negative view taken of the Yukawa theory in the Oppenheimer–Serber letter (Oppenheimer and Serber 1937) and Serber's later account of its history (Serber 1983), in which he refers to Millikan's view that his own experiments had already shown that there was no absorption anomaly, let alone one that could be explained by the theory of an obscure Japanese physicist [30].

While we are on the subject of Millikan, this may be a good place to remark on his campaign to have the name of the new particle recognized as *mesotron* [31]. The name was proposed in print in a letter dated 30 September 1938 to *Nature* (Anderson and Neddermeyer 1938). On 7 December 1938, Millikan wrote a letter to the *Physical Review* (Millikan 1939):

> After reading Professor Bohr's address at the British Association last September in which he tentatively suggested the name "yukon" for the newly discovered particle, I wrote to him incidentally mentioning the fact that Anderson and Neddermeyer had suggested the name "mesotron" (intermediate particle) as the most appropriate name. I have just received Bohr's reply to this letter in which he says: "I take pleasure in telling you that everyone at a small conference on cosmic-ray problems, including Auger, Blackett, Heisenberg and Rossi, which we have just held in Copenhagen, was in complete agreement with Anderson's proposal of the name 'mesotron' for the penetrating cosmic ray particle."

And so it remained, at least for a while [32]. Still Millikan had to come back to the same question about five years later, when he wrote to the Soviet physicist A. Alichanow that he was "particularly pleased to find you, contrary to the British and Indian scientists, writing 'mesotron' and not 'meson' " [33]. On 5 November 1946 he reported "the history of the word 'mesotron' " to Robert Brode at the University of California at Berkeley [34]:

> I have no idea who started the use of "meson". A couple of years ago I wrote to Bethe, about the only man in this country who was using "meson", and asked him if he did not think it would be desirable if we got together and tried to get some common usage.

With respect to Hans Bethe's suggestion that "it might be well to keep the name 'mesotron' for the experimental thing and 'meson' for the theoretical", Millikan did not think that was wise or practical. He also reported in his letter to Brode that he

> spoke to [W.F.G.] Swann about this recently in Philadelphia and he feels very vigorously about it that the use of "meson" is a very unfortunate one, not only because it violates all historical and etymological properties but is also so close in name to a word that has come in French to be used as a word for a house of ill fame, that he will not tolerate its use at all.

Soon, after all these quibbles, the situation would be clarified experimentally by the discovery of the π-meson and its subsequent decay into the μ-meson, the latter being identified with the old cosmic-ray "mesotron".

8.4 The mesotron lifetime measurements of Rossi and Rasetti (1938–41)

As we have seen, by the end of 1938 a number of cosmic-ray specialists had considered the effect of mesotron decay on the variation of the intensity of the penetrating component as a function of depth in the atmosphere, or under various absorbers. They all had agreed that the effects could be explained with a mesotron lifetime at rest of about $(2-4) \times 10^{-6}$ s [35].

A Symposium on Cosmic Rays was held at the University of Chicago from 27 to 30 June 1939. Organized by Arthur H. Compton, it was attended by sixty-odd physicists, including Bothe, Hess, Neddermeyer, Anderson, Auger, Heisenberg, Blackett, Swann and Rossi [36]. In his report on the disintegration of mesotrons, Rossi summarized the relevant experimental information available at that time, which he said was consistent with $\tau_0 \approx 3 \times 10^{-6}$ s, but not consistent with τ_0 less than 1×10^{-6} s (Rossi 1939). However, he asserted that there was "no evidence that mesotrons which have been stopped actually disintegrate into an electron and a neutrino". In fact, he argued: "The present experimental evidence is against rather than in favour of the emission of electrons by mesotrons which have effectively stopped." [37]

Bruno Rossi, who was born in Venice, Italy in 1905, had become a leading cosmic-ray experimentalist in the 1930s. In October 1938, he had left Italy, fleeing from anti-Semitic persecution, and went first to Niels Bohr's Copenhagen Institute. Bohr, in the same month, organized a meeting and invited other cosmic-ray experts, including Blackett; the latter then brought Rossi to Manchester on a fellowship. One of Rossi's main achievements had been the development of the coincidence-counting technique using vacuum tube circuits; at Manchester, he worked with the Hungarian physicist Lajos Janóssy (also a refugee from his home country) in extending this method to include the use of anti-coincidences as well. (In this way one could, e.g., know that an incoming particle, whose path was given by the coincidental discharge of two vertically aligned counters, had not emerged from an absorber placed beneath them.)

Rossi and Blackett were both interested in what could be learned about meson decay. As Rossi said later in his autobiography [38]:

> Blackett and I carefully examined the experimental results that were quoted...(variation of the anomalous absorption with height, with zenith angle, with atmospheric temperature). We reached the conclusion that none of these results (with the possible exception of some questionable data concerning the zenith angle variation) could be regarded as proof of mesotron instability.

These critical remarks notwithstanding, it appears hard to explain why the different experiments gave nearly the same lifetime, especially when that result disagreed with theoretical expectation. In any case, a somewhat critical spirit certainly motivated Rossi's talk at the Chicago Symposium and also led to his next series of experiments. They were carried out on Mt Evans in Colorado, suggested by Arthur Compton as a good site for making measurements of the absorption anomaly [39].

That very summer of 1939, Rossi and two of Compton's associates compared the absorption of air and carbon, working at the summit and at several lower attitudes (Rossi *et al.* 1939, 1940). In his autobiography, Rossi said that "no similar measurements of comparable accuracy had ever been performed before" and concluded: "We had thus achieved the first unambiguous demonstration of the anomalous absorption of mesotrons in the atmosphere, therefore proving their radioactive decay in flight." [40] The measurement again yielded a decay time $\tau_0 \approx 2 \times 10^{-6}$ s [41].

Rossi apparently felt that the decay in flight of mesotrons was now established by the Mt Evans experiments [42]. Still, no one had seen or measured the effect of the decay of a mesotron brought essentially to rest. The question, which later took on major importance, was the competition between capture of a mesotron by a nucleus and its decay. For the moment, Rossi continued the absorption experiments after moving to Cornell University (Rossi and Hall 1941; Rossi *et al.* 1942), confirming the earlier result. These experiments also demonstrated that the mean lifetime of the mesotron was proportional to its momentum, in accordance with the special theory of relativity (as had been suggested earlier by Bhabha 1938a) [43].

The first successful measurement of the lifetime of slow or "stopping" mesotrons we owe to another Italian refugee, Franco Rasetti [44]. He used a fourfold coincidence system of counters (a counter telescope) to select a "beam" of mesotrons which passed through a 10 cm block of iron, establishing that they were absorbed by "a battery of anticoincidence counters" (Rasetti 1941a). A fraction of the stopped mesotrons emitted a secondary particle (an electron), whose emission was delayed by a few microseconds, as measured by a delay-timing circuit. Thus the decay electron was finally "seen" and the existence of the decay was established. By a refinement of the method he found $\tau_0 = 1.5 \pm 0.3$ μs (Rasetti 1941b).

A more accurate measurement of the same mean lifetime, with an improved timing circuit, was then begun by Rossi at Cornell (Rossi and Nereson 1942; Nereson and Rossi 1943). This gave for the lifetime of stopping mesons $\tau_0 = 2.15 \pm 0.07$ μs. Nereson and Rossi found, further-more, that "only half of the absorbed mesotrons undergo disintegration, in agreement with the results of Rasetti" [45].

8.5 Meson-decay and β-decay (1938–41)

Yukawa's theory of heavy quanta (the meson theory) was intended to be, as we have stated repeatedly, a fundamental unified theory of nuclear forces. Thus, its justification would be found in its prediction of nuclear forces of the right range, strength, spin-dependence and exchange character. As a unified theory, it also had to account for the observed β-decay phenomena, thus the meson itself had to be β-active.

One of the major consequences of the theory was that mesons would be produced in collision processes of sufficient energy, at that time available only in the cosmic rays. Either in the atmosphere itself, or in other material contained in experimental arrangements, mesons were expected to be produced by γ-rays, electrons, or other cosmic ray particles; they would then be scattered and absorbed with typical nuclear cross-sections. Furthermore, they should decay at a rate consistent with known rates of β-decay, corrected for the relativistic time dilation of clocks in motion. Once the existence of charged cosmic-ray particles of mass $(100-200)m_e$ had been established, their detailed properties of production, scattering, absorption and decay became tests of Yukawa-type theories; hence their determination became an experimental programme for the cosmic-ray physicists, while the comparison with the meson theories became an important task for the experts in theoretical high-energy nuclear physics.

In his first paper proposing the meson theory (Yukawa 1935), Yukawa related the Fermi coupling constant G_F to the meson-decay constant g' by the equation

$$G_F = 4\pi g g'/\mu^2, \tag{8.3}$$

where μ denotes the mass of the meson and g its coupling constant to the nucleon. On the basis of this relationship — suitably modified in the alternative versions of the theory that followed (e.g., for the spin-1 theory) — Yukawa's theoretical estimates of the meson lifetime were generally 10–20 times smaller (Yukawa *et al.* 1938a, b) than the generally agreed upon experimental values, as we have seen.

8.5.1 *Developments in the West, especially in the USA*

Besides Yukawa and his collaborators, physicists in the West soon turned to the problem of calculating the meson decay time from various schemes developed at that time. One of the earliest estimates, by the Nordheims at Duke University, yielded a lifetime, on the basis of β-decay, of about 10^{-8} s (Nordheim and Nordheim 1938). Noting that this value was even smaller than that of the Yukawa group, Lothar Nordheim found the reason to lie in different assumptions for the Fermi constant. This, he said, "depends quite appreciably on the group of elements which are taken for comparison" [46]. Thus, using "the value deduced for light positron emitters", which he claimed to be correct for this purpose, Nordheim obtained $\tau_0 = 1.6 \times 10^{-9}$ s, about 1000 times smaller than the experimentally observed value. He also realized that the Konopinski–Uhlenbeck version of β-decay theory would only make matters worse and concluded: "A real improvement can only be expected by a complete reformulation of the theory".

Robert Serber picked up the problem of Nordheim. He pointed out, in particular, that, in the spin-1 theory (Yukawa *et al.* 1938b), the most general β-decay interaction has contributions both from the direct four-fermion coupling and from the two-step process involving the meson as an intermediary. The latter process alone gives a β-decay lifetime proportional to the seventh power of the energy release, rather than to the fifth power "as required by experimental evidence" (Serber 1939). He continued:

> The only escape is to deny the mesotron any role in β-decay, and return to direct emission of the light particles by the heavy ones... One is thus forced to

From left to right: H.A. Bethe (born 1906), L. Nordheim (1899–1992), I. Rabi (1898–1988) and E.U. Condon (1902–1974). Photograph reproduced by permission of AIP Emilio Segrè Visual Archives.

give up any theoretical connection between β-decay and mesotron decay; both can take place, but they must be supposed completely independent processes.

On the other hand, Serber referred to the fact that the results of Yukawa *et al.* for the Proca meson "seem to give both a satisfactory β-theory and a relation between β-lifetimes and the meson lifetime". However, their conclusions, he claimed, rested on the following procedure: Yukawa *et al.* (1938b) included both direct and indirect terms for β-decay as well as for meson-decay; these terms had different coupling constants and exhibited different energy-dependences. The choices of these coupling constants to accommodate the data, made by the Yukawa group, said Serber, "seem inadequately motivated" and hence "the discrepancy noted by Nordheim [in Nordheim (1939)]... is purely fictitious" [47].

Another response to Nordheim came from a Copenhagen group, namely Christian Møller, Léon Rosenfeld and Stefan Rozental (Møller *et al.* 1939). They suggested that a theory containing a mixture of meson fields for spin 0 and spin 1 (Møller and Rosenfeld 1939a) would result in a theory of β-decay involving

> a certain amount of arbitrariness, which a comparison with existing experiments is not yet able to remove. Anyhow it may be expected that it will enable us not only to avoid the discrepancy pointed out by Nordheim, but also to account for such considerable variations of the form of the beta-spectrum and the value of the beta-decay constant from elements to element, as are already indicated by the present experimental data.

About the same time Serber wrote his response to Nordheim, Bethe submitted a new and simplified version of the meson theory of nuclear forces (Bethe 1940a, b). His aim was to construct a classical version of the spin-1 theory, resembling classical electrodynamics, in order to study how various assumptions led to different types of nuclear forces. In the second part of the paper, he considered in detail the theory of the deuteron. Since the charge-independence symmetry of nuclear forces permitted only a completely neutral meson theory or the charge-independent theory of Kemmer (the "symmetrical" theory of Kemmer (1938b)), which involved both charged and neutral mesons, only these theories were considered. Finding that the symmetrical vector meson theory gave wrong results for the quadrupole moment of the deuteron, and rejecting the mixture theory of Møller and Rosenfeld "because of its intrinsic complication", Bethe remained finally in doubt [48]:

> The question remains open whether the neutral theory is correct because it gives a quantitative agreement for the deuteron, or the symmetrical theory, which is qualitatively preferable but in violent quantitative disagreement both with the theory of the deuteron and with the β-decay.

Of course, the purely neutral meson theory had the disadvantage that it contributed nothing to the understanding of β-decay or to the magnetic moments of the nucleons. Also, it was the *charged* mesotrons that had been observed so far, not neutral ones.

Together with Lothar Nordheim, Bethe then proceeded to apply the different versions of the spin-1 meson theory to the problem of radioactive decay (Bethe and Nordheim 1940). They showed that this theory resulted in a scheme equivalent to Fermi's β-decay theory "with the Gamow–Teller selection rules (Gamow and Teller 1936) as required by experiment" [49]. Since the detailed study of the rate of β-decay showed that the corresponding meson-decay time should be about 10^{-8} s, the authors concluded [50]:

> One has to admit, therefore, that Yukawa's theory in its currently used form does not give a quantitative account of the meson decay. Therefore, in order to obtain a sufficiently rapid nuclear β-decay, it seems necessary to introduce again a direct interaction between heavy and light particles such as the original Fermi interaction...If this is done, there is, of course, not much point in introducing an additional β-decay through the meson field, and we may just as well use the neutral meson theory which does not yield any β-decay at all. However, we must, at the present time, leave the question open whether the introduction of a direct interaction between light and heavy particles is really

necessary, or whether Yukawa's theory is still in a state where quantitative predictions cannot be made even in cases where the usual divergence difficulties do not enter directly.

8.5.2 *Developments in Japan*

Let us return to Japan, where at the same time as in Europe and in the United States the problem of meson decay occupied much of the interest of the meson physicists. Following their pioneering work (Yukawa *et al.* 1938a, b), Yukawa and Sakata published letters which considered the comparison between the observed mesotron decay time and that calculated from their vector-field theory, fitting the coupling constants to the nuclear forces and β-decay (Yukawa and Sakata 1939a, b). Although they found reasonably good agreement especially if the meson mass were taken on the low side, i.e., nearer $100m_e$ than $200m_e$, Yukawa wrote shortly afterwards to Heisenberg to say that they had used too small a value for the Fermi constant; consequently, "the theoretical lifetime is always a little too short in comparison with $(2–4) \times 10^{-6}$ s, which was estimated by several authors from the cosmic-ray data according to your suggestion" [51].

Having received these letters, the meson-decay calculations were continued by Sakata and Taketani, who had earlier speculated that the β-decay theory could be improved by adding a new "Bose-neutrino" to accompany the Fermi neutrino in the β-process (Sakata and Taketani 1938). About this collaboration Taketani recalled many years later [52]:

> After finishing the fourth paper (Yukawa *et al.* 1938), I was working to solve this lifetime problem with Sakata. The difficulty seemed to disappear if two neutrinos, a fermion and a boson, were emitted in the β-decay. However, my scientific activity was forced to be interrupted on 13 September 1938, since the thought police arrested me for the reason that I joined a group publishing a liberal journal called *World Culture*.

Some physics papers were smuggled into jail, including Euler and Heisenberg's cosmic-ray report (Euler and Heisenberg 1938), and:

> I again became interested in the lifetime problem and sent a letter to Sakata, in which I suggested separating the meson decay from the nuclear β-decay. This model was worked out successfully by Sakata.

Indeed, Sakata later published two papers picking up on the idea mentioned by Taketani (Sakata 1940, 1941a).

Independently of Sakata, and far away from him in Germany, Sin-Itiro Tomonaga also suggested a similar scheme [53]. In his *German Diary*, he noted on 17 November 1938 that Fermi's theory of β-decay would fit the lifetime data only if one assumed that the overlap integrals of the nuclear wavefunctions were of the order of unity for the light nuclei, but much smaller for heavier nuclei. This assumption, on the other hand, made the theoretical lifetime of the (cosmic-ray) meson much too short compared with the experimental results. After further detailed study of the nuclear β-decay papers, Tomonaga consulted Heisenberg (on 12 December 1938), who told him that many problems had still to be clarified in the conventional theory of β-decay. Encouraged by this response, Tomonaga proceeded to modify, even to reverse Yukawa's concept of these processes. He now assumed that the process of the *nuclear* β-decay is due to the fundamental Fermi interaction, while the (cosmic) ray meson first decays into a (virtual) nucleon–antinucleon (either proton–antineutron or neutron–antiproton) pair and then the virtual neutron (antineutron) decays again according to the original Fermi interaction. By this two-step process (especially the decay of the virtual neutron) the lifetime of the meson would eventually turn out to be much longer than in the previous calculation by Yukawa *et al.* Heisenberg fully agreed with this. In carrying out the actual calculation, Tomonaga encountered divergences when integrating over the virtual nuclear–antinucleon states, which he and Heisenberg could not overcome at that time (in late 1938 or early 1939), and so he finally gave up.

Two years later, Sakata used the same idea. He found, compared with the Yukawa-type calculations, that a meson lifetime resulted which is larger by a factor about $(\hbar c/g^2)^2$, which is about 100, in reasonable agreement with experiment [54]. Sakata avoided the divergence problem by using a cut-off of the "loop-integral"; the same cut-off had provided good results earlier in the calculations of the magnetic moments of nucleons in which a similar divergent integral appeared (Kemmer 1938a).

Sakata published yet another paper on the subject (Sakata 1941b). It contained a general discussion of processes for the four kinds of meson theories, namely vector and pseudovector (which are two different theories of spin-1 mesons) and scalar and pseudoscalar (two different theories of spin-0 mesons). The reason for again considering different types of meson theories involved the problems mentioned by Bethe concerning the low energy nuclear forces, especially in the theory of the deuteron, as well as the lifetime discrepancy and the failure to observe any evidence of the strong interaction of the cosmic-ray mesotrons. These four kinds of meson theories had been described earlier by Kemmer (1938a), and mixtures of spin-0 and spin-1 meson fields had been proposed (Møller and Rosenfeld 1939a,b).

Sakata's paper had a note added in proof that showed a possible way out of some of the difficulty faced by the meson theory [55]:

> Recently it has been shown by Rarita and Schwinger (1941a), p 439, that the nuclear forces derived from the pseudoscalar meson theory agree in sign and spin dependence with the sign and magnitude of the singlet–triplet difference and quadrupole moment of the deuteron system, if we take the influence of the non-central force into account correctly. In this connection it may also be remarked that the experimentally observed burst frequency of the mesons is in good agreement with the calculation based on the pseudoscalar theory (Christy and Kusaka 1941a,b). See also J.R. Oppenheimer (1941).

However, an Oppenheimer student, E.C. Nelson at Berkeley, claimed that the observed lifetimes implied that, for a pseudoscalar meson as an intermediary, β-decay lifetimes would be much too long, by a factor 10^6 (Nelson 1941) [56].

8.6 Meson decay versus meson capture (1939–42)

As the postulated heavy quantum of the strong nuclear interaction, mesons were expected to show evidence of strong interaction with nuclei; thus they should have been produced copiously in high-energy nuclear interactions, scattered strongly from nuclei and captured with high probability by nuclei, provided that they made a sufficiently close approach or stayed long enough near a nucleus. An appreciable fraction of meson interactions should also have resulted in nuclear disruption.

However, experimental observation revealed very little evidence to support the expected strong interactions of the mesotrons. Without detailed discussion here, we may mention a cloud-chamber study of Ralph P. Shutt of Bartol Research Foundation (Swarthmore, Pennsylvania), using two different thicknesses of lead plates without a magnetic field. In addition to finding the expected multiple small-angle electromagnetic scattering, Shutt said [57]:

> All theories based upon a mesotron spin of 1 predict a nuclear scattering cross section per neutron or proton between 10^{-25} and 10^{-26} cm^2, values 100–1000 times too large to agree with the experiments... As Williams (1939) and others have shown, because of their intensity and short range, the nuclear forces must scatter into large angles considerably wider than those given by the electrical theory, and this type of scattering would, therefore, be easily recognized.

Some theories not of the Yukawa type (e.g., Marshak and Weisskopf 1941) gave smaller cross-sections, but had other problems. Shutt's own result for the total cross-section of mesotrons (per nucleon) was $\sigma = 4.5 \times 10^{-28}$ cm^2 ± 50%. However, he cautioned that the fraction of tracks undergoing large angle — or, as he said, "anomalous" — scattering was about 2%, about the same as the estimated fraction of protons in the hard component, hence (and he put it in italics) "the possibility cannot be excluded that a great fraction or even the total of the anomalous large angle scattering observed is associated with the proton component and not with the mesotron component" [58].

One of the experimental difficulties was distinguishing the cloud chamber tracks of a proton and a mesotron from each other [59]. This was possible only for particles near the ends of their ranges, where one could make use of observations of range, as well as ionization and multiple scattering.

When continuing his experiments with T.H. Johnson — using the same chamber filled with argon (plus saturated water and alcohol vapour), but with magnetic field — he obtained a stereographic photograph showing a "stopping" negatively charged mesotron (Johnson and Shutt 1942). No electron was seen emerging from the end of the track, which did not surprise the authors, because a review of the half-dozen or so published pictures of stopping mesotrons did not reveal any either. Hence Johnson and Shutt concluded: "It appears that disintegration electrons have only been found from positive mesotrons." [60]

Later that year, the Shutt group, testing a new high-pressure cloud chamber, were able to publish a second photograph of a stopping positive mesotron with a very clear positron secondary emerging (Shutt *et al.* 1942). Noting that "until now a single photograph by E.J. Williams (Williams and Roberts 1940) has constituted the only indisputable cloud chamber evidence for mesotron decay", they estimated that they might expect to find such an event once in about 100 pictures in their new chamber.

We consider now what the theory had to say about the relative probability of meson capture versus decay under various conditions. Yukawa and a student, Taisuke Okayama, made a first investigation of that problem (Yukawa and Okayama 1939). However, they neglected to consider the effect of the Coulomb field of the nucleus, which plays an important role in repelling positive mesotrons and attracting negative ones. This creates an enormous difference in the ratio of captured to decaying mesotrons for the two signs of charge, as Tomonaga and G. Araki pointed out a year later in a letter to the *Physical Review* (Tomonaga and Araki 1940).

Yukawa and Okayama had obtained the result that the capture probability in a dense medium should always be larger than the decay probability. Thus for lead they found a capture probability for slow mesotrons of about 10^8 s^{-1}, to be compared with a decay probability of 10^6 s^{-1} (i.e., the reciprocal decay mean lifetime). Also, they had written [61]:

> It should be noticed that the time required to stop the mesotron is still smaller as can easily be calculated... Thus in the case of Pb, it is 6×10^{-11} for the initial energy 10^7 eV and 2×10^{-10} s for 10^8 eV, so that the majority of mesotrons are captured by nuclei after having stopped completely... On the contrary... in gaseous medium... we have only to consider spontaneous disintegration.

Tomonaga and Araki now modified these results as follows [62]:

> The effect of the Coulomb force on the capture of mesons can roughly be taken into account by multiplying the capture probability, which was derived by various authors on the assumption of free mesons, by the factor

$$(2\pi\alpha Ze/v)(1 - e^{\pm 2\pi\alpha Ze/v})^{-1}$$

> for negative and positive mesons respectively, where Z is the atomic number of the material and v is the velocity of the incident meson.

Giving a few numerical results, they concluded:

> Since the probability for negative mesons being captured is seen always to be larger than the probability of disintegration... the negative mesons will be much more likely to be captured by nuclei than to disintegrate spontaneously, not only in dense materials but also in gases. On the other hand, practically all positive mesons will disintegrate spontaneously because of the extremely small capture probability due to the existence of the potential barrier. Practically all positive mesons, which come to rest, should therefore be necessarily accompanied by a disintegration electron at the end of their range.

In view of the last statement, Tomonaga and Araki asserted that, if their theory were right, experiments reported which did not find any disintegration electrons from stopping mesotrons—they explicitly cited Montgomery *et al.* (1939)—"seem hardly to be understood, unless we assume that the slow mesons they observed are not identical with the ordinary cosmic-ray mesons and have much smaller lifetime". Later experiments on meson decay (see, e.g., Rasetti (1941a), and also the work discussed above of Shutt *et al.*) found that about half of the stopping mesons (presumably the positive ones) did decay [63].

8.7 Preliminary conclusions and post-1942 development

By December 1941, with much of the world at war, most research in pure science ceased. However, some theoretical physics continued (e.g., in Ireland and Japan), as did some experiments on cosmic rays, notably in Canada, France, England and Italy. Important Italian experiments begun during the war would lead in 1945 to the remarkable discovery that slow negative mesotrons were captured in iron, but not in carbon (Conversi *et al.*, 1945, 1947a, b). Since this result contradicted the prediction of Tomonaga and Araki for mesons, one had to conclude that the cosmic-ray mesotrons seen at sea level were different from the mesons postulated by Yukawa as the quanta of nuclear forces. That conclusion was, of course, consistent with the lack of evidence of strong interaction of mesotrons, and with the clear indication of a lifetime discrepancy between Yukawa's theory and experiment amounting, as we have seen, to a factor of about 100.

The final blow to meson–mesotron identity would result from cosmic-ray observations by the nuclear emulsion group in Bristol, led by Cecil F. Powell and Giuseppe P.S. Occhialini [64]. Plates exposed at mountain altitude showed tracks of mesons stopping and decaying into other particles of intermediate mass. The latter also stopped and decayed into a visible electron. This was interpreted as follows: a positive Yukawa meson stops and decays into a neutral particle plus a cosmic-ray mesotron, which then stops and decays into a positron and one or more neutral species (Lattes *et al.* 1947a, b).

The existence of a spin-0 meson decaying into a lighter spin-$\frac{1}{2}$ meson, which subsequently decays into an electron and two neutral species, was anticipated by the two-meson theory of Sakata and Inoue (1942, 1946), and will be discussed in detail in chapter 11. We only wish to note here that meson decay played a major role in a classical confrontation between theory and experiment, typical of those encounters that are studied by historians, philosophers and social critics of science. The situation could not be more dramatic. After Yukawa's meson theory of nuclear forces had been ignored for over two years, the discovery that the penetrating component of the cosmic rays consisted of particles that seemed in their mass and instability remarkably similar to those theoretically predicted by Yukawa called forth efforts, theoretical and experimental, to prove their identity. One could cite this as an example of the "social construction" of a concept, or perhaps even of a paradigm, in the sense of Thomas Kuhn. Alternatively, the process can be viewed as one of scientists responsibly testing the conjecture that the cosmic ray mesotron was the same as Yukawa's postulated meson.

In our account in this chapter of one aspect of this theoretical–experimental confrontation, that concerning the meson–mesotron decay, we saw that the theory had room for variation in its formulation, which made possible a certain range of predictions capable of accommodating differing experimental outcomes. However, at some extremes this effort took on an evident artificiality. During the period in which the theorists began to be confident in their knowledge of what the theory could or could not accommodate, experimental techniques were refined and believable experimental limits were established, to the point that the existence of an anomaly became apparent. Together with other discrepancies in the identification of the mesotron and the Yukawa particle (namely, the mass and the interaction probability), physicists began to entertain serious doubts about the basic concepts.

On the theoretical side, these doubts gave rise to new proposals favouring the separate treatment of β-decay and meson decay, as well as two-meson theories, proposing separate treatment of the cosmic-ray mesotron and Yukawa's nuclear force meson. Together with greatly improved experimental techniques — first the accurate electronic measurements of the decay lifetime and ultimately, after several years, the photographic emulsion technique, which exhibited the complete decay chain — the problem was eventually resolved.

Notes to text

[1] Apart from a short reference to meson decay in Mukherjee (1974), we are not aware of any historical treatment of this subject. The Heisenberg-Yukawa correspondence has been discussed by Konuma and Rechenberg (1985).

[2] Stueckelberg (1937a, p 42).

[3] See Heisenberg to Pauli, 12 June 1937 ([WPSC], p 523): "It seems reasonable to me that there are very many different masses [of cosmic-ray particles]...Only very few of them will be stable."

[4] Yukawa (1935, p 57).

[5] Yukawa *et al.* (1938a, p 337).

[6] [WPSC], p 563.

[7] [WPSC], p 565. Actually, Wentzel's published value for τ_0 was "as an estimate, at least 10^{-7} s" (Wentzel 1938a, p 276).

[8] Heisenberg to Pauli, 14 April 1938. [WPSC], p 566.

[9] [WPSC], p 571. The papers referred to are presumably Kemmer (1938b) and Bhabha (1938c).

[10] Taketani (1985, p 289).

[11] A major concern of the thesis was, as can be seen from the title, that of the Hoffmann bursts. See chapter 4.

[12] E.g., he had stimulated the work of Heinrich Maass in Munich on the absorption of single cosmic-ray particles by a counter telescope (Maass 1936).

[13] See the description of the impact of Kulenkampff's ideas in Bagge (1985).

Note, however, that Erich Bagge's account cannot be correct in all details.

[14] Euler and Heisenberg (1938, p 41).

[15] Euler and Heisenberg (1938, pp 42–3).

[16] Letter of 16 June 1938 [YHAL].

[17] This question has presumably to do with a supposed disagreement between Fermi's theory and experiments on the shape of the low-energy part of the electron spectrum in β-decay. Because of that, many physicists then preferred to describe the situation by the Konopinski–Uhlenbeck modification of Fermi's theory (Konopinski and Uhlenbeck 1935).

[18] Yukawa to Heisenberg, 15 July 1938 (copy in [YHAL])

[19] Yukawa to Heisenberg, 6 August 1938 (copy in [YHAL])

[20] Heisenberg to Yukawa, 14 September 1938 [YHAL].

[21] Blackett to Heisenberg, 10 September 1938 [WHA].

[22] The variation of absorption with density is also related to the "temperature effect", about which Blackett later wrote a letter to the *Physical Review* (Blackett 1938b).

[23] Blackett to Heisenberg, note 21.

[24] Blackett and Rossi had just met in Copenhagen, where a group of cosmic-ray experts including Heisenberg had held a topical conference (25–29 October 1938).

[25] Ehrenfest to Heisenberg, 20 November 1938 [WHA].

[26] The lifetime *at rest* would be about ten times shorter both for experiment and for theory.

[27] Proca to Yukawa, 15 November 1938 [YHAL]. Proca repeats the words of the Yukawa group's article (p 721) in referring to "asymmetry": "Of course, the Fermi theory also produces a momentum spectrum that is somewhat asymmetric between electron and neutrino".

[28] Yukawa to Proca, 12 December 1938 [YHAL].

[29] Oppenheimer to Millikan, 1 January 1939 [RMA]. Millikan has given the content of his own letter on the same page.

[30] In September 1925, Millikan and G. Harvey Cameron had done a series of experiments in deep snow-fed lakes at high altitudes in the California mountains. Millikan claimed that these measurements showed for the first time that the cosmic rays were of extraterrestrial origin and he began to call them *cosmic rays*. Millikan's claim led to a famous controversy, with Europeans claiming the discovery for Victor F. Hess, see, e.g., Xu and Brown (1987). For our purposes here, the point is that, although Millikan and Cameron measured absorption in sufficiently large thicknesses of air and water, they did not find the relatively large absorption anomaly that others reported later. After the Chicago Conference of 1939, measurements similar to those of Millikan and Cameron were done in mountain lakes by a Caltech group (Neher and Stever 1940), from which the authors inferred that mesotrons decay with $\tau_0 \approx 2.8 \times 10^{-6}$ s for $m = 160 m_e$.

[31] See Rechenberg and Brown (1990, notes 4 and 72).

[32] Thus, for example, Arthur H. Compton wrote in the foreword of the proceedings of the Chicago Conference on cosmic-ray physics, held in June 1939: "An editorial problem has arisen with regard to the designation of the particle of mass intermediate between the electron and the proton. In the original papers and discussion [at the conference] no less than six different names were used. A vote indicated about equal choice between *meson* and *mesotron* with no considerable support for *mesoton, barytron, yukon* or *heavy electron*. Except where the authors have indicated a distinct preference to the contrary, we have chosen to use the term mesotron (Compton 1939)."

[33] Millikan to Alichanow, 14 February 1945 [RMA].

[34] Millikan to Brode, 5 November 1946 [RMA].

[35] In addition to the publications already cited we mention that in Zürich, Henri Rathgeber found from the *"Barometereffekt"* about 0.7×10^{-6} s (Rathgeber 1938); and in Philadelphia, Thomas Johnson and Martin Pomerantz obtained $(2.4\text{–}2.5) \times 10^{-6}$ s (Johnson and Pomerantz 1939). The latter followed the Euler–Heisenberg suggestion and carried out a water–air experiment.

[36] This symposium was dominated by experimental topics. Thus Neddermeyer and Anderson reviewed developments in the understanding of cosmic-ray particles, including the discovery of the mesotron and the determination of its mass and decay time (Neddermeyer and Anderson 1939).

[37] Rossi (1939, p 296).

[38] Rossi (1990, pp 44–5).

[39] Compton also helped Rossi to get a research associateship at the University of Chicago for the next two years. See also Rossi (1983, 1964).

[40] Rossi (1990, p 51).

[41] For completeness, we mention another approach to deducing the mesotron lifetime of the same period (Pomerantz 1940; Bernardini *et al.* 1940). This compared the intensity ratio of the hard and soft components of the cosmic rays, assuming that most of the soft (i.e., shower) component arose from the decay products of the mesotron. From this assumption a lifetime of about 6 μs can be derived. The actual situation, however, is much more involved: indeed, we know that most of the showers are produced by γ-rays from neutral meson decay, discovered in 1950.

[42] We have to mention, in this context, Fermi's study of whether the absorption anomaly could be explained in terms of the polarization effect limiting the extension of the electric field of a fast charged particle in a condensed medium (Fermi 1939, 1940). Fermi's conclusion, confirmed by others (Halpern and Hall 1940), was that, although there was a marked reduction in ionization energy loss for relativistic particles (the density effect), the effect at Rossi's energies was much too small to explain the anomaly without also having the mesotron decay.

[43] Of course, a particle as light as a mesotron never comes to rest; but a small velocity gives a negligible correction to its lifetime at rest.

[44] Rasetti left Italy and accepted a professorship at Laval University, Quebec, in 1939. He had been a co-worker of Fermi in Rome since 1927.

[45] Nereson and Rossi (1943, p 201).

[46] See Nordheim (1939, p 506).

[47] Serber seems to suggest that Yukawa and his collaborators could get *any* value for the mesotron lifetime, while still fitting β-decay. However, too short a mesotron lifetime would make the indirect process dominant in β-decay, thus giving the wrong dependence on energy release in β-decay.

[48] Bethe (1940b, p 390).

[49] Bethe and Nordheim (1940, p 998). The same result had been shown earlier by the Brazilian physicist Mario Schönberg (1939).

[50] Bethe and Nordheim (1940, p 1004).

[51] Yukawa to Heisenberg, 25 April 1939 [YHAL]. This interesting letter refers to Nordheim (1939), saying that he had used the correct Fermi constant and had correctly "concluded that the theoretical lifetime was too short except for the exaggeration due to the omission of a factor 16 . . . ". Yukawa also informed Heisenberg that he had been invited to attend the Eighth Solvay Conference in Brussels (never held because of the outbreak of war) and that he had a new address (having moved from Osaka University to accept a chair at Kyoto Imperial University).

[52] Taketani (1985, p 289).

[53] This information from Tomonaga's *German Diary* is contained in a paper, presented by H. Ezawa at the Heisenberg Conference at Leipzig (Ezawa 1992).

We thank Professor Ezawa also for private communications on Tomonaga's stay at Leipzig.

[54] Sakata referred to Tomonaga's idea in his second paper (Sakata 1941a, footnote 9). We should mention that this treatment does not abandon the connection between nuclear β-decay and meson decay as, e.g., proposed by Serber (1939).

[55] Sakata (1941b, p 291).

[56] After the nuclear force meson had been observed in 1947, it was shown indeed to be pseudoscalar.

[57] Shutt (1942, p 7).

[58] Shutt (1942, p 13).

[59] Shutt's chamber had no magnetic field.

[60] Johnson and Shutt (1942, p 381).

[61] Yukawa and Okayama (1939, p 388).

[62] Tomonaga and Araki (1940, p 91). The factor is the so-called "Sommerfeld factor", equal to $|\psi(0)|^2$, where $\psi(0)$ is the value of the continuum Coulomb wavefunction of the meson at the nucleus. The quantity α denotes the fine-structure constant (about 1/137).

[63] Tomonaga and Araki also predicted a large difference between the probabilities of capture in flight for positives and negatives (e.g., in Pb 0.001 for positives versus 0.1 for negatives at 10^8 eV). Responding to an inquiry of Oreste Piccioni, Fermi wrote in 1947 that this result, which assumed a point nucleus, was, in his opinion, certainly wrong and that, for the energy mentioned, there should be "no appreciable difference in capture rates" (Fermi to Piccioni, 8 June 1947).

[64] See Powell *et al.* (1959) for a full account of the history of the nuclear emulsion technique.

Chapter 9

The Meson Theory and Yukawa Circumnavigate the Globe

9.1 Introduction

In an essay entitled "Thirty years of meson theory", Yukawa wrote of the period when the meson's lifetime and its interaction with matter were becoming problematic [1]:

> I was full of confidence at the outset when the meson had not yet been discovered; but after it was found and its properties began to be understood, curious to say, I gradually lost my self-assurance. It was partly a matter of age: in my twenties I had no doubt at all, but in my thirties I could not help wavering. Confucius said in his *Analects*, "At forty I had no doubt"; on the contrary, in my twenties I had no doubt, but in my thirties, great doubt.

When Yukawa turned thirty on 23 January 1937, he was essentially unknown, but within the next two years his meson theory began to dominate nearly all aspects of nuclear and cosmic-ray physics. The year 1939 would be triumphant for the young physicist, beginning with a step up the academic ladder at home. According to Sakata [2]:

> Just at that time, Yukawa was offered a new job. He was invited to Kyoto to succeed Professor Tamaki who had died the previous year. Tanikawa and I decided to follow Yukawa to Kyoto, but Kobayashi and Taketani remained at Osaka University. As a result our group was divided and we met with some troubling conditions. In the older universities, like Kyoto, there was still a feudal system, which did not exist in the newer universities and which was an unconscious pressure against the free atmosphere of research. The difference of environment was refreshing to Yukawa when he had moved to Osaka from Kyoto, and it was discouraging to us when we moved to Kyoto from Osaka.

These remarks of Yukawa and Sakata suggest some slackening in the pace of research on meson theory in Japan in 1939, and perhaps the separation and movement, together with Taketani's imprisonment in 1938, did reduce the vigour of their activities. Nevertheless, the work continued. All of the Yukawa group, except for Taketani, presented papers at the Annual Meeting of the PMSJ on 3 April 1939. M. Kobayashi and T. Okayama spoke on "Radioactive processes relating to mesotrons" and on "Scattering of mesotrons in the Coulomb field"; Sakata and Y. Tanikawa on "Creation processes relating to mesotrons"; Yukawa and Sakata on "Mass and lifetime of mesotrons"; and Yukawa on "The applicability of field theory" [3].

On the manuscript of the last talk, Yukawa wrote the marginal remark [4]:

> ... our theoretical understanding of mesotrons has been worked out to the limit and there remain only very difficult problems. It may be impossible in the framework of conventional quantum mechanics to find a way out of these difficulties. In this situation, it may be timely to expect some suggestions from all of you.

Before turning to the "very difficult problems" Yukawa and Sakata looked again at the vexing question of meson decay, sending two letters (of identical content) to the British *Nature* and to the *Proceedings of the PMSJ*, in which they tried to improve the calculations already presented in *Interaction III* and *IV* (Yukawa and Sakata 1939a, b). The earlier calculations had given a lifetime at rest of $\tau_0 = 1.3 \times 10^{-7}$ s for a meson of mass $200m_e$, about a factor ten shorter than the value of Euler and Heisenberg (1938) from cosmic-ray data. The Japanese authors now claimed that the discrepancy was due to "inaccuracies in theoretical estimation rather than to experimental errors", noting especially a rapid increase of τ_0 with decreasing meson mass μ. Of course, μ also entered into the determination of the meson-nucleon coupling constant, albeit less critically, and this again entered into the determination of Yukawa's weak coupling constant from β-decay. Overall, Yukawa and Sakata found an increase by a factor of 20 in the meson lifetime by a decrease of the mass to $100m_e$. They concluded: "Thus, the agreement between theory and experiment is very satisfactory, if we assume a value of μ intermediate between $100m_e$ and $200m_e$". [5]

Other than the work on slow-meson capture in nuclei (Yukawa and Okayama 1939, mentioned in chapter 8) and a Japanese textbook, *Discussions on New Particles*, published in paperback in August 1939, much of Yukawa's activity after 5 April 1939 was related to an invitation that he received on that date, to attend the Eighth Solvay Conference in Brussels, scheduled for 22–29 October 1939 [6]. Soon after this, new invitations arrived both from Heisenberg and from Pauli to participate actively in conferences that each was organizing around this time, one in Marienbad (of the German Physical Society), the other in Zürich.

In the previous chapter, we have discussed the lively work on mesons that took place between 1937 and 1941, emphasizing the comparisons between theory and experiment. We turn now to the inner development of the theory itself during this period, concentrating on two main efforts. The present chapter is devoted to applications and modifications of the theory of the vector-meson field discussed in chapter 7. The modifications were proposed in the light of problems posed by the observed nuclear and cosmic-ray phenomena. In contrast to this more phenomenological approach, the next chapter will deal with the general theoretical properties of elementary particles, taking into account the restrictions on meson fields that are demanded by the principles of relativistic quantum field theory.

When the vector-meson theory emerged in 1937–38 as the first fundamental theory promising to deal with the known phenomena of nuclear and cosmic-ray physics, Yukawa's ideas finally achieved the attention of the world-wide scientific community. In addition to the Japanese and European physicists, the American theorists climbed aboard, partially on the initiative of European colleagues or émigrés. By the summer of 1939, when Yukawa commenced his trip to Europe, hardly anyone working on either low- or high-energy nuclear forces could avoid being infected by meson fever.

After the first wave of recognition, occurring at several European conferences which took place in late 1937 and early 1938 (section 2), and the formation of new centres of meson research in Switzerland and Germany (section 3), the theory slowly gained support across the Atlantic, as Americans began to interpret experiments in terms of the meson approach (sections 4 and 5). Ultimately, the meetings in which Yukawa was to make his European debut were cancelled because of the imminent outbreak of war, and the Japanese scholar returned home via the United States, where he took the opportunity to meet and discuss the meson theory with American physicists (section 6).

9.2 European conferences take note of the mesotron and the meson theory (1937–38)

By the end of 1937, there was clear evidence for the existence of the new cosmic-ray particles that Anderson had mentioned at the end of his Nobel address in December 1936 (see chapter 6), suggesting that they "will provide interesting material for future study" [7]. The first rough mass values (about $(100–200)m_e$) were also obtained in 1937 (Street and Stevenson 1937b; Nishina *et al.* 1937). However, the scatter of mass values seemed to suggest more than one particle of intermediate mass. In a note sent in December of that year, Neddermeyer considered the possibility that there were several kinds of particles, their masses being multiples of a certain mass (Neddermeyer 1938). This speculation was prompted by experiments reporting a variety of masses: $350m_e$ (Corson and Brode 1938a) and $(120 \pm 30)m_e$ (Rulhig and Crane 1938). An experiment in Liverpool measured $(220 \pm 50, 190 \pm 60, 160 \pm 30)m_e$, as well as $430m_e < m < 800m_e$ (Williams and Pickup 1938). Further results came from the French cosmic-ray groups (Auger 1938; Ehrenfest 1938).

To help clarify the mass situation, Dale Corson and Robert Brode at Berkeley carefully studied the relationship between the specific ionization and the mass of cosmic-ray particles, as observed by themselves and others (Corson and Brode 1938b). They found that their measurements agreed well with Hans Bethe's formula given in volume 24/1 of the *Handbuch der Physik* (Bethe 1933, p 518). They also noted [8]:

It seems that all heavy tracks reported thus far, with the exception of Rulhig and Crane's, are not in serious disagreement with a unique value lying within the limits $(200 \pm 50)m_e$. The disagreement between the different values given for the mass of the heavy particle may be due to errors in determining the radius of the curvature in the magnetic field. When the curvature is small it may be influenced considerably by scattering and turbulence. The range can be determined with a relatively high degree of accuracy. In nearly every case one observes a minimum range, thus placing an upper limit on the mass.

In agreement with this conclusion, Neddermeyer and Anderson (1938) reported a measurement on a mesotron brought to rest in the gas of their chamber, from which they found the mass to be "about 240 electron-masses".

The new particles (whether possessed of a unique mass or not) certainly attracted the attention of European theorists. First, Lev Landau and George Rumer of Moscow (Landau and Rumer 1937), studying the production of electromagnetic cascade showers, and in order to explain the frequency of showers at sea level and under the earth, invoked a charged primary object heavier than the electron, to initiate the shower through *Bremsstrahlung*. "If we take for [its mass] the value of some tens of electron masses proposed by Anderson and Neddermeyer, we get fair agreement with the observed order of magnitude", they concluded in a letter dated 1 September 1937. This idea was pursued at physics conferences in late 1937.

Hans Euler mentioned the mesotron in his review on cosmic radiation at the September meeting of the German Physical Society [9]. A few weeks later, at the Paris International Meeting of early October, Patrick Blackett, analysing the particles observed in cosmic rays also referred to "Anderson's suggestion of the existence of a heavy electron", but declared it "rather too simple to explain the experimental evidence" [10]. At the same meeting, as Jenö Barnóthy recalled: "Anderson and Neddermeyer presented a paper on the discovery of the [mesotron]. It raised heated discussion. Some, Bohr among them, preferred to believe in a breakdown of the electromagnetic field." [11] In a third important meeting of autumn 1937 in Bologna (the 29th Riunione of the Italian Physical Society, celebrating the 200th anniversary of Luigi Galvani), Bruno Rossi of Padua called attention to the Neddermeyer–Anderson particle [12].

No mention was made in the above conferences of a possible connection between the "heavy electron" and Yukawa's U-particle. However, that did occur, beginning with conferences held in the spring of 1938. At the discussion meeting on cosmic rays of the Royal Society of London on 26 May, Blackett (as well as other experimentalists) summarized evidence concerning what he termed "the birth, life and death of the heavy electron", and Bhabha and Heitler told of their work on Yukawa's theory of nuclear forces and how it might explain the behaviour of heavy electrons [13]. Another occasion to discuss these topics arose at the Warsaw Conference of 1938, sponsored by the International Union of Physics and the Polish Intellectual Co-operation Committee. Yukawa's theory played an important

Second row, from right to left: O. Klein (1894–1977), E. Wigner (1902–1995), Mrs and Mr N. Bohr. First row: G. Gamow (second from left) and L. Rosenfeld (fourth from left).

role in talks given by Oskar Klein, Hendrik Kramers and Léon Brillouin.

Klein, noting that the "mesoton" (as he termed the Yukawa particle) had mass about two orders of magnitude larger than the electron and thus a Compton wavelength about 100 times smaller, argued that [14]:

> The logical consistency of this enlargement of the field concept would seem to require the removal of the self-energy difficulty of the electron at least down to distances approaching the radius of the new particle. Considering the order of magnitude and range of nuclear forces — the nuclear binding energies being comparable with the rest mass energy of the electron and the range of forces of the order of the electronic radius — it would not seem unreasonable to assume that a theory explaining nuclear attractions would also account for the rest mass of the electron, the attractive forces required as a compensation of the Coulomb repulsion being of a similar nature to the nuclear forces. A necessary condition for such an explanation... is that the new forces belonging to the heavy-electron field are determined by means of the elementary electric charge in the same way as the electromagnetic forces, so that no other independent constant than the mass of the new particle will appear in the theory.

With this motivation, Klein proposed a new unified field theory, merging ideas he had developed as early as 1926 in an attempt to unify Einstein's general relativity with Maxwell's electrodynamics (Klein 1926, 1927). This involved introducing a new auxiliary fourth space coordinate x_0, canonically conjugate (apart from a constant factor) to the electric charge, the quantization of charge leading to a "periodicity in x_0 corresponding to the length $l_0 = (he/c)(2\chi)^{1/2}$ where χ is the Einstein gravitational constant". Klein then promised [15]:

> The theory outlined, which may be derived from a variation principle containing as well the spinor as the tensor field quantities, will describe an interaction between the proton, the neutron, the electron and the neutrino through the intermediate of the charged and uncharged fields giving as it would seem a quantitative formulation of the considerations of Yukawa, Kemmer and Bhabha on nuclear fields of force. Especially the theory will contain no new physical constants other than the mass of the mesoton, and the nuclear interactions will, like the electromagnetic forces in the first approximation neglecting all direct gravitational effects, depend upon e^2.

In these speculations about a unified field theory, Klein clearly assigned a fundamental role to the meson [16]. The same might be said of another contribution to the conference, not directly, but through a messenger. In the discussion his talk on the limitation of the applicability of the present theory, Kramers (1939) called upon L. de Kronig of Groningen University to summarize some of the ideas of Heisenberg connected with the idea of a new fundamental length, related in part to the existence of "explosive showers". (See our presentation in chapter 4.) Heisenberg himself was not permitted by the German authorities to attend the Polish conference. After this, Rosenfeld was asked to summarize a letter from Heisenberg to Bohr on the same subject. There he had argued that the usual quantum mechanics may be inapplicable at distances small compared with the universal length r_0 (or equivalently, when the invariant momentum transfer exceeds $\hbar/r_0$). Heisenberg had concluded: "The quantitative applicability of quantum mechanics ceases for all cases which introduce changes of energy and momentum sufficiently large to allow of the creation of the mesoton" [17].

Finally, Léon Brillouin devoted the last four sections of his report on elementary particles to the "mesoton theory", including the results obtained by the Yukawa school and the workers in Britain and on the continent [18]. In the extended discussion that followed, Eugene Wigner, Samuel Goudsmit, Francis Perrin and George Gamow all addressed various aspects of the meson theory.

These conferences showed how strongly European theorists had been affected by the Yukawa theory from the beginning of 1938. American physicists still held back; no conferences on nuclear forces or cosmic rays were held in the USA in 1938. Nevertheless, by the middle of 1938 it may be fairly said that the mesotron theory, its consequences and its problems, had come to dominate nuclear and cosmic-ray physics before the outbreak of the Second World War.

9.3 The vector meson theory in Switzerland and Germany (1938)

On 9 March 1938, Kemmer wrote from London an eight-page letter to Wentzel in Zürich to give a "short report" on work by Fröhlich, Heitler,

Bhabha and himself [19]. He described his own paper on the "mathematics" of meson theories, his joint paper with Fröhlich and Heitler on applications of the vector field theory, Heitler's work on meson scattering and the results of Bhabha, all submitted to the *Proceedings of the Royal Society* A (see chapter 7).

From his solo paper considering four possible choices for the meson field (Kemmer 1938b), he described the deduction of the nucleon–nucleon force, and, comparing with the singlet S state and the triplet S (ground) state of the deuteron, he concluded: "Only Proca is thus possible; naturally, it is not at all trivial that the sign [of the force] is correct." He then exhibited the "right" n–p potential,

$$V = (\tau_x^N \tau_x^P + \tau_y^N \tau_y^P)[c\kappa/(4\pi)]\{|g|^2 + |f|^2[(\boldsymbol{\sigma}^N \cdot \boldsymbol{\sigma}^P)$$
$$-(\boldsymbol{\sigma}^N \cdot \mathrm{grad})(\boldsymbol{\sigma}^P \cdot \mathrm{grad})]\}e^{-\kappa r}/r, \tag{9.1}$$

and also noted:

> In [my] section 4, something is said about the occurrence of the universal length κ^{-1} in perturbation theory in general. The f- and g-interactions both (because of the longitudinal part) lead to Heisenberg showers. Only the [scalar and pseudoscalar cases [20]] do not give rise to showers; in the Proca case they cannot be avoided.

After mentioning the high degree of divergence in the self-energy calculation, two orders higher than in the electromagnetic case, Kemmer turned to the three-man paper with Heitler and Fröhlich, outlining its main results and conclusions, including a detailed discussion of the deuteron S terms, the (fourth-order) p–p interaction and the derivation of the formula for the anomalous magnetic moments. Then he reported on Heitler's calculations of various meson processes (using the symbols $Y^\pm$, for "Yukon"), namely the processes:

(1) $Y^+ + N = P + h\nu$

(2) $Y^+ + \begin{Bmatrix} N \\ P \end{Bmatrix} = \begin{Bmatrix} N \\ P \end{Bmatrix} + Y^+$

(3) $Y^+ + N = P + Y^+ + Y^-$
(4) $Y^+ + N_{\mathrm{bound}} = P$ (depending on binding energy)
(5) $h\nu + P = N + Y^+$

and he explained:

> It is important that for $E \approx m_0 c^2$ [m_0 being the meson mass] the processes (3) and (2) are already of the same order of magnitude. (Again $1/\kappa$ occurs and shower theory!) Also here Heitler sticks to the belief in cut-offs. The fact that, e.g., the cross section of process (2) grows as $[E/(m_0 c^2)]^2$ seems to be satisfactory to him; he just leaves out everything above κ. Then he can calculate with non-relativistic heavy particles and the evaluation is so simple! In the energy region up to $m_0 c^2$ he believes he can describe the behaviour of cosmic rays by the processes treated. Primary Yukons, he claims, need not exist. I cannot judge whether one really gains much if one considers only the region $E < m_0 c^2$. In any case, one can hardly speak of the possibility that the Yukon

energy can be transferred, according to this theory, to electrons, as Blackett wants to have it.

Kemmer also wrote a few words about Bhabha's paper and his relativistic treatment of the p–n scattering, which yielded a "mad increase" of the cross section. "One must comment on this by admitting that the expansion fails" was his conclusion. Finally, he spoke of his interest in the "neutretto", the neutral meson, needed to obtain attractive p–p forces in second order, and announced that a future paper would soon appear (Kemmer 1938c). This letter, although sent with some delay [21], informed Wentzel of the important work being done in England, and the Zürich professor was now able to finish writing a review on the subject for *Naturwissenschaften*, which would be the first published summary of this recent research (Wentzel 1938a). The report had five sections:

(1) The penetrating component of cosmic rays.
(2) Theories of the proton–neutron forces and of β-decay.
(3) Neutral Yukawa particles.
(4) Are there other unstable particles?
(5) Heisenberg's explosion-like showers.

Since there is little in this report that we have not already addressed earlier, we shall mention only a few points. Following the authors in Britain (not having yet seen *Interaction III* from Japan), Wentzel explained how the need for nuclear forces of the Majorana type led to the preference for a vector field obeying generalized Maxwell equations, because it gave both Majorana and Heisenberg forces. Wentzel concluded section 2 by presenting the Yukawa β-decay theory and its implications for meson decay [22]:

> It shows that one must ascribe to the free ϵ-particle [i.e., the meson] at rest a lifetime of at least 10^{-7} s. This lifetime is so long that one cannot expect with certainty to observe directly the spontaneous decay of cosmic-ray particles in a Wilson chamber. In denser matter the Yukawa particle will not decay spontaneously at all, since it will be beforehand absorbed by nuclear processes or be degenerated in some other way.

Section 3 of the review discussed the arguments on nuclear forces that favoured the existence of a neutral meson. Wentzel had previously tried to extend the Fermi-field theory to account for the charge-independence of nuclear forces. (See chapter 3.) Now he referred to the new papers soon to appear in the *Proceedings of the Royal Society*, especially Kemmer's charge-independent theory. Section 4 remarked that Yukawa's theory, as shown by experiment, was incomplete as regards the interaction with the light particles (i.e., electrons and neutrinos). For β-decay it was equivalent to Fermi's theory, which predicted, e.g., a symmetrical electron spectrum for light elements and high decay energy (at which charge and mass were unimportant). However, the experiments gave an asymmetrical spectrum. To meet this objection, Wentzel referred to earlier work of his (Wentzel

G. Wentzel (1898–1978) in the 1930s. Photograph from S.A. Goudsmit and reproduced by permission of AIP Emilio Segrè Visual Archives.

1937a, b) suggesting "heavy boson protons and neutrons" as intermediate states in β-decay, as a possible solution to this "difficulty" [23].

In the final section, discussing explosion-like showers, Wentzel argued that Heisenberg's expectation of their existence was based upon the elementary nature of the Fermi β-decay theory. Since this became a two-step process in the Yukawa formulation, Wentzel claimed that Heisenberg's theory had lost its foundation. However, Wentzel considered it possible that "in the framework of the theories presented here, there are explosions of another kind, in which besides protons and neutrons *only Yukawa particles* (positive, negative and neutral) take part, e.g., simultaneous emission of several "heavy electrons" [24]. Thus he opened new possibilities for Heisenberg-type showers involving the new particles.

When Wentzel quoted the new British papers of Kemmer *et al.*, he remarked that "similar results have been obtained by E.C.G. Stueckelberg, according to a letter" [25]. Stueckelberg, who was the first European to mention Yukawa's theory (see chapter 6), published a three-part article in the spring of 1938, with the title (in English translation): "The interaction forces in electrodynamics and in the field theory of nuclear forces" (Stueckelberg 1938a). There he derived a systematic scheme for treating quantum fields of the tensor type, e.g., Maxwell, Yukawa and Proca fields. As the classical prototype of all of these, he considered the retarded interaction of two electric charges.

In part I, Stueckelberg treated the case of the massive scalar field, i.e., the case treated by Pauli and Weisskopf, and later by Yukawa, deriving the advanced and retarded solutions of the wave equation in the many-time formalism of Dirac *et al.* (1932) — that is, he used what later (in the 1940s) was called the "interaction picture". He concluded [26]:

> The advantage of the present method lies in that it allows the calculation of the retarded, and naturally also the non-retarded interaction terms, without explicit reference to the quantum structure of the fields and without specific assumptions about the representation of the matter. In agreement with Kemmer it appears possible to me that a complex Proca field (whose particles are "heavy electrons" and "heavy antielectrons") and a Proca field (whose particles are uncharged) completely describes the nuclear forces.

Kemmer had informed Stueckelberg about the papers in press in the *Proceedings of the Royal Society*, but the latter sent his own treatment of the vector-meson field to the Swiss journal in early April, still before the British papers had appeared in print. In it, he produced essentially the same theory as the British authors and the Japanese *Interaction III*. However, he incorporated it into a unified relativistic scheme that he had been pursuing already for several years (Stueckelberg 1936a, b, c). (See chapter 6 for a discussion of these papers.) A noteworthy feature of his treatment was that he stressed the conservation of what he called "heavy charge", attributed only to the nucleons, which today we call nucleon or baryon number.

Discussing the "possible fields in nature", Stueckelberg took into account those with and without electric and/or "heavy charge". Among those with heavy charge, Stueckelberg also considered the heavy "Bose" proton and neutron suggested by Wentzel to serve as intermediates in β-decay and thus to explain a supposed "stronger asymmetry" than that predicted by the original Fermi theory. In a talk at the Delémont meeting of the Swiss Physical Society on 7 May 1938, he discussed some high-energy effects occurring in cosmic rays, concerning the absorption of charged or neutral Yukawa particles by bound nuclear particles, using the analogy of the behaviour of light quanta. That is, he estimated the cross-sections for the "photo-effect" and the "Compton effect" of mesons, using dimensional arguments (Stueckelberg 1938b). These were the only quantitative results that he drew from his theory: he predicted a preference for the photo-effect in the case of low-energy Yukawa particles and for the Compton effect at higher energies. Because of the high energy release on capture of a meson (at least $200m_ec^2$), this would result in "heating" of the nucleus (in the sense of Bohr) and several "nuclear constituents should leave the nucleus".

With Wentzel's involvement and Stueckelberg's independent work,

Switzerland became a centre for research on the Yukawa particle. Although Pauli did not publish a paper on mesons at that time, he corresponded on the subject with Kemmer and also with Heisenberg [27]. At the beginning of 1938, he wrote to the latter: "The Yukawa theory seems to suffer from the same infinities as the earlier theories!" [28]. Heisenberg responded by sending Pauli the manuscript of a new work that he had recently submitted to the *Annalen der Physik* for an issue honouring the eightieth birthday of Max Planck: "On the universal length occurring in the theory of elementary particles" (Heisenberg 1938a).

Recalling an earlier paper (Heisenberg 1936) in which a universal length had been inferred from the Fermi-field theory and used to explain the existence of "explosive showers" or "explosions" in the cosmic rays (see chapter 4), Heisenberg argued that the field theory of nuclear forces still (i.e., after the meson theory) required a universal length $r_0 = e^2/(mc^2) = 2.81 \times 10^{-13}$ cm. This was still justified, he now argued, by the divergences present in all known quantum field theories, and supported by the close agreement of r_0 with the characteristic length introduced by Yukawa's meson, namely its Compton wavelength. Since the meson's coupling strength $g^2/(\hbar c)$ was of order unity, Heisenberg suggested that it would also mean the production of "explosions" containing a number of mesons.

Pauli wrote back on 10 March 1938, saying that he had discussed the paper with his student Markus Fierz and with Wentzel [29]. They had agreed that some arguments of Heisenberg relating the universal length to the possibility of position measurements of the nucleons were sloppy (*schlampig*). However, Pauli agreed that the Yukawa theory gave "results more favourable for your point of view". He went on:

> Namely, if one describes the "heavy electrons" by a vector field instead of a scalar field, then the higher approximations always yield multiple processes (explosions) whose frequency is larger by the factor $[E/(\mu c^2)]^n$ than the simple processes, i.e., that $\hbar/(\mu c)$ appears itself as a characteristic length.

Because "no assumption needs to be made about the strength of coupling of the 'Yukons' with the light particles", Pauli argued that the theory of β-decay could still be separate from that of nuclear forces. Hence, he said, "*I hold it to be possible (even probable) that the explosions would practically contain only heavy particles, thus no electrons at all.*" (Original emphasis).

In the following months, the correspondence dealt with the problem of the meson's decay lifetime, the occurrence and composition of explosive showers, a new Japanese paper (*Interaction III*), and Heisenberg's reformulation of his critique of quantum field theory (Heisenberg 1938b) [30]. On 15 August 1938, Pauli informed Heisenberg about a new work of his Zürich colleague (Wentzel 1938b), that tried to account for what appeared to be a surprisingly large occurrence of hard cosmic-ray showers [31]. According to Wentzel, the small angular spread of these showers (less than 10°) required the introduction of a special "form factor" that would cut off the production of showers with a large angular spread. He justified this truncation of the theory by saying, "...it is well known from other applications that Yukawa's theory, like quantum electrodynamics, must be assumed to break down in problems involving very large energy" [32].

9.4 New experiments and calculations on mesotron production (1938–39)

During 1938, there were many experimental studies of cosmic rays. In the United States, especially, the groups of Millikan in Pasadena and of Compton in Chicago carried out their often competing programmes, concentrating on the origin and composition of the penetrating component. Other American and European groups were also active in this field [33]. Analysis of earlier data (Barnóthy and Forró 1937; Maass 1936) brought new suggestions, e.g., for "the *neutral counterpart of the heavy electron*, for which we propose the name *neutretto*" (Arley and Heitler (1938), original emphasis) [34]. Francis R. Shonka of Compton's team looked for the production of charged secondaries by neutral cosmic-ray particles, as several others had done since 1934. However, unlike the others who worked at sea level, Shonka's experiment was done at the Mount Evans Observatory (Colorado) at 14200 feet altitude. He also used a great thickness of lead (19–23 cm), and found a value of 1.06 ± 0.02 for the ratio of charged particles above and below the absorber, concluding that (Shonka 1939): "This high penetrating power suggests their identification with the neutrettos...postulated by Heitler." [35]

A theoretical paper that tried to account for the experimental observations of the hard component [36] by photoproduction of mesons was produced by two members of Yukawa's school in Osaka (Kobayasi and Okayama 1939). They used the vector-field theory (*Interaction IV*; Nordheim and Nordheim 1938), noting that the earlier scalar theory calculation [37] yielded cross-sections rapidly falling with energy, which did not fit the data. The vector theory, however gave rising cross-sections (as shown also by Heitler (1938a)). For the total photoproduction cross-section of positive or negative mesons (of velocity v and mass m_U) from a nucleus containing N neutrons and Z protons, near the threshold for the process, the Japanese authors obtained

$$\sigma \approx 6\pi(N+Z)[ge/(m_U c^2)]^2 (v/c) \qquad (9.2)$$

(e and g being the electric and meson couplings), and for large photon energies,

$$\sigma \approx \pi(N+Z)[ge/(m_U c^2)]^2. \qquad (9.3)$$

With regard to the latter expression, they noted: "It is remarkable that this expression neither increases unlimitedly with the energy of the incident photon, nor decreases as $1/E_0$, which was the case in the scalar theory" [38]. Thus the cross-section increases from threshold until it reaches the limiting

UNIVERSITÉ DE LYON
FACULTÉ DES SCIENCES

TÉL. PARMENTIER 05-45

INSTITUT DE PHYSIQUE ATOMIQUE
1, RUE RAULIN
LYON

LYON, le 31 décembre 1938

Dear Yukawa,

I want first to thank you for your kind forwarding of your interesting reprints and to congratulate you to the important work that you have initiated.

I learned about your idea about one year ago, when I left Russia, and I got at once very enthusiastic about it. I have thought very much about the problem and I must say, that I like your idea the more, the more I think about it.

I have, however, a few objections, which do by no means affect your principal idea, but only the technical treatment of the problem. My criticism concerns mainly the treatment of the problem by Kemmer and Heitler, but also your own calculations on nuclear forces. The main point is, that I feel quite sure that any field theory leads automatically to proton neutron and proton-proton forces of the same order of magnitude and that the previous treatment - which leads only to proton-neutron forces, by a perturbation calculation - is wrong. I have tried recently to discuss in detail the limits of applicability of the perturbation method and I think that I can show now, that the second order perturbation energy is identical with the exact enery eigenvalue if we consider a uncharged field (e.g. the electromagnetic field). The perturbation method leads, however, to wrong results for a charged field (e.g. heavy electrons), because it neglects the fact, that say a proton, will always be able to emitt virtually a great number of negative heavy electrons - the charge of which is of course neutralised by a still greater number of virtually emitted positive quanta. Thus, a proton will be only 50% of the time a proton, and 50% of the time a neutron. and p-p forces will be "equal p-n forces.

The second point is, that the exact theory does not lead to forces of the Heisenberg type and that I think, that the theory will have to use only one interaction constant. It seems to be very unlikely, that there should exist nonrelativistic forces between spins. I shall be glad to send you soon a reprint of a preliminary note on this subject, which is to appear in Journ. de Physique.

I shall be very happy to hear about your further work and I remain with my very best wishes

yours sincerely

Guido Beck

A letter of G. Beck (1903–1993) to Yukawa. Reproduced by permission of Yukawa Hall Archival Library.

value given in equation (9.2). Results of a similar order of magnitude were obtained for the inverse process of meson capture with photon emission.

Finally, Kobayasi and Okayama addressed the question of whether [39]

the hard component consists of heavy quanta created by photons in the atmosphere. As was pointed out by Bhabha, Heitler, Nordheim and Nordheim, it is necessary for the explanation of the intensity of the hard component at sea level by this assumption that a photon of high energy has a chance of about one in ten to create a heavy quantum of comparative energy before it is absorbed by the production of an electron pair.

They estimated that their prediction was one in 100, rather than one in ten, and argued:

> But, we believe this discrepancy to be due to the roughness of our calculations...Moreover, a considerable fraction of the hard component could be attributed to heavy quanta produced by the [Heisenberg] explosion process. We think, therefore that our result is satisfactory for the explanation of the cosmic-ray hard component at the present stage of the theory.

An unconventional approach to meson theory at about this time was embodied in an argument of Guido Beck, given in a letter to *Nature*, that a neutron or proton spent about half its time dissociated into a "heavy electron" and the other type of nucleon (Beck 1938a). Shortly afterwards, he submitted a full paper (Beck 1938b) and also wrote about his idea to Yukawa [40]:

> I learned about your idea years ago, when I left Russia, and I got at once very enthusiastic about it. I have thought very much about the problem and I must say I like your idea the more, the more I think about it. I have, however, a few objections, which do by no means affect your principal idea, but only the technical treatment of the problem.

Beck then criticized the treatments of Kemmer, Heitler and Yukawa, claiming that any field theory would give charge-independent nuclear forces. Previous calculations (without neutral mesons) had not led to it because their perturbation theory method "neglects the fact, that say a proton, will always be able to emit virtually a great number of negative heavy electrons" besides a greater number of positive quanta". Thus, a proton will only be 50% of the time a proton and 50% of the time a neutron, and p–p forces will be equal to p–n forces".

Yukawa answered Beck, reminding him that on his visit to Japan in 1937 he had accepted "the spirit of my first paper on the theory of the mesotron". Yukawa found Beck's argument "quite plausible if we accept the present formulation of the theory" [41].

9.5 Progress in theory until summer 1939: the Americans enter the scene (1938–39)

We have remarked previously (chapter 7) on the relative neglect of Yukawa's theory by American theoreticians in its early stages. While their European colleagues discussed it in their reviews [42], the Americans were not sure whether to take it seriously. Indeed, Robert Serber's talk in December 1937 claimed that the meson theory of nuclear forces had defects which could not be remedied (Serber 1938). Bethe, however, in an April 1938 meeting in Washington, pointed out that including neutral "barytrons" and using vector-meson fields might remove two of Serber's difficulties, namely, the explanations of charge-independence and the saturation of the nuclear forces [43]. Also Oppenheimer's students at Berkeley softened a bit their negative views on meson theory. Thus Lamb and Schiff (1938), using the scalar theory, obtained repulsive n–p forces in the deuteron ground state (as was to be expected), but they derived the correct order of magnitude for the p and n magnetic moments by adding an additional six-vector field [44].

9.5.1 *The Nordheims and the Sommerfeld Festschrift*
The next American paper on meson theory was by G. and L.W. Nordheim, two German refugees at Duke University in Durham, North Carolina [45]. One of their results gave a photoproduction cross-section of mesons that was much too small to account for the observed number of mesons in the atmosphere. The Nordheims concluded:

> Therefore no simple picture of barytron production in terms similar to radiation theory can be given. The failure, however, indicates only the inapplicability of perturbation calculations, but does not constitute an actual disproof of the link between nuclear forces and cosmic radiation.

In a note added in proof, they referred to a "more exact evaluation" of the meson's interaction strength, yielding $G^2/(\hbar c) \approx 0.3$, on which they commented [46]:

> The cross sections of the production processes are thereby increased... so that they approach already those required by observation. Furthermore, the convergence of the perturbation calculations becomes doubtful, even for the scalar barytron field. This means that the energy dependence of the cross sections cannot be determined any more. Even if the possibility for a better estimate of the production probabilities seems thus to be still more remote, the prospect for establishing the discussed connection between nuclear forces and the hard component of the cosmic radiation is certainly much better now.

The 1 December issue of the *Physical Review* contained four papers on mesons. In addition to one by Wentzel (1938b), there were articles by Heitler, Stueckelberg and Otto Laporte of Ann Arbor, Michigan, a former student of Arnold Sommerfeld. Using a time-dependent perturbation theory, Laporte found that the elastic scattering of mesons gave rise, already in first approximation, to a polarization effect, in contrast to the electron scattering case [47].

Heitler, on the other hand, discussed nuclear disintegration by mesons in the vector-meson theory, emphasizing [48]:

> The problem of the interaction between heavy electrons (and electrons or light quanta) and nuclear particles for high energies seems to be a very fundamental one, leading beyond the limits of the present quantum theory... It seems that experiments on the behaviour of the hard cosmic-ray component could, for the first time in physics, open an insight in the laws of physics which are beyond the validity of the quantum theory.

Finally, Stueckelberg tried to give a unified treatment of the interaction with matter of "radiation" — in which term he included both the electromagnetic field and the massive vector field — by introducing the static interaction of both fields into the system's Hamiltonian through a "contact" interaction. This yielded a singular behaviour for the static interaction, already in the second order of perturbation theory (i.e., in order $[e^2/(\hbar c)]^2$ or $[G^2/(\hbar c)]^2$).

The number of papers from abroad in the cited issue of the *Physical Review* of 1 December was unusually large because the issue was dedicated to Sommerfeld on his seventieth birthday. In 1938, the Nazi government launched further restrictions on the German and Austrian Jews and, as Heisenberg informed Pauli in a letter of 15 July, the *Annalen der Physik* (in which an issue to celebrate Sommerfeld had been planned) would not accept papers from "non-Aryans". In response Pauli, with Laporte's help, got the *Physical Review* to publish 18 contributions from American and non-American authors who were former students or associates of Sommerfeld [49].

Edward Condon, in concluding his article in this *Festschrift*, indirectly referred to the darkness that had fallen upon Germany [50]:

> Everyone of my generation grew up in atomic physics by way of [Sommerfeld's] great *Atombau* und *Spektrallinien*, a large group have profited by the stimulation of his lectures on his American visits, a fortunate few have derived boundless stimulation from the opportunity of working in his Institut für Theoretische Physik in the former brighter days.

Although the brighter days had passed, the brilliant students of Sommerfeld who were refugees abroad in Europe and the USA, were deeply involved in helping to establish the meson theory of nuclear forces.

9.5.2 *The American–European interchange*

By late 1938, the meson theory was replacing the hitherto dominant Fermi-field theory of nuclear forces, and some physicists began to think about the possible effects of the exchange of charged mesons in the nucleus upon the electric and magnetic properties of the nucleus. For example, it was realized that size effects and deviations from the Coulomb law could affect even atomic spectra. This would especially be the case for the atomic S-levels, whose wavefunctions do not vanish at the origin. Since 1934, there had been reports of small deviations in hydrogen and deuterium from the predictions of the Dirac theory for a point nucleus. In 1938, Robley C. Williams of Cornell reported accurate measurements on the $n = 3$ to $n = 2$ levels in hydrogen and helium that confirmed discrepancies in the fine structure predicted by Dirac's relativistic electron theory (Williams 1938).

Simon Pasternack then pointed out that nuclear size effects would be too small to account for Williams' results, but they could be accounted for by a simple shift (of unknown origin) of the 2S levels, upwards in energy by 0.03 cm^{-1} (Pasternack 1938) [51]. Fröhlich, Heitler and Kahn at Bristol suggested that the deviation, rather than indicating a breakdown of the Dirac theory, could be due to a departure from the Coulomb law at small distances. They argued [52]:

Such a discussion has now been made possible by the recent theories of the proton and neutron based on the existence of the mesotron... This theory has led to the conception that the proton spends a certain fraction of the time in a dissociated state as a neutron and a positive mesotron distributed around the neutron in a volume with linear dimensions of the order of the electronic radius. It is obvious that this would lead... to a decrease of the Coulomb attraction and hence to an additional repulsion, as required by the experiments.

The authors point out that their effect is related to the self-energy of the electron in the complicated field of the proton, that the effect becomes noticeable in the potential at about one-fifth of the classical electron radius and that, e.g., the "spreading" of the nuclear charge in deuterium gives a shift of energy by a factor 10^4 too small to account for the observed effect. However, they cautioned [53]:

> The results of our theory should not, however, be taken too literally. Yukawa's theory, like any quantized field theory, breaks down for small wavelengths, and no result can be relied upon for distances considerably smaller than the electronic radius. But this is just the region where our departure takes place.

In calculating the energy shift, the Bristol group had to manipulate the difference of two diverging integrals, which is always a tricky procedure. Willis Lamb criticized their paper, saying that "no mesotron theory which is not radically different from those considered at present can possibly give such a short range repulsion between a proton and an electron" (Lamb 1939). He showed that, if α is the probability that a proton is dissociated, then the proton's potential $V(r)$ satisfies

$$V(r) < -(1 - \alpha)(e^2/r), \tag{9.4}$$

so that there can be no repulsion, because α cannot exceed unity.

Although the Bristol group (Fröhlich *et al.* 1939b) countered that field theory might permit a negative charge density very near the origin, because of virtual pair production, and hence give an α larger than one, Lamb argued back that, in fact, the existing field theories did not allow such a negative charge density. He concluded his letter with this caution [54]:

> For the present, at least, it seems that one may draw qualitative conclusions from a divergent field theory, if at all, only if it is found that one's results are not seriously affected by the introduction of a reasonable converging device.

Concerning the suggestion of a modified Coulomb law, Kemmer had already written to Pauli [55]:

> Heitler's authority must be doubted with respect to the paper of Fröhlich, Heitler and Kahn, which will appear soon, and according to which the virtual presence of mesons changes the Coulomb field (e.g., in a hydrogen atom) such that for small distances a violent repulsion occurs. I do not understand this physically at all; hence one should attribute it most probably to the account of cut-off physics.

Pauli's response was uncharacteristically mild [56]:

> The paper of Heitler, Kahn and Fröhlich we have now discussed more carefully here in the seminar, and I shall soon write my opinion about it to Heitler in

detail. I consider it to be mathematically correct (in the approximation of perturbation theory considered) *but physically it seems to me* that the result lies outside the validity of the present theory.

A more quantitative attempt at nuclear forces within meson theory was made by Møller and Rosenfeld in Copenhagen [57]. So far, they wrote, only the first approximation to the force had been considered by Kemmer and others, and so:

> It would thus seem desirable to discuss more closely the reliability of such results, and for this purpose a possible method of attack is suggested by an analogous situation in quantum electrodynamics, where a suitable canonical transformation allows us to separate, from the expression of the total energy of a system consisting of electrons and an electromagnetic field, a term depending only on the coordinates of the electrons and representing the Coulomb energy.

This method could be applied equally to a system of nuclear particles and a meson field, providing static forces which are valid also in regions smaller than the nuclear force range.

The main motivation for the work of Møller and Rosenfeld was what they called the "main defect" of the meson theory; namely, the occurrence of "a term of dipole interaction energy [a $1/r^3$ term] which is so strongly singular for infinitesimally small [distances] that it would not allow the existence of stationary states for a system of such particles" [58]. They therefore proposed adding "besides the four-vector wavefunction a further pseudoscalar wavefunction", which would give a static interaction "just capable of cancelling the dipole interaction term without affecting the others". They claimed that the new term would also modify the predicted β-ray spectrum to give a better agreement with experiment.

9.5.3 *Neutral meson and mixed-meson theories*

In the same year, 1939, Bethe discussed some consequences of the meson theory for nuclear forces, referring to the British papers of 1938, but surprisingly, not at all to those of Yukawa *et al.* He remarked that the meson theory, which he took to be purely neutral, gave both a spin-independent force

$$U = g_1^2 \exp(-\lambda r)/r \tag{9.5}$$

and a spin-dependent force

$$V = V_1 + V_2, \tag{9.6}$$

with

$$V_1 = \frac{2}{3}g_2^2 \, \boldsymbol{\sigma}_1 \cdot \boldsymbol{\sigma}_2 \, \exp(-\lambda r)/r, \tag{9.6a}$$

$$V_2 = -g_2^2[3(\boldsymbol{\sigma}_1 \cdot \boldsymbol{r})(\boldsymbol{\sigma}_2 \cdot \boldsymbol{r})/r^2 - \boldsymbol{\sigma}_1 \cdot \boldsymbol{\sigma}_2] \, \exp(-\lambda r)/[r^3(1 + \lambda r + \lambda^2 r^2/3)], \tag{9.6b}$$

in the usual notation. Although the authors quoted by Bethe (as well as the Yukawa team) had argued that V_2, which depends on the vector r but whose average over all directions gives zero, "has no influence on spherically symmetrical states such as the ground state of the deuteron", Bethe said that was false and argued that: "In fact, V_2, as it stands, will give an infinite binding energy for the ground state of the deuteron, for it represents an inverse cube potential which is attractive for a certain linear combination of an S and a D state, and for such a potential the Schrödinger equation does not possess a lowest eigenvalue." [59]

Bethe suggested cutting off the potential, although, as he said: "This is not very satisfactory from the aesthetic point of view". Still, the resulting nuclear forces were "superior to the old forces in *predicting a quadrupole moment for the deuteron* because the ground state of this nucleus will now be a mixture of an S and a D state rather than a pure S state. Such a quadrupole moment has been observed by Kellogg, Rabi, Ramsey and Zacharias" [60]. Bethe also found the prediction of the neutral meson theory to be superior to that of Kemmer's symmetric meson theory as regards the magnitude and the sign of the quadrupole moment (i.e., whether "pill-box" or "cigar" shape), but he admitted that the experimental results were still in doubt [61]. Of course, the neutral theory was silent on the questions of β-decay and the magnetic moments of the nucleons.

In their correspondence, Pauli and Heisenberg were strongly critical of Bethe's suggestion. Thus, Pauli wrote, the American physicists [62]

> rated it very highly. To me, however, this assumption [of only neutral mesons] appears to be most nonsensical (*höchst unsinnig*); also, his result, according to which there should be no Majorana force between proton and neutron (and instead only spin-dependence) appears to be extremely suspicious.

Heisenberg, having just returned from the United States, replied [63]:

> I have finished with Bethe's paper on the deuteron and the neutral mesons. Bethe gave a talk on it in Lafayette [Indiana]. His result that the neutral mesons are more suited than the Kemmer forces to explain the quadrupole moment of the deuteron rests entirely on his senseless way of "cutting-off". If one keeps the relation between the tensor force and Yukawa's potential, according to which the tensor force is caused essentially by the term $(\boldsymbol{\sigma}_1 \cdot \nabla)(\boldsymbol{\sigma}_2 \cdot \nabla)V(r_{12})$, and then inserts for $V(r_{12})$ something like a Gaussian curve, probably the opposite follows from what Bethe claims. Bethe actually admitted this and will continue his calculations on the problem.

As noted above, Møller and Rosenfeld avoided the use of arbitrary cut-offs in their mixed vector–pseudoscalar meson theory. In a letter to *Nature*, dated 3 July 1939, they wrote (Møller and Rosenfeld 1939b):

> In view of this unsatisfactory feature of the vector meson theory [the $1/r^3$ term], we should like to point out that the question of the quadrupole moment of the deuteron, as well as the whole problem of consistency of the meson there of nuclear forces in its present provisional form, takes a quite different aspect when as already suggested in a previous note [Møller and Rosenfeld 1939a], pseudoscalar meson fields are introduced besides the vector fields in such a way

as to cancel the strongly singular terms in the static interaction energy of the heavy particles.

Møller and Rosenfeld published a detailed paper in the following year. Unlike Bethe, they preferred to use Kemmer's symmetric theory (Kemmer 1938b, 1939) because of its unifying character, and they also noted that there were effectively four types of meson fields possible that satisfied relativistic invariance and had positive definite energy. Using their method of canonical transformation, at the end of their part I, they found, with $\phi(r) = \mathrm{e}^{-\lambda r}/(4\pi r)$ [64]:

The most general form of static interaction resulting from an arbitrary mixture of the four types of meson fields is

$$\frac{1}{2}\sum_{I,K}(\tau^I\tau^K)\left\{\begin{array}{c} G_1 + G_2(\boldsymbol{\sigma}^i\boldsymbol{\sigma}^k) \\ +\dfrac{1}{\kappa^2}\,G_3(\boldsymbol{\sigma}^i\mathrm{grad}^i)(\boldsymbol{\sigma}^k\ \mathrm{grad}^k) \end{array}\right\}\phi(r^{ik}) \qquad [(9.7)]$$

with coefficients given by

$$\begin{aligned} G_1 &= g_1^2 - f_1'^2 \\ G_2 &= g_2^2 - g_1'^2 \qquad\qquad\qquad\qquad [(9.8)] \\ G_3 &= g_2^2 - f_2'^2 - g_1'^2 - g_2'^2. \end{aligned}$$

The various f and g terms are the couplings to the nucleons of the four field types. Letting $G_3 = 0$, to make the "dipole" term in the force vanish, and imposing conditions on the forces from those known for the deuteron, they found that they could not have only one type of meson field [65]. Having a mixture of two will work, provided that they are the vector and pseudoscalar fields, and it requires that $g_2^2 = f_2^2$ [66].

In 1939, indeed, many theorists began to work on the theory of mesotrons, or "meson theory" as it became commonly known, especially in British journals [67]. The spectrum of this work ranged from speculative proposals — such as that of Proca and Goudsmit (1939), who set up a six-dimensional theory of elementary particles (with spin and charge as coordinates, in addition to space-time) — to more practical ones. Regarding β-decay, for example, the Brazilian physicist Mario Schönberg showed "...that the meson theory of disintegration leads to the Gamow–Teller selection rule, not only for the theory with spin 1 that gives correct neutron–proton force, but even more for one of the forms of the theory with 0 spin" [68]. In India, Majumdar and Kothari (1939) addressed the high-energy interaction of mesons and nuclei, as seen in the cosmic rays, which was beginning to be, perhaps, as important a problem for the theory as the decay lifetime discussed in the previous chapter.

9.6 Yukawa's trip to Europe and America (1939)

At the Seventh Solvay Conference in 1933, whose subject was nuclear physics, Heisenberg had presented his neutron–proton theory of nuclear forces and Pauli had described his neutrino hypothesis; these subjects had been forerunners of the meson theory. The title chosen for the postponed 1936 conference was originally "Cosmic Rays and Nuclear Structure". The rapid progress since then had produced the more ambitious title in 1938, heralding a new field of physics: "Elementary Particles and their Mutual Interactions".

Although not asked to serve as a *rapporteur*, Yukawa was to join a select group of non-lecturing participants, including Carl Anderson, Enrico Fermi, Alexandru Proca, Walther Bothe, Arthur Compton, Bruno Rossi, George Gamow, Irène and Frédéric Joliot-Curie, Lise Meitner and Merle Tuve, as well as the meson theorists Homi Bhabha, Christian Møller and Léon Rosenfeld (the last group of three being "scientific secretaries").

Yukawa was excited to receive the invitation, and immediately talked it over with Sakata and with Seichi Kikuchi at Osaka and with Professors Arakatsu and Kimura at Kyoto (on 6 April, according to Yukawa's diary). Two days later, he went to Tokyo to discuss it with Nishina. On 17 April, he received more details about the conference programme and participants, and on 19 April, he was interviewed and a story was printed in the evening edition of the newspaper *Asahi Shinbun*. On 22 April, he received a letter signed by Nishina, informing him that the National Committee on Physics of the Research Council of Japan had appointed him a "delegate" to the Solvay Conference. On 25 April Yukawa sent a letter to Brussels in French, accepting the invitation. The trip would be financed by Riken. Meanwhile, he received two further invitations from Europe.

The first came from Heisenberg, who knew of Yukawa's invitation to Brussels, and it asked him "in the name of the German Physical Society to deliver a shorter talk on his theory at the Marienbad meeting". (The meeting was to be held in the Sudetenland, which had been ceded to Germany by the Czechs, under threat, the previous year.) The letter added [69]:

> A certain difficulty in this invitation seems to me the problem which language you can use. We prefer, most of all, of course, that you speak in German. If you cannot do so, the lecture can also be given in English. Foreign language talks have so far been given in the so-called smaller lectures and your talk should in this case not exceed 20 minutes.

Heisenberg mentioned that the society would contribute a "suitable" amount to the travel expenses, and concluded: "Personally I would be very happy to get to know you this autumn and to talk with you about many problems".

That letter arrived on 4 May, but a day earlier Yukawa had already received a letter from Wolfgang Pauli, inviting him to lecture at an international meeting in Zürich at the beginning of September. Yukawa accepted both invitations, to Marienbad and to Zürich [70]. He made his preparations during the weeks that followed, also moving to Kyoto to take up his chair at the Imperial University on 26 May. At the end of June, he boarded the steamer Yasukuni Maru in Kobe, from which he wrote to

Heisenberg [71]: "I hope that I shall have the honour of visiting your university soon after my arrival in Berlin. It is to me a great pleasure that I can accept your kind invitation to the meeting at Marienbad and shall deliver a short lecture there."

In June 1939, Heisenberg was on a tour of the USA, where he received another letter from Yukawa (also dated 30 June); to this he replied while crossing the Atlantic back to Germany, saying that he would not be in Leipzig during August but that he would meet Yukawa at the conferences in Zürich and Marienbad [72]. Yukawa arrived in Naples on 2 August, visited Pompeii and went on to Rome, arriving on 7 August in Berlin, where he worked on his lecture for Marienbad and "kept in practice of speaking German" [73]. He contacted the secretary of the German Physical Society, W. Grotrian, who welcomed him in a letter, also requesting a short summary of the Marienbad lecture [74]. Yukawa provided the following text (original in German) [75]:

> The Present Status of the Theory of the Mesotron
> by Hideki Yukawa
> Dr, Professor of Theoretical Physics
> at the Imperial University of Kyoto
>
> The most important properties of the mesotron are mentioned, which follow immediately from the basic assumptions of the present theory. Then it is discussed briefly how in this theory the forces between the constituents of nuclei, the β-decay and the phenomena connected with the penetrating component of cosmic radiation can be explained. Finally the difficulties are discussed — both those which this theory shares with all other quantum field theories and those which are peculiar to it.

The preliminary programme for Marienbad (Deutscher Physiker- und Mathematikertag), dated 5 June, scheduled Yukawa's lecture in the session on "Nuclear Physics", on 28 September. The session's organizers were Otto Hahn, Heisenberg and Gerhard Hoffmann. Yukawa drafted most of his lecture in German and divided it into these sections [76]:

(1) Introduction (fundamental properties of mesotrons).
(2) Problem of the deuteron. Neutral mesotrons and their decay.
(3) Mesotrons as the penetrating component of the cosmic radiation.
(4) Applicability of the present mesotron theory.

On 18 August, Tomonaga, who was taking a two-year sabbatical leave from Riken to study with Heisenberg, came to Berlin to fetch his friend to stay with him in Leipzig, where Yukawa could make use of the library to look up some references. Most of the university staff were on their summer vacations. He then returned to Berlin and worked on his Zürich lecture. On 25 August, however, he received word from the "association of Japanese residents in Berlin" and soon afterwards from the Japanese Embassy to "flee to Hamburg and to take the steamer Yasukuni Maru because of the increasingly strained situation in Europe" (Yukawa 1971). In the evening, Yukawa left for Hamburg and boarded the ship, the same one that had taken him to Naples, which sailed for Bergen, also carrying Tomonaga, arriving there on 28 August.

On 1 September, the *Blitzkrieg* attack on Poland began; on the same day, from aboard ship in Bergen, Yukawa wrote to Heisenberg:

> I received your letter from *"Europa"* with great pleasure and have been waiting for the Meeting in Zürich. I am now in Bergen starting from Hamburg according to the advice of the Japanese Embassy. Our ship will go back again to Hamburg, when the circumstances in Europe become better. If this were not the case, the ship will go directly to Japan according to the order of the Japanese Embassy and it will be very difficult for me to remain in Europe apart from other passengers, so that I shall be able neither to see you nor to attend the Meeting in Marienbad. I beg your pardon in advance for my inability to carry out the promise in the worst case, although I still expect that it will not happen. In either case, I shall never forget your kindness and I believe that the science will develop steadily through the cooperation of scientists of countries all over the world. I remain always,
> yours very sincerely,
> H. Yukawa

The Marienbad meeting was, of course, cancelled, as was the Zürich international conference which was to have preceded it on 4–17 September.

The Japanese ship left Bergen on 4 September and reached New York ten days later. The two physicists aboard visited the World's Fair then in progress; then Tomonaga returned to the ship, which headed for the Panama Canal. Yukawa, however, instead travelled across the United States by train, visiting various physicists, taking ship again in San Francisco on October 13. At Columbia University, Yukawa met Fermi, Isadire Rabi and Otto Stern; he also saw John Dunning and Harold Urey. ("Urey seemed to be absorbed in separating the isotopes of C and N.") At Princeton he met John Wheeler and visited Einstein at his home. ("I had the impression that Einstein still believed that the present quantum theory was incomplete and there should exist some correct continuous theory to replace the present one... Wheeler gave an account of the Bohr–Wheeler theory of nuclear fission. He was very kind, and still a young man." [77])

In Washington, at the Carnegie Institution, he saw Gamow and Teller, both of George Washington University, and Merle Tuve, at whose suggestion "we all sat on the lawn, set a blackboard against a tree and discussed the meson theory". Then to Boston, where he met K.T. Bainbridge and J.C. Street at Harvard and saw the cyclotron and the Van de Graaff machines at MIT. On 30 September, he arrived at Chicago and met Rossi. ("He had just returned from his experiment on Mt Evans (4300m) in Colorado.") On a trip to the University of Michigan at Ann Arbor, he gave a thirty-minute talk to members of the Physics Department in the evening, after meeting Otto Laporte.

Back in Chicago, he met Arthur Compton and discussed the recent "supposition that the primary cosmic rays may include protons". Yukawa gave a talk to the cosmic-ray physicists and then took the train to Los Angeles. Arriving on 6 October, he immediately went to Caltech and met

Neddermeyer and also Anderson, who invited him to dinner. Millikan was in Australia on a cosmic-ray expedition. ("Anderson has produced brilliant achievements in physics, but there is no pretentiousness about him.") He then travelled to San Francisco and gave a colloquium talk at the University of California in Berkeley. On 13 October Yukawa sailed on the Kamakura Maru, reaching Yokohama on the 28 October.

Notes to text

[1] Yukawa *et al.* (1965); translation of p 13 by R. Kawabe.
[2] Sakata in Yukawa *et al.* (1965); translation by N. Eguchi, p 31.
[3] [YHAL] E15 010 T25 has Japanese abstracts of these talks. The meeting was held at Kyoto University shortly before Yukawa moved there.
[4] [YHAL] E15 20 T25. H. Yukawa, "General remarks on mesotrons", manuscript of a talk at the 1939 Annual Meeting of the PMSJ, 5 pp, marginal note on p 1.
[5] Yukawa and Sakata (1939a, p 139).
[6] Yukawa's diary, [YHAL], from which most of the following is derived.
[7] Neddermeyer and Anderson (1937) and Street and Stevenson (1937a).
[8] Corson and Brode (1938b, p 777).
[9] Euler (1937, p 518): "If the assumption of Anderson and Neddermeyer should turn out to be correct, one could introduce a new type of particle into a future theory as follows, according to Heisenberg: just as in quantum mechanics there exist many discrete energy states for the nucleus, of which some are stable and others unstable, a future quantum theory of fields might well yield several field quanta of different rest energy, among them some stable ones like protons and electrons, others being unstable like those of Neddermeyer and Anderson, and eventually still many others".
[10] See the report "Congrès du Palais de la Découverte. International Meeting in Paris – Physics" *Nature* **140**, pp 710–12, especially p 711.
[11] J. Barnóthy to L.M. Brown, 10 October 1978.
[12] Rossi (1938a, especially pp 63–4). At the Galvani Conference, Gregor Wentzel, of Zürich, reported on the production of showers in cosmic rays (Wentzel 1938a, especially p 278 and note 32 on p 279).
[13] See E.J. Williams (1938), Bhabha (1938c) and Heitler (1938c).
[14] Klein (1939, pp 77–8). See our discussion in chapter 1 regarding the stability and the mass of the electron, according to H.A. Lorentz.
[15] Klein (1939, p 80).
[16] The importance of Klein's ideas was recognized only decades later, in connection with the gauge theories appearing in the current Standard Model. However, Klein tried to merge strong, weak and electromagnetic forces (in a grand unified theory), whereas the Standard Model has separate electroweak and strong nuclear (quantum chromodynamics) interactions.
[17] Discussion section of Kramers (1939, p 110).
[18] Brillouin (1939, sections 19–22).
[19] Kemmer to Wentzel, 9 March 1938 (in German) [AHQP]. The letter begins by stating that "not much has happened since Christmas", since Kemmer was assisting in the beginners' courses, "which at first takes some time".
[20] These are excluded by the sign of the n–p forces.
[21] Kemmer excused himself for the delay in the letter. The reason he gave was that he was trying to get a tenured job as "demonstrator" at London, a position with a small increase in his stipend and, of course, less time for research. Trying other

possibilities for employment, he reported: "I was twice in Cambridge to give a lecture; there one is not very enthusiastic about the new professor. Nevertheless, wait and see. When Pauli is there, I will go as well".

[22] Wentzel (1938a, p 276).
[23] Actually, the asymmetry was only an experimental artefact.
[24] Wentzel (1938a, p 278).
[25] Wentzel (1938a, endnote 22).
[26] Stueckelberg (1938a, p 244). See also Stueckelberg (1938c, 1939a).
[27] See, e.g., [WPSC2], pp 522–7, for correspondence in June 1937 on the possible existence of the mesotron.
[28] Pauli to Heisenberg, 22 February 1938. [WPSC2], p 522.
[29] Pauli to Heisenberg, 10 March 1938. [WPSC2], pp 556–8.
[30] This paper contained results that Heisenberg prepared for the 1938 Warsaw Conference. Because Heisenberg was not allowed by the German government to attend, for political reasons, Rosenfeld read selections from his manuscript.
[31] [WPSC], pp 593–4.
[32] Wentzel (1938b, p 870). These "penetrating showers" were only understood later as nucleon cascades.
[33] E.g., in the USA: Neddermeyer and Anderson (1938) (showing a mesotron brought to rest in a cloud chamber), Swann and Ramsey (1938), V.C. Wilson (1938, 1939a, b) and Schein and Wilson (1938). In Europe: J.G. Wilson (1938, 1939), Geiger and Heyden (1938) and Maier-Leibnitz (1938).
[34] N. Arley visited Bristol from Copenhagen, where Christian Møller wrote his first note on meson theory (Møller 1938), suggesting a real vector-field description of the neutretto, similar to Kemmer (1938c) (which had not yet appeared in print).
[35] Evidence for a neutral component of the penetrating cosmic rays had been presented by Maass (1936), Barnóthy and Forró (1937, 1939) and others. Clay and von Gemert then measured the absorption in lead of the hard component at various depths in the earth and found the decrease in lead absorption to be consistent with the absorption of charged particles. They concluded: "We see, therefore, that the hypothesis according to which the coincidences, found in the deep layers of the earth, are produced by neutrinos [or by neutrettos, i.e., neutral mesotrons] must be given up" (Clay and von Gemert 1939, p 503).
[36] Schein and Wilson (1938) and Bowen *et al.* (1938a, b, c, d).
[37] This was done for scalar theory in *Interaction IV* because "the vectorial theory for the U-field is rather complicated and will be constructed by Kobayasi in another place" (Yukawa *et al.* 1938b, p 738).
[38] Kobayasi and Okayama (1939, pp 11–12).
[39] *Ibid.*
[40] Beck to Yukawa, 31 December 1938. [YHAL]
[41] Yukawa to Beck, 15 March 1939. [YHAL]
[42] Besides Wentzel (1938a), section 7 of Heitler (1938c, pp 381–8) deals entirely with Yukawa's theory, as does Peierls (1939).
[43] Bethe (1938). Only a little earlier, Bethe and his student Robert Marshak had proposed a new electron-pair theory of nuclear forces (Bethe and Marshak 1938).
[44] Lamb and Schiff did not know yet of the vector meson theory of the Yukawa school. They mention Bhabha (1938b) and his "four-vector dynaton; however, in the nonrelativistic limit for the heavy particles, this is completely indistinguishable from the scalar dynaton formalism of Yukawa" (their note 13). Of course, that is not true.
[45] Nordheim and Nordheim (1938), discussed in chapter 8. See also Nordheim (1938).
[46] Nordheim and Nordheim (1938, p 265).

[47] Laporte (1938b). See also Laporte (1938a).

[48] Heitler (1938b, p 876).

[49] In addition to those mentioned, their names included W.V. Marston, L. Pauling, C. Eckart and E.U. Condon from the USA; L. Brillouin from France; P. Ewald, F. London and R. Peierls from the UK; and A. Rubinowicz from Poland. Former Germans, then in the USA, were H. Bethe, A. Landé, K. Herzfeld and E. Teller. (See also chapter 7 for German refugees in Britain; see also Rechenberg (1988a).)

[50] Condon (1938, p 940).

[51] The effect (the Lamb shift) was accurately measured in 1947. It results from higher order radiative corrections in QED and it stimulated the development of the technique of renormalization, by which it was calculated.

[52] Fröhlich *et al.* (1939a, p 270).

[53] *Ibid.*

[54] Lamb (1940). Of course, Lamb was proven to be right. After 1947, renormalized QED gave accurate results for the shift of energy levels in a Coulomb field, i.e., what became known (after Lamb's experiment) as the Lamb shift.

[55] Kemmer to Pauli, 20 April 1939; [WPSC2], pp 629–32, especially pp 631–2.

[56] Pauli to Kemmer, 13 June 1939. [WPSC2], pp 665–6.

[57] Møller and Rosenfeld (1939a). Note the use of the term "meson" in their title, which was common in Copenhagen.

[58] *Ibid.*, p 242.

[59] This is, of course, just the "dipole term" that is called in note 46 the principal defect of the meson theory.

[60] Kellogg *et al.* (1939). All quotations are from Bethe (1939, pp 1261–2), with original emphasis.

[61] In the symmetrical theory, the factor $\tau_1 \cdot \tau_2$, which is -3 in the triplet state of the deuteron, changes the sign of the quadrupole moment. The tensor force was later analysed by Rarita and Schwinger (1941a,b).

[62] Pauli to Heisenberg, 7 August 1939; [WPSC2], pp 672–3.

[63] Heisenberg to Pauli, 14 August 1939; [WPSC2], pp 672–3. Bethe (1940a, b) also discussed the Kemmer symmetric theory.

[64] Møller and Rosenfeld (1939b, especially p 476).

[65] Møller and Rosenfeld (1940, p 26).

[66] *ibid.*, p 33.

[67] We return once more to the question of naming the meson. Bhabha (1939a) used both names in his title, but added a footnote: "The name 'mesotron' has been suggested by Anderson and Neddermeyer...for the new particle found in cosmic radiation...It is felt that the 'tr' in this word is redundant, since it does not belong to the Greek root 'meso' for middle; the 'tr' in neutron and electron belong, of course, to the roots 'neutr' and 'electra.'...It would therefore be more logical and also shorter to call the new particle a meson instead of a mesotron."

[68] I.e., for the pseudoscalar theory (Schönberg (1939), original emphasis.)

[69] Heisenberg to Yukawa. 13 April 1939.

[70] There followed an amusing exchange of letters with Heisenberg. Yukawa wrote that he would "endeavour to deliver the lecture in German instead of English under the title, 'Die gegenwärtige Zustand der Theorie des Mesotrons' or something like that for 30–40 minutes" (Yukawa to Heisenberg, 9 May 1939). Heisenberg then wrote that he could have 30 minutes but should leave some time for discussion. He added, "We are very happy and are grateful to you that you will talk in German. Since I know how difficult it must be for a Japanese to master so many European languages, I may perhaps advise you to read your lecture from a manuscript. Perhaps you will forgive me for adding, due to my experiences with so many Japanese friends, the following advice. The German

language differs from the Japanese especially by the fact that the vowels are often pronounced as long vowels. Perhaps you have the opportunity in Japan to get used a bit to this property of the German language with the help of a German acquaintance." (Heisenberg to Yukawa, 1 June 1939).

[71] Yukawa to Heisenberg, 30 June 1939. Details of Yukawa's visit to Europe have been treated by Konuma and Rechenberg (1985).

[72] See note 71 for Yukawa's letter. Heisenberg to Yukawa, 5 August 1939.

[73] "YHAL Resources: Hideki Yukawa (III)", edited by YHAL (Kyoto 1988), pp 193–6, especially p 193. There the detailed itinerary is given, as derived from H. Yukawa, "A tour to Europe and the USA 1939" in Yukawa (1971, pp 251–66).

[74] Yukawa to Grotrian, 14 August 1939; Grotrian to Yukawa, 17 August 1939.

[75] H. Yukawa, "Der gegenwärtige Stand der Theorie des Mesotrons" [YHAL].

[76] There are several, mostly unfinished, manuscripts of the Marienbad lecture: [YHAL] FO8150, FO8152 and FO8160. There is a practically finished typescript version of the Zurich lecture [YHAL] FO8 P19, 9 pp, which was published later as Yukawa (1941).

[77] Quotations here and in the following are from Yukawa (1971) (see note 73).

Chapter 10

General Properties of Elementary Particles

10.1 Introduction: what is an elementary particle?

Just before the turn of the century, the atom of the Greeks lost its role as the smallest constituent of matter: the atom was split into electrons and ions, and atomic fragments, the α-particles, emerged from radioactive atoms. Three decades later, the most elementary objects were thought to be the electron, the proton and (though not universally acknowledged) the photon. In the thirties, new particles made their appearance: the positron (the first theoretically predicted particle), the neutrino, the neutron and then the mesotron.

What qualified these particles as more elementary than other atomic-sized systems? In the first place, they were obtained by fragmenting atoms and did not themselves seem capable of further fragmentation. Protons and neutrons were seen as the irreducible *heavy* parts of matter, while the α-particle, although it had not been "smashed", was conveniently represented as composed of protons and neutrons, as were the other nuclei. However, from the point of view of *weak* interactions, the neutron was not indivisible, decaying into a proton, an electron and a neutrino. Thus it was possible for an "elementary particle" to be unstable. Maurice Goldhaber has recalled the impact of this realization [1]:

> I remember being quite shocked when it dawned on me [in 1934] that the neutron, an "elementary particle" as I had by that time already learned to speak of it, might decay by β-emission with a half-life that I could roughly estimate...to be about half an hour or shorter.

As we have discussed in earlier chapters, elementary particles were assumed to be those that participated directly in the elementary interactions of quantum field theory, i.e., QED and the Fermi interaction. In chapter 2, we described the attempt to derive also the strong nuclear interaction from the Fermi-field. It was the failure of this approach that led Yukawa to propose the meson theory of nuclear forces. With his U-particle, an object strongly coupled to the heavy particles (today's hadrons) and weakly coupled to light particles (today's leptons), Yukawa introduced, for the first time, separate *strong* and *weak* couplings, while at the same time unifying them through the U-particle.

Relativistic quantum field theory in the 1930s was beset with difficulties and controversial. Plagued with infinities, such as those of the mass and charge in QED, the situation was even worse in the field theories of the nuclear forces. Disregarding the pessimism of pioneers like Bohr, Heisenberg and Pauli, other particle theorists tried to cure the difficulties by new formulations. While electromagnetic interactions were strongly limited by gauge invariance, nuclear interactions were available in richer variety. The meson (or mesotron) that carried the force had properties that experiment had not entirely pinned down, as regards its mass, spin, decay lifetime and other interactions. Thus different types of quantum fields — scalar, vector, tensor — and a variety of permissible interactions needed investigation. Meson physics became a kind of theoretical laboratory, and from its study in the 1930s emerged many of the concepts and tools that played a decisive role in the elementary particle physics of the future.

The first step towards a fundamental understanding of relativistic quantum field theory consisted in a theoretical justification of the well-known empirical relation between spin and statistics, achieved by Pauli and his student Markus Fierz for free particles, beginning in 1937 (section 2). To account for the puzzling lack of evidence for strong interactions between cosmic-ray mesons and nuclei, aside from copious meson production, Bhabha and Heisenberg independently tried a new approach, namely classical field theory, which they used especially to calculate the multiple production of mesons (section 3). Both of these enterprises were prominent in the joint report which Pauli and Heisenberg prepared for the international Solvay Conference, scheduled for autumn 1939 (section 4).

Although the Solvay Conference was cancelled because of the outbreak of war in Europe, Pauli continued this work, on which he eventually published two papers in American journals. We discuss the Heisenberg–Pauli report in section 5. In section 6 we consider the attempt to determine the spin of the mesotron from its observed electromagnetic interaction, whose theory is summarized in Pauli's American review article on relativistic quantum fields.

10.2 The description of particles of any spin and the spin-statistics theorem (1938–39)

After the completion of non-relativistic quantum mechanics in 1927, Heisenberg and Pauli — initially with the support of Pascual Jordan — started on an ambitious programme to develop a fully relativistic quantum field theory [2]. Although they succeeded in the first step of laying down a general framework (Heisenberg and Pauli 1929, 1930), they discovered serious difficulties in doing so, which led them to a series of investigations, carried out with their respective associates in Leipzig and Zürich. While Heisenberg addressed the difficult problem of interactions, Pauli felt that there was still much to be done with the simpler source-free theory. One result of the Zürich programme was the theory of the complex scalar wavefield (Pauli and Weisskopf 1934), to which we have already referred.

Pauli was especially interested in demonstrating that a consistent quantization of a relativistic quantum field was possible without the necessity for Dirac's holes, as in the positron theory. (Delighted with his success, he took to calling the new theory the "anti-Dirac theory".) The important quantum condition in the scalar theory was applied to the commutator between the field operator and its canonical conjugate (in the positron theory, to the *anti*-commutator). In the non-relativistic case, Jordan and Oskar Klein had shown that these commutator relations led to particles obeying Einstein–Bose statistics (Jordan and Klein 1927). The generalization of this idea, that is, the question of whether integral values of spin always required Bose–Einstein statistics and half-integral values Fermi–Dirac statistics, impressed Pauli as an important one. He found an able collaborator in this effort in Markus Fierz, a Swiss who had received his doctorate at Zürich University under the direction of Gregor Wentzel.

In September 1938, as his *Habilitation* thesis, Fierz submitted a paper concerning free particles of arbitrary spin, in the abstract of which he noted [3]:

> It has been shown that particles with integer spin must always have Bose statistics, particles with half-integer spin Fermi–Dirac statistics. The wavefields with spin smaller than or equal to one are, however, already distinguished in the force-free case, in that for them alone the charge density and the energy density are uniquely determined and gauge invariant quantities, while this is the case for those of higher spin only for the total charge and the total energy.

The equations that Fierz wrote down had been given "already in essence" (*im wesentlichen schon*) by Dirac (1936). However, he noted that "the physical meaning of these equations is not clear in Dirac's work and, moreover, one finds there representations of which some can give rise to misunderstandings, and some are wrong". As for the note of Yukawa and Sakata (1937b), Fierz considered it to be "totally unsuccessful" (*ganz missglückt*) [4].

"Fierz has a long paper in press", Pauli informed Dirac in November 1938, "where he can show that no difficulties arise by the quantization of these equations, so long as no interaction between the particles (or with other particles' electromagnetic fields) is taken into account. In the last time, however, we investigated more closely the question of this interaction and came to quite different results." He then pointed out three difficulties that might occur in cases of higher spins: (i) Dirac's statement that one has simply to substitute $P_\mu - (e/c)A_\mu$ for P_μ is not correct if the spin of the particle is greater than 1 (P_μ being the four-momentum and A_μ the four-potential); (ii) the equations must describe at least two different particles which can make transitions into each other; and (iii) at least one kind of particle has negative energy (in all of the cases that they had investigated). Thus he summarized: "So we came to the conclusion that no *elementary particle* (at least with non vanishing zero [*sic*] rest mass) *with a spin greater than 1 can exist.*" [5]

From left to right: M. Fierz (born 1912), W. Pauli (1900–1958) and H. Jensen (1907–1973). Photograph by P Ehrenfest, Jr. in 1934 and reproduced by permission of AIP Emilio Segrè Visual Archives.

While Fierz worked with well-established tensor and spinor analysis, Frederick Josef Belinfante, a student of Hendrik Kramers in Leyden, suggested a new set of quantities, which he called "undors" to describe either integral spin or half-integral spin fields (Belinfante 1939a). These are essentially outer products of Dirac spinors, a single Dirac spinor being an "undor of first rank". He developed the different theories, especially that of the vector mesons, in terms of such undors, noting that the second-rank undor could be decomposed into a scalar, a pseudoscalar, a vector, an axial vector and a symmetrical tensor (Belinfante 1939b, c, d). The new scheme added little new physics, perhaps, but it gave a consolidated view, bringing together a number of known results.

In particular, Belinfante emphasized the general condition of charge conjugation, such that "to one description of Dirac particles, mesons, neutrettos [i.e., neutral mesons] and the electromagnetic field by undor wave functions there is an equivalent *charge-conjugated description*... This suggests a kind of symmetry between the two ways of describing physical situations. By way of hypothesis one might assume that such a symmetry is a *fundamental property of nature*. We shall call this property the '*charge invariance*' of the physical world." [6]

Belinfante further claimed that this postulate can be used to distinguish physically meaningful quantum observables from others; in addition, he said

[7]:

> We shall show here that the postulate of charge-invariance implies directly that photons and neutrettos *must* be neutral, that Dirac electrons must obey Fermi–Dirac statistics and that mesons *must* obey Einstein–Bose statistics. The interesting fact is that this statistical behaviour of particles and quanta follows much more directly from the postulate of charge invariance then [*sic!*] from postulates concerning the positive character of the total energy of free particles or quanta.

At about the same time as Belinfante, Kemmer also tried to set up a scheme for treating wavefields in a more unified way but avoiding the use of "spinor calculus". Since late 1938 he had exchanged letters on this subject with Pauli, who encouraged him and supplied some advice [8]. Kemmer later discovered that an American mathematician, R.J. Duffin, had proposed a similar formalism (Duffin 1939). Kemmer's stated physical motivation is to emphasize the particle character of the (charged) meson by employing a formalism similar to Dirac's electron equation, rather than the Proca equation, which is a generalized Maxwell equation, thus "theoretical work has laid stress on the wave aspect of the meson practically throughout" [9]. The Kemmer–Duffin formalism applies to mesons of spin 0 and 1, and provides a unified treatment of them.

10.3 The classical approach to meson interaction: Bhabha and Heisenberg (1939)

In May and August of 1939, J.G. Wilson reported on a series of accurate measurements of the absorption and scattering of mesons in metal plates placed in a cloud chamber, making use of using Blackett's large magnet (Wilson 1939, 1940). For "low energies" ($< 7 \times 10^8$ eV), he found that "mesotrons suffer no other energy loss in dense materials comparable in magnitude to that due to ionization" [10]. Of the particles with higher energy, most were consistent with energy loss by ionization only, but there were nine "more absorbable particles", of which eight were positively charged. The single negative "lies close to the main mesotron distribution, and may well belong to it", one positive stops, and may represent absorption, while it "seems most probable that the remaining seven positive particles are protons", which agreed with the estimate of the number of protons by Blackett [11].

However, Wilson observed that [12]:

> in at least three of these cases the outgoing particle [below the plate] cannot still be a proton, for it shows normal ionization at an energy for which a proton would ionize heavily...We therefore consider the possible existence of a process by which a proton interacting with a nuclear field can produce a mesotron. If this is the correct interpretation, the cross-section for this process must be very large, as compared with very small cross-sections for the processes already considered.

The quotation clearly shows the dilemma: large cross-sections for fast protons, possibly producing mesons in some cases, while mesons themselves showed little evidence of absorption.

The author strengthened his conclusion in the second paper, based on accurate measurements of the mean angle of meson scattering in lead, copper and gold (Wilson 1940). These results followed [13]:

> The evidence for the existence of large-angle scattering due to short-range forces between mesotrons and nuclear heavy particles (protons and neutrons) is discussed. The cross-section for this type of scattering is estimated to be of the order 10^{-28} cm^2, and this value is in agreement with that given by Bhabha for a "classical" meson theory. There is no experimental evidence for the large increase of scattering due to short-range forces at low mesotron energies given in the quantum mechanical treatment due to Heitler. For the available mesotrons of lowest energy, the cross-section is found to be less than 10^{-27} cm^2. This result is not compatible with the present development of mesotron theory, and may be interpreted as indicating a failure in the treatment of the charge-exchange which leads to the interaction between charged mesotrons and heavy particles.

Although the experiment of Wilson thus contradicted the perturbation approach of Heitler (1938a), it seemed to agree with the result of a new classical approach, first advanced by Homi Bhabha (1939a). In a letter to *Nature*, the Indian physicist considered the phenomenon of "explosive showers" which Heisenberg had connected with the existence of a fundamental length and argued that it set a limit on the applicability of quantum field theory [14]. Bhabha pointed out that the existence of multiple processes in QED, the "infra-red" catastrophe, in no way places a limit on the applicability of QED to less probable processes by the methods of perturbation theory. He went on [15]:

> For example, in the collision of two protons with energy very large compared to mc^2, the probability becomes large for the simultaneous emission of a large number of mesons, which is the analogue of the "infra-red catastrophe" for quanta of finite rest mass, and hence quantum mechanics is none the less competent to deal with it. It can be similarly shown that *we can calculate the production of large explosions to a high degree of accuracy by treating the meson field quantities classically*, that is, as non-quantized magnitudes, for since mesons satisfy Einstein–Bose statistics, the meson field becomes a classical one just in the case where we are dealing with a large number of mesons.

Thus Bhabha revived Heisenberg's idea of classical meson theory (Heisenberg 1936), applying it (Bhabha 1939b) to explain the low meson scattering cross-section, as referred to by Wilson. It was also put to use in cosmic-ray problems by Russian authors (Iwanenko and Sokolov 1940), who found that the interaction cross-sections of mesons did not increase with energy, due to a damping effect [16].

The detailed paper (Bhabha 1939b) was communicated to the Royal Society by Dirac in April 1939. Its main motivation was to avoid the divergence difficulty which arose in the perturbation treatment, which had "led Heitler (1938a) and others to doubt the correctness of the fundamental equation even for mesons of energy comparable with their rest mass" [17]. Rather than using quantized fields, Bhabha proposed to treat all observables in the Hamiltonian for mesons and nucleons in interaction as commuting, and thus obtain classical equations, "since it is possible either to solve them exactly, or at least to give approximate solutions the errors of which can be strictly estimated" [18]. To simplify the problem further, he considered only neutral mesons, treated neutrons and protons as equivalent, and put the second coupling constant of the vector meson theory $g_2 = 0$. He noted that g_2/g_1 could be considered a second characteristic length, the first being $\lambda = \hbar/\mu$, with μ the meson mass. He asserted that the paper would demonstrate that [19]

> the finite rest mass of the meson introduces no essential difference in the behaviour of mesons at high energies [compared with QED] and hence that the fundamental length required by Heisenberg cannot be identified with λ. The type of explosions investigated by Heisenberg, if they exist, cannot therefore be connected with the rest mass of the meson but would be due either to the interaction constant g_2 or to the fact that the meson field carries electric charge.

In the classical theory, neither μ nor $\hbar$ appear explicitly, but only their ratio λ, so the equation for the meson field U_ν becomes:

$$\frac{\partial}{\partial x_\mu} \frac{\partial}{\partial x^\mu} U_\nu - \lambda^2 U_\nu = 4\pi R_\nu + 4\pi \frac{\partial}{\partial x_\mu} S_{\mu\nu}, \tag{10.1}$$

where R_ν is a four-vector describing the effect of the nuclear charge on the meson field and $S_{\mu\nu}$ is an antisymmetrical tensor describing the effect of the neutron dipole moment. The Green function solution of this equation is obtained and has only a δ-function contact singularity. This "shows at once that the worst singularities in the meson field are identical with those of the electromagnetic field, the additional singularities [which are discontinuities on the light cone] being of lesser order" [20].

Bhabha then calculated the meson–neutron scattering cross-sections for transverse and longitudinally polarized mesons of low energy, these being the generalization of the Thomson scattering formula for low-energy photons. (For $\lambda \to 0$, the limits are correctly reached: zero for the longitudinal and Thomson for the transverse.) The result for high energy becomes independent of the neutron mass and nuclear charge, and Bhabha noted "that a behaviour of this sort cannot be approximated to by a series in ascending powers of g^2" [21]. He argued that the formulae, with validity claimed to 10^9 eV, showed "that the scattering of mesons is a small effect, the cross-section being of the order $8\pi g^4/(3M^2) \approx 10^{-28}$ cm^2, though scattering may be through large angles when it takes place" [22]. Except for the oversimplifications introduced, one might conclude that the vector meson theory, treated non-perturbatively, *would account for the small observed meson interaction cross-sections.*

In spite of the apparent success of Bhabha's latest work, Heisenberg did not approve of it, although he had himself advocated the use of classical theory, as early as 1936 for the Fermi-field theory and in 1939 for the Yukawa theory. Heisenberg sent a new manuscript to Pauli, asking for his comments, describing it as in part "a correction to Bhabha's note, which I hold to be false. (You also, as I heard from Fierz.)" [23].

In this manuscript, submitted a few weeks later to the *Zeitschrift für Physik*, Heisenberg noted that although various authors, including Bhabha, Heitler and the Yukawa group (in *Interaction IV*) had concluded that the Yukawa theory must yield processes in which many mesons are simultaneously produced, Bhabha's new work (1939a) had the opposite conclusion (Heisenberg 1939b). Heisenberg's paper first considered the general properties of multiple processes, then the specific case of Yukawa theory and finally comparisons with experiment. One of his goals was to see what would be the new features of a future divergence-free quantum field theory that yielded multiple processes.

The property that he suggested for a future field theory was that it could not, in general, be expanded in products of field operators. He cited an example given earlier by Born (1933) with the Lagrangian:

$$L = l^{-4}\{1 + l^4[(\partial\phi/\partial t)^2 - (\text{grad } \phi)^2]\}^{1/2}, \qquad (10.2)$$

where $\phi(x)$ is a scalar field of dimension reciprocal centimetres and l has the dimension centimetres (other quantities, such as energy, are in suitable units). For small energy of interaction ϵ, L can be expanded in powers of $\phi(x)$ and higher powers (which would give infinite self-energy) can be neglected, but for $\epsilon \gg 1/l$ this is impossible. In the latter case, Heisenberg advocated the use of a "semi-classical" method, similar to that employed in the treatment of the infra-red problem in QED (Bloch and Nordsieck 1937). He quoted the results of these authors for the *Bremsstrahlung* spectrum (in which the mean number of photons emitted is infinite due to their masslessness) and simply replaced the frequency k_0, which is equal to the wavenumber k in the massless case, by $(k^2 + \lambda^2)^{1/2}$ in the meson case, which gives a finite mean number of mesons.

Applying this idea to the Yukawa case, Heisenberg made use of the same simplifying assumption that Bhabha had adopted, namely neutral mesons only. These he described by the four-vector field $(u_0, \boldsymbol{u})$, interacting with protons (equivalent to Bhabha's assumption of only neutrons) represented by the Dirac spinor ψ. The spin-1 meson Lagrangian then becomes

$$L = \frac{1}{2}(f^2 - g^2) + \frac{1}{2}\lambda^2(u_0^2 - \boldsymbol{u}^2) + i\psi^*\partial\psi/\partial t - i\psi^*\alpha_k\partial\psi/\partial x_k + \psi^*\beta\psi K.$$
$$(10.3)$$

(f, g) denotes the six-vector associated with the vector-meson field, i.e. $g = \text{curl } \boldsymbol{u} - l\psi^*\beta\boldsymbol{\sigma}\psi$; $f = -\partial\boldsymbol{u}/\partial t - \text{grad } u_0 - il\psi^*\beta\boldsymbol{\alpha}\psi$. λ and K are, respectively, the meson and the proton masses, expressed as reciprocal lengths $(mc/\hbar)$. The constant length l has the value

$$l = g_2[4\pi/(\hbar c)]^{1/2}\lambda^{-1},$$

thus empirically its order of magnitude is $l \approx 1/\lambda$. Heisenberg ignored the g_1 interaction for his purpose, which was to study the high-energy interactions.

He then applied this model Lagrangian to calculate the production of mesons in the collision of a proton with any other elementary particle, causing it to release a *"Bremsstrahlung"* of mesons, the scattering of a meson by a meson (e.g., a virtual meson in a nucleus) and the scattering of a meson by a proton (the analogue of Compton scattering). In all this he includes the nonlinear interactions of outgoing mesons with each other, modelled by a sixth-order term, as an example.

For the empirically important case of the meson–proton scattering, he found a much smaller result than that yielded by perturbation theory, concluding [24]

> that the usual quantum theoretical perturbation calculations lead to entirely wrong results for the scattering of mesotrons by protons; that correspondence-like considerations for the classical wave theory allow one to approach the region in which the present quantum mechanics fails; further, that the cross-section for mesotron–proton scattering is much smaller than has so far been claimed. This also seems to follow necessarily from the experiments, because the observed large range of mesotrons is not reconcilable with the large cross-section given earlier.

Consequently, the classical meson theories of Bhabha and Heisenberg, while differing in some principles and results, both seemed to remove the serious discrepancy of the meson theory between the small interaction cross-sections of the mesons observed in the cosmic rays and their supposed role as the mediators of the strong nuclear binding force.

10.4 International physics conferences in the fateful year 1939

Three international conferences in which meson theory was to be given a prominent role were planned for the months of June, September and October. In addition, Yukawa had been invited to address a meeting of German physicists and mathematicians at Marienbad in the newly Germanized Sudetenland in September [25]. Of these, only the cosmic-ray conference of 27–30 June, held at the University of Chicago, actually took place, as we shall now describe. The others were all cancelled because of the outbreak of war in Europe. However, we will also give their proposed programmes, because they show the state of the art and the problems envisioned at the time of their planning.

The Chicago symposium on cosmic rays was the first American conference in which the meson became prominent [26]. Its principal organizer, Arthur H. Compton of Chicago, wrote in the foreword to the proceedings (Compton 1939):

An editorial problem has arisen with regard to the designation of the particle of mass intermediate between the electron and the proton. In the original papers and discussion no less than six names were used. A vote indicated about equal choice between *meson* and *mesotron* with no considerable support to *mesoton*, *barytron*, *yukon* or *heavy electron*. Except where the authors have indicated a distinct preference to the contrary, we have chosen to use the term *mesotron*.

Experimental topics dominated the symposium and there was not much discussion of the fundamental field theories of mesons, but several of the experimental talks referred to calculations of cross-sections (J. Clay, M. Schein and P.S. Gill) or to the "barometer effect" derived by Blackett from the Euler–Heisenberg analysis (Compton and Gill). Three talks, however, had the meson at their centre. First, Neddermeyer and Anderson (1939) gave a detailed historical review of the particle content of the cosmic rays, especially of the discovery of the mesotron, and they gave the latest status of mass and decay-time problems. Secondly, Heisenberg (1939c) presented his

Der Gegenwärtige Stand der Theorie
des Mesotrons
von Hideki Yukawa
(Ein Vortrag gehalten am 28. Sept, 1939.
auf der Tagung der Deutschen Physikalischen
Gesellschaft in Marienbad)

§ 1. Einleitung

Seit einigen Jahrzehnten war es die Hauptaufgabe der Physik, die elementare Bausteine der Materie aufzusuchen und ihre Wechselwirkungen untereinander auszustudieren. Eine Aufgabe, die niemals völlig gelöst werden möchte! Doch scheint es uns als ob wir kürzlich der Lösung sehr näher gekommen sein, vornehmlich durch die Entdeckung des Mesotrons oder des Mesons von von Amerikanischen Physikern in der kosmischen Ultrastrahlung. Diesbezüglich habe ich darauf aufmerksam gemacht worden, daß Herr Prof. Kunze schon in 1933, die Nebelkammeraufnahme gewonnen hatte, die eine Spur des Teilchens von mittler Masse enthalten hatte.

Yukawa's draft of the Marienbad table. Reproduced by permission of Yukawa Hall Archival Library.

new theory of explosive showers based on classical meson theory. Finally, Rossi (1939) criticized the existing evidence on the mean life of meson disintegration, as we have discussed in chapter 8.

The published proceedings contain also comments made at the symposium. After Montgomery and Montgomery (1939) had discussed the behaviour of high-energy electrons in the cosmic rays, Robert Oppenheimer rose to make a lengthy comment on bursts. Addressing Heisenberg's argument that the frequency of large bursts required a new explosive mechanism, he stated that there were "a number of points having to do with cascade theory and with the production of energetic secondary electrons by the penetrating component whose consideration tends to weaken this argument". He attributed bursts observed under a large thickness of matter to high-energy knock-on electrons produced by mesotrons and he concluded: "It is sure that in the production of bursts, cascade multiplication plays in the end a decisive part." (Oppenheimer 1939).

We do not know whether Oppenheimer presented all the remarks contained in the printed discussion; he may have expanded them later, based on his paper with Serber and Hartland Snyder, submitted in early November 1939 (Oppenheimer *et al.* 1940). In any case, Heisenberg's biographer, David Cassidy, has constructed from this and other remarks a "Chicago confrontation" and an "onslaught" on Heisenberg's ideas (Cassidy 1981). The published documents, at least, do not seem to support a harsh encounter. (The sharpest statement opposing Heisenberg's idea occurred, in fact, in a "note added in proof" by the Montgomerys) [27]. We feel that the situation in Chicago was far less polarized and dramatic for Heisenberg than Cassidy has asserted. Heisenberg saw no reason to alter his opinion about bursts drastically in the years that followed.

An "International Meeting on Physics" was planned for Zürich on 4–10 September. The nuclear physics section [28] was to contain eight main reports [28]. Those of Bohr, Otto Hahn and Frédéric Joliot would deal with recently discovered uranium fission, whereas those concerning problems of elementary particles were: Chadwick ("Some aspects of β-transformations"), Euler ("Das neue Elementarteilchen in der kosmischen Strahlung"), Heisenberg ("Der Folgerungen der Yukawaschen Theorie für das Verhalten sehr energiereicher Teilchen") and Rasetti ("Einige Experimente über die harte Komponente der Höhenstrahlung"). Shorter contributions were advertised from Eva Barnóthy[-Forró] ("Absorption der Ultrastrahlung"), Gilberto Bernardini ("Les recherches sur la nature des mesotrons, etc.") and Yukawa ("Some problems of the mesotron").

Finally we turn to the Eighth Conseil de Physique of the Institut International de Physique Solvay, which was to be held in Brussels on 22–29 October 1939. Letters inviting participation were sent in March 1939, saying that the conference would be on "questions concerning the elementary particles and their interactions" [29]. The following authors agreed to make reports as follows.

(1) W. Heisenberg: General problems. Particle equations. Cases of positive and negative electrons. Materialization. Limits of application of the quantum theory.
(2) P. Blackett: The heavy electron (mesotron) from the experimental point of view.
(3) W. Heitler: The mesotron from the theoretical point of view.
(4) E. Fermi: The neutrino, experiment and theory.
(5) H. Bethe: The interactions of electron, proton and neutron. Indications derived from the study of simple nuclei and from deviations. Heisenberg and Majorana forces.
(6) Showers from the experimental point of view.
(7) L. de Broglie: The photon.
(8) C. von Weizsäcker. Astronomical limits concerning particle properties.
(9) F. Bloch: The magnetic moments of protons and neutrons.

Besides the *rapporteurs*, a group of non-lecturing participants was invited, including Carl Anderson, Alexandru Proca, Walther Bothe, Arthur Compton, George Gamow, Irène and Frédéric Joliot-Curie, Lise Meitner, Merle Tuve and Hideki Yukawa; in addition, the meson theorists Bhabha, Møller and Rosenfeld served as scientific secretaries. Copies of the main reports were to have been distributed several weeks in advance.

10.5 The 1939 Solvay report of Heisenberg and Pauli, and Pauli's publications based on it

On 20 April 1939 Heisenberg wrote to Pauli that he had been asked to provide a report for the Solvay Conference on the subjects: "General problems, limitations of the present theory, concept of elementary particles" [30]. A few days later he wrote that he had "noticed that a large part [of the subject] deals with problems which you know in more detail than I" [31]. He therefore asked Pauli to take over the "General properties of elementary particles" as part 1, while he would prepare a part 2 on particular forms of interaction (including QED and Yukawa theory) and a part 3 on the limitations of the present theory.

Soon Pauli agreed to do so, in spite, as he wrote, "of his laziness to compose such a report", but he asked that he restrict himself to non-interacting quantum fields and give his part the title, "Relativistic wave equations of free particles and their quantization" [32]. Pauli's expertise in these matters had been developed over several years with his students and collaborators in Zürich. We have already noted parts of it: the theory of the charged scalar field, with Victor Weisskopf (Pauli and Weisskopf 1934), and the later work done especially with Markus Fierz [33].

The subjects discussed in the foregoing papers by Fierz and Pauli formed the substance of Pauli's contribution to his joint report with Heisenberg to the Solvay Conference. In a letter to Heisenberg, Pauli outlined it as follows [34]:

(1) Spin $\frac{1}{2}$. Dirac's hole theory.
(2) Spin-0 and 1. Scalar and vector theory. (Energy-momentum tensor; current vector; quantization; Fourier decomposition.)
(3) Most general theory of particles of spin 0 and 1. (Dual cases; pseudoscalar and pseudovector.) Particular formulations (Belinfante, de Broglie, Kemmer).
(4) Reduction of theories by the requirements of reality. (Neutral particles.) (For spin-$\frac{1}{2}$ carried out by Majorana.)
(5) Higher spin; quantization and statistics.

A month later, he organized this material into two chapters, the first dealing with general considerations, the second with individual field theories [35].

Much of Pauli's report (Pauli 1939) was used in an article he published about two years later in the *Reviews of Modern Physics* (Pauli 1941), which followed quite closely the original plan. This article also included newer results obtained after mid-1939 and a new section of "Applications" that dealt with the interaction of particles of spin-0, $\frac{1}{2}$ and 1 with the electromagnetic field, replacing his original "Remarks on gravitational waves and gravitational quanta". We shall discuss this article below.

Heisenberg's part of the report also consisted of two chapters, one entitled "The interaction of elementary particles", the other "Limitations of the present theory" (Heisenberg 1939d). We have already touched on its basic ingredients in discussing Heisenberg's analysis of high-energy cosmic-ray phenomena and explosive showers; but his Solvay report stressed general principles. In the introduction to the first chapter, he separated QFTs into two classes: in class 1 the interaction term has only a dimensionless coupling constant $Z \ll 1$; in class 2, the constant Z has the dimension of some power of a length. Even for a class 1 theory like QED, large field strengths introduce terms (such as pair production) which can effectively transform it into class 2.

Meson theory is from the outset a class 2 theory, containing significant nonlinear effects, such as those Heisenberg had discussed in his recent shower paper (Heisenberg 1939b). In the second chapter of his Solvay report, he summarized the problems of the high-energy behaviour of the existing QFTs, which in his opinion were all connected with a new fundamental length, whose value was between the classical electron radius $e^2/(mc^2) \approx 2.8 \times 10^{-13}$ cm and a tenth of this value. He concluded that Yukawa's theory gave a correct picture of nature only in the static limit. Dynamical phenomena, such as high-energy collisions of elementary particles, would be treated in a future theory for which only a classical correspondence limit existed at that time.

Heisenberg soon dropped out of elementary particle theory, due to his involvement in the German wartime Uranium Project, but Pauli was able to continue to work on aspects of his Solvay report. Its main contents appeared, with improvements, in the form of two articles (Pauli 1940, 1941). Both papers had great relevance for fundamental theories of nuclear forces, as theorists began more and more to explore new types of fields in an effort to obtain improved agreement with the accumulating experimental data.

Pauli (1940) was an elegant paper, in which the conjectured connection between spin and statistics was proved for "the relativistically invariant wave equation for free particles" [36]. For arbitrary integral spin, he demonstrated that the sign of the charge is indefinite (as was known for spin 0 and 1) and for arbitrary half-integral spin, that the sign of the energy is indefinite (as was known for spin $\frac{1}{2}$). From these results and the condition now known as *local commutativity* (meaning that "measurements at two space points with a space-like distance can never disturb each other") Pauli concluded: "*For integral spin the quantization according to the exclusion principle is not possible.*" Although "it is formally possible to quantize the theory for half-integral spins according to Bose–Einstein statistics . . . *the energy of the system would not be positive*" [37].

The second paper (Pauli 1941) amounted to a compact modern treatise on theories of elementary particles, with the following contents:

Part I. Transformation properties of the field equation and conservation laws
1. Units and notation
2. The variation principle and the energy-momentum tensor; gauge transformation and current vector
Part II. Special fields
1. The wavefields of particles without spin
2. Wave fields for particles with spin 1
3. Dirac's positron theory (spin $\frac{1}{2}$)
4. A special synthesis of the theories for spin 1 and spin 0
5. Applications

We note that in part I the theory is developed first for *c-number* fields without interaction. Electromagnetic interaction is then introduced for complex fields by the use of the gauge principle, i.e., the replacement of the partial derivative $\partial/\partial x_k$ by the covariant derivative $D_k = \partial/\partial x_k - ie\phi_k$ (and its complex conjugate), ϕ_k being the electromagnetic potential. Pauli also considered the possibility of introducing additional terms depending directly on the field strengths "for the description of particles which have a magnetic moment" [38].

In part II, the special non-interacting fields for spin 0, $\frac{1}{2}$ and 1 were considered in fully relativistic form, together with their quantization. Then Pauli introduced the interaction *with an external electromagnetic field* via the gauge principle, the spin-1 case, especially, receiving extended treatment. As in the original Solvay report, the new paper contained a section on "a special synthesis of the theories for spin 1 and spin 0" which referred not to the Møller–Rosenfeld theory, but rather to the Kemmer–Duffin formalism.

Replacing a section on gravitational waves and spin 2 in the original Solvay report, Pauli's paper (1941) provided a section on "applications"

consisting of a brief discussion of the cross-sections of various electromagnetic processes for *mesotrons* of different spins and magnetic moments,
accompanied by six tables giving the relevant formulae (some of which we
have discussed above.) Like other physicists moving from Europe to England
or the USA, Pauli, who was spending most of his time in Princeton,
apparently felt that his theoretical work should be supplemented by
applications to the open experimental problems of the day [39]. The usual
cautions regarding the questionable validity of perturbation methods as
applied to mesons were put forward at the end of the article.

10.6 The spin of the meson from its electromagnetic effects (1939–41)

The thorough study of the general properties of elementary particles,
especially the study of the possible variety of quantized relativistic free fields
(as described above), led to insights into some problems that had previously
baffled meson physicists. Now they looked into the electromagnetic
couplings of fields of different spin, hoping that they could use experimental
phenomena, such as burst frequency, to determine the spin of the cosmic-ray
meson.

We have already mentioned (in chapter 9) the papers of Otto Laporte
(1938a,b) on the elastic scattering of a spin-1 meson in the Coulomb field,
which gave results similar to the well-known Rutherford scattering, except
that a polarization effect that appears in the second approximation for spin-$\frac{1}{2}$
is present already in the first Born approximation in the spin-1 case. (Clearly,
there is no such effect for spin 0.) Any such observable difference in
electromagnetic behaviour could be used, in principle, to determine the spin
of the mesotron, a point that was stressed after 1939. Peierls, in a review of
1939 on the meson, devoted only a brief paragraph to electromagnetic effects
(in connection with the subject of energy loss) and he quoted only the then
unpublished work of Massey and Corben (1939) and Bhabha and
Carmichael (private communication) "who have shown that the probability
of the production of fast knock-on electrons is very much greater [than for
spin $\frac{1}{2}$] if the meson has a spin of one unit" [40].

10.6.1 *Meson scattering by electrons*

The effect mainly considered was the production of "knock-on" electrons
through largely forward momentum transfer by a fast meson; the electron so
produced could initiate a large cascade shower or "burst" (see chapter 4).
Radiation processes, such as *Bremsstrahlung*, strongly dependent on the spin
[41], are proportional to the inverse square of the mass of the radiating
charged particle, so they would tend to be much smaller for a meson than for
an electron. Most of the authors treating radiative effects relied on an
experimental review presented at the Chicago conference (Schein and Gill
1939) to compare the theory with observation.

In order of submission, the first such comparison is by Bhabha *et al.* (1939), which referred to the experiments on bursts of the last two authors, Carmichael and Chou (1939), carried out in a London tube station. The three authors concluded that "the frequency–size curve of bursts provides evidence that the meson has a spin of one unit in agreement with what is believed from nuclear considerations" [42].

The next contribution to the problem came from Oppenheimer *et al.* (1940), who were very concerned about the validity of perturbation calculations that gave large electromagnetic, as well as nuclear, cross-sections. The authors remarked at the outset that [43]:

> Because of the large nuclear coupling and high mesotron mass it has usually been assumed that other nuclear processes would alone be important for high energy transfers [to the soft component]. In fact, calculations of nuclear effects, such as mesotron absorption and scattering by nuclei...lead to cross sections so large, for high energies, that they completely contradict the high penetrating power of the mesotrons. It has been emphasized especially [Heisenberg 1939b] that the prediction of these large cross-sections rests on the essentially erroneous treatment of the interactions as small; despite several attempts no reliable estimate of them has been given, and this problem probably goes beyond the framework of present theory...Under these circumstances we have thought it profitable to re-examine the electromagnetic effects a little more closely.

The authors then gave the cross-sections for knock-on electrons and for *Bremsstrahlung* that would apply "if mesotrons satisfied the Dirac equation" and concluded that the first cross-section has the wrong energy-dependence and the second the wrong Z-dependence. However, they stated that both of those cross-sections were of doubtful applicability anyhow [44]:

> For Yukawa's theory requires an integral mesotron spin, and the spin dependence of nuclear forces shows that this spin must be one. The theory which has been developed to describe these particles makes the mesotron electromagnetic current density depend on derivatives of the mesotron field, and associates, with transitions in which the direction of the mesotron spin changes, current distributions more singular than those of a Dirac electron...These singular currents of course interact very strongly with high frequency radiation fields. They radically alter the high-energy cross-sections, giving much larger values [than the Dirac case]. At the same time they introduce couplings so large that the question of the perturbation theoretic estimate of the cross-section requires reexamination.

These remarks reflect the paradoxical situation that high cross-sections consistent with the spin-1 theory would not necessarily confirm that the meson's spin was 1, because the theory itself was internally inconsistent. Neither would lower cross-sections refute spin 1, for the same reason. (Nevertheless, a tacit agreement developed among cosmic-ray theorists that low burst production would imply either spin 0 or $\frac{1}{2}$.) Independent experimental test of the theory was impossible because the only source of high-energy spin-1 particles, if indeed they existed at all, would have been the cosmic rays.

E.J. Williams used his well-known method of impact parameters (Williams 1933, Weizsäcker 1934) to calculate R, the average number of electrons accompanying a cosmic-ray meson due to the knock-on process, and R', the number above a certain energy, and found reasonable agreement with his experimental paper (Williams 1939, published earlier). However, he cautioned [45]:

> The absolute values of R and R' given above depend, of course, on the assumption that the knock-on electrons produced by a meson arise from a pure Coulomb interaction...This assumption may be seriously vitiated in the first place by the operation of a spin interaction, and secondly by a departure of the electric field between a meson and an electron from the Coulomb form. As regards the latter, the closest distance of approach in collisions in which the knock-on electron acquires energy ϵmc^2 is of the order of $137\epsilon^{-1/2}$, so that when ϵmc^2 exceeds 10^{10} eV departures from a Coulomb field might be expected.

This question of validity was taken up again by Tomonaga (1940). Accepting the importance of attempts to determine the spin of the meson experimentally from its electromagnetic behaviour, he made the additional point [46]:

> However, it appears to us also interesting from the theoretical standpoint; for, if one makes of the spin a literally "classical representation, somewhat as though it resembles a magnetic needle", it is not easy to see that the transition from spin $\frac{1}{2}$ to spin 1 — the magnetic moment of the mesotron is exactly the same as the electron's [for the same mass] — should bring about such an important difference.

Although one might expect only about a factor of two or so difference, one obtains more than a factor of eight. Since the spin is a purely quantum mechanical effect, it is not possible to represent this effect as a classically visualizable (*anschaulich*) one. Tomonaga asked, therefore, how one might adapt Williams' collision parameter method to cases in which there is a change in spin direction of the mesotron. He did this by replacing, in first approximation, the point electric charge by an electric dipole (noting that a similar idea could be found in Oppenheimer *et al.* (1940). He then applied the method to the ionization of atoms and to the production of electron pairs by mesotrons.

10.6.2 *Electromagnetic properties of nucleons*

An additional motivation for studying the electromagnetic properties of mesons was given by Corben and Schwinger (1940). They pointed out that nuclear electromagnetic properties, e.g., the neutron's magnetic moment, depend both on the nuclear and on the electromagnetic couplings of the meson [47]:

> Data obtained experimentally from electromagnetic nuclear properties do not therefore determine independently the mesotron — heavy particle coupling and the mesotron — radiation interaction, and it would be of advantage to study each of these forms of interaction directly. Information concerning the influence of the electromagnetic field on mesotrons may be obtained by the

investigation of recoil electrons resulting from collisions with the mesotrons forming the penetrating component of cosmic radiation.

Corben and Schwinger considered possible spins σ $(0, \frac{1}{2}, 1)$, with magnetic moment μ (arbitrary except for zero spin), and found that "only for $\sigma = \frac{1}{2}$, $\mu \neq 1$ (especially $\mu = 0$) and $\sigma = 1$, $\mu = 1$ [the Proca case] is the cross section of the correct magnitude and form... to account for the observed burst phenomena at energies greater than 2×10^{10} eV". Although, "on the basis of cosmic-ray evidence", one could not exclude the possibility of a spin-$\frac{1}{2}$ meson with magnetic moment different from unity, they argued that "such evidence as is available from nuclear phenomena indicate that this is not likely" [48].

The Russian theorist Igor Tamm sounded a critical note [49]:

> The theory of mesons of spin 1 leads in its present form to divergent or otherwise unreasonable results not only in the majority of nuclear problems, but also in the treatment of the interaction of mesons with the electromagnetic field. The simplest instance is the scattering of mesons by a point charge.

Exploring whether the problem lay with the use of the Born approximation, Tamm studied the exact solutions, finding that "the regular solutions of the Proca equation in a field of a point charge do not form a complete set of functions and the problem of the Coulomb scattering of mesons has no solution". Thus: "the Coulomb scattering of mesons with energies $E \geq \hbar c/r_0$ must substantially depend on the size r_0 of elementary particles" and one must take this size into account.

A very complete calculation of radiative processes (light scattering, *Bremsstrahlung* and pair production) of fast mesons (Booth and Wilson 1940) was carried out in Cambridge, England, using Kemmer's matrix formalism for spin 1, which allowed the use of some simple methods developed for Dirac matrices. Directly calculating the *Bremsstrahlung* differential cross-section, the authors found that the result was "extremely complicated", so much so that they did not even quote it in their paper, giving instead the cross-section integrated over angles, obtained by E.J. Williams' method of impact parameters. Regarding their results they noted [50]:

> In most discussions it has hitherto been assumed that, on account of the large mass of the meson, the radiation loss from mesons would be about 40000 times smaller than from electrons. This naive argument... turns out to be unfounded. Our calculations show that... the energy loss from mesons due to radiation varies as E_0^2, where E_0 is the energy of the meson; the energy loss for electrons... only varies as E_0.

Although "the highly singular nature of the interaction... renders an exact theory impossible", they argued that "the history of the quantum theory has shown that physical theories are often valid far beyond the limits within which they can be proved to be consistent" [51]. Finally, however, after considerable back-and-forth arguing, Booth and Wilson reached the conclusion: "The 'heavy electrons' found in cosmic rays are usually identified with mesons, but there is as yet no really convincing evidence

Robert F. Christy (born 1916), left, and S. Kusaka (1915–1947). Photographs reproduced by permission of AIP Emilio Segrè Visual Archives (Kusaka from Physics Today Collection).

that these cosmic-ray particles have spin 1 rather than spin 0 or $\frac{1}{2}$". They have here tacitly assumed that the nuclear forces unambiguously select spin 1.

We finish this chapter by discussing the calculations of radiative interaction of mesons by Oppenheimer's students Robert F. Christy and Shuichi Kusaka (1941a), and Oppenheimer's inferences from their work. The former authors noted that the worst singular behaviour of the cross-sections comes from high *Bremsstrahlung* frequencies and that this argument "is incorrect because [it] fails to distinguish between a Coulomb field and the actual nuclear field near the nuclear radius". Assuming a spin of 1, they used a nuclear form factor, described as follows [52]:

A closer [than the Coulomb] approximation to the nuclear potential is $(Ze/r)(1 - e^{-r/d})$, where d is the nuclear radius and is taken to be $5\hbar Z^{1/3}/(6\mu c) = 1.82 \times 10^{-13} Z^{1/3}$ cm for $\mu = 177$ electron masses. This essentially sets an upper limit of about $\mu c/Z^{1/3}$ on the momentum transfer to the nucleus and gives

$$\sigma = B'\alpha Z^{5/3} e^4 E/(\mu c^2)^3,$$

where E [is the γ-ray or initial mesotron energy] and $B' = \pi/18$ and $\pi/6$, respectively, for *Bremsstrahlung* and pair creation.

Christy and Kusaka also calculated the knock-on electron cross-section in a second paper (1941b) and used this to make a comparison of predicted burst frequency with the summary given at the 1939 Chicago symposium on cosmic rays (Schein and Gill 1939). They found good agreement for spin-0 mesons and about twice as many events for spin $\frac{1}{2}$ and normal magnetic moment (with about 50% uncertainty for both cases). However, for spin 1, they obtained too many bursts by a factor of 20.

In a letter to the editor, Oppenheimer called attention to the papers and conclusion of Christie and Kusaka (in the same issue of the *Physical Review*) and remarked that, because about half of the energy is shower-producing [53],

> this would seem fully to have confirmed Yukawa's prediction that mesotrons can disintegrate into electrons and neutrinos; and this would in our opinion make a half-integral value of the mesotron spin improbable. These arguments would then establish that the mesotron was described by a scalar or pseudoscalar field. It may be remarked in this connection that the neutron proton forces derived from a charged pseudoscalar field [Rarita and Schwinger 1941a] by "classical" or perturbation-theoretic approximation agree in sign and spin dependence, though not of course in their singular dependence upon distance, with the sign and magnitude of the singlet triplet difference and the quadripole [*sic*!] moment of the deuteron system, whereas the corresponding theory for charged mesotrons of unit spin gives a quadripole moment of wrong sign. The results of CK can thus not be regarded as adding a further difficulty to this in itself highly unsatisfactory theory of nuclear forces.

Evidently the spin of the meson had not become clear from these considerations of electromagnetic interaction. However, the calculations were beginning to cast serious doubt upon the vector-meson theory that had been in favour for several years.

Notes to text

[1] Goldhaber (1979, p 88).
[2] For a historical survey of this programme to 1958, see Rechenberg (1993a).
[3] Fierz (1939, p 3).
[4] *ibid.*, p 4.
[5] Pauli to Dirac, 11 November 1938. [WPSC2], pp 607–8. The detailed paper is Fierz and Pauli (1939a).
[6] Belinfante (1939c, pp 881–2). Original emphasis. The charge conjugation operation was introduced by Pauli in 1936.
[7] Belinfante (1939c, p 882). See also Fierz and Pauli (1939a, b) and Pauli and Belinfante (1940).
[8] [WPSC], e.g., Pauli to Kemmer, 14 October 1938 and an exchange of letters in 1939, from 6 April to 2 July.
[9] Kemmer (1939, p 92).

[10] Wilson (1939, p 525). Meson decay in flight is important only in gaseous absorbers.
[11] *ibid.*, p 527.
[12] *ibid.*, pp 527–8.
[13] Wilson (1940, p 84).
[14] Bhabha (1939a). Both "meson" and "mesotron" appear in the letter's title, but Bhabha argues that the "tr" is redundant and "It would therefore be more logical and also shorter to call the new particle a meson instead of a mesotron".
[15] *Ibid.*, p 277.
[16] See also for the classical theory of the meson: Iwanenko (1939, 1940) and Stueckelberg (1939b).
[17] Bhabha (1939b, p 385).
[18] *Ibid.*
[19] *Ibid.*, p 385, with original emphasis.
[20] *Ibid*, p 387.
[21] *Ibid*, p 398.
[22] *Ibid*, p 399.
[23] Heisenberg to Pauli, 20 April 1939, [WPSC2], p 629. Heisenberg is evidently referring to the letter to *Nature* (Bhabha 1939a) and not to Bhabha (1939b), which had just been submitted.
[24] Heisenberg (1939b, p 82).
[25] Deutsche Physiker- und Mathematikertag, 24–30 September. See Yukawa's *Einladung*, dated 5 June 1939 [YHAL] and section 6 of chapter 9.
[26] The programme of the symposium had these general topics: I. The intensity of cosmic rays. II. Time variations of cosmic rays. III. Composition of cosmic rays. IV. Production of secondary radiation. The printed talks or summaries of them are in *Reviews of Modern Physics* **11**, Nos. 3 and 4, July–October 1939. Heisenberg attended this conference as part of an American tour: New York (22–26 June), Chicago (27 June–1 July), Purdue University (1–22 July) and again New York (23 July–1 August).
[27] Montgomery and Montgomery (1939, footnote on p 261). Also Montgomery and Montgomery (1941).
[28] Other sections were to be held on solid state physics, technical physics, television and high-frequency technologies.
[29] This and the following information were taken from the letter of the Commission Administrative to Yukawa, 11 March 1939 [YHAL].
[30] Heisenberg to Pauli, 20 April 1939; [WPSC], p 629.
[31] Heisenberg to Pauli, 23 April 1939; [WPSC], pp 634–5.
[32] Pauli to Heisenberg, 27 April 1939; [WPSC], pp 636–9
[33] There had been continuing correspondence between Heisenberg and Pauli on the basic problems of field theory since their joint work on QED (1929, 1930). However, Pauli had not put his name on any paper involving interacting quantum fields after 1930, because of the unresolved divergence difficulties. He left that to Heisenberg and his collaborators, especially Hans Euler. (See Rechenberg (1993).)
[34] Pauli to Heisenberg, 11 May 1939; [WPSC2], pp 652–3.
[35] Pauli to Heisenberg, 10 June 1939; [WPSC2], pp 662–3. The Solvay manuscript of Pauli has been published recently in [WPSC3], pp 833–901.
[36] Pauli (1940, abstract). Note the restriction to free fields and to wave equations in the abstract (not *quantized fields*, although the quantization of the fields is discussed in the paper). A footnote states: "The relation of the present discussion of the connection between spin and statistics, and the somewhat less general one of Belinfante, based on the concept of charge invariance, has been cleared up in [Pauli and Belinfante 1940]."
[37] Pauli (1940, pp 721–2), original emphasis.

[38] Pauli (1941, p 208).
[39] We will further describe Pauli's activities during the war in the next chapter.
[40] Peierls (1939, p 90).
[41] This can be seen, e.g., in Richtmeyer (1940), where it is shown that radiative effects rise quadratically with the energy for spin 1; this gives a quadratic divergence in the self-energy of a spin-1 particle, as opposed to the logarithmic divergence obtained for spin $\frac{1}{2}$.
[42] The paper with Bhabha in the *Proceedings of the Indian Academy of Science* was received on 5 October. It notes, "This work was carried out in Cambridge in June, but external circumstances followed by the outbreak of war have delayed its publication."
[43] Oppenheimer *et al.* (1940 p 77).
[44] *Ibid.*, p 78.
[45] Williams (1940, p 191)
[46] Tomonaga (1940, p 400).
[47] Corben and Schwinger (1940, p 953).
[48] Corben and Schwinger (1940, p 954).
[49] Tamm (1940, p 952).
[50] Booth and Wilson (1940, pp 485–6)
[51] *Ibid.*, p 486
[52] Christy and Kusaka (1941a, p 406).
[53] Oppenheimer (1941, p 462). In conclusion we merely mention here some additional works dealing with this subject, none of which change the general picture that has been presented: Iwanenko and Sokolov (1940), Ma and Yu (1942), Ma (1942), Kusaka (1943), Kobayasi (1941) and Chakrabarty and Majumdar (1944).

Part D

Meson Physics from 1939 to 1950: the Meson Puzzle Resolved

This is the concluding part (aside from an epilogue); it covers meson physics during the war years, mainly theoretical, and during the immediate postwar period, when experiments, first with the cosmic rays and then with accelerators, revealed the existence of the strongly interacting meson postulated by Yukawa, until then invisible.

The phrase "war years" had, of course, a rather different meaning for those locked in battle in Europe from September 1939, punctuated by the German invasion of Russia in June 1941, and those whose "war" began with the attack on Pearl Harbor in December 1941 (although Japan had been at war with China since at least 1937). Normal physics research, especially experimental, could be carried on only in a limited way where bombs were falling; nevertheless, some significant work was done by physicists who had been assigned to teaching responsibilities. For example, George D. Rochester, Cecil F. Powell and Lajos Janóssy continued their cosmic-ray researches at Manchester, as did Marcello Conversi, Ettore Pancini and Oreste Piccioni in Rome. The same was true of groups partly involved in war work, especially in Germany (Heisenberg), Japan (Nishina) and USA (groups at Berkeley, Princeton and Cornell University). Important theoretical developments took place in the Soviet Union and in the countries occupied by Germany, and also under active war conditions. The issues include Heisenberg's S-matrix theory and the two-meson theory of Takesi Inoue and Shoichi Sakata.

We shall, somewhat arbitrarily, divide the wartime research on a geographical basis, devoting chapter 11 to the work done in the United States, Britain, Ireland, Switzerland and Holland (the "West"), and chapter 12 to research done in Germany, Italy, Japan and the Soviet Union (the "East"). In chapter 13 we discuss the experimental studies in Rome, begun under wartime conditions, which proved the lack of strong nuclear interaction of the cosmic-ray meson. Then we turn to the Bristol experiments in 1947, the discovery of the pion. In the same year, strange particles were first seen at Manchester, and the following year the Berkeley synchrocyclotron produced its first mesons. Together, these experiments mark a typical node in the history of particle physics, solving the puzzles of one era, while posing new enigmas for the era to follow.

Chapter 11

Meson Theory During the War (West)

11.1 Introduction. Research in nuclear physics

In December 1938, Otto Hahn and Fritz Strassmann of Berlin's Kaiser Wilhelm Institut für Chemie noticed for the first time that uranium nuclei bombarded by slow neutrons could break into two nearly equally large fragments. A few months later, several experimental groups, especially Irène Curie and Frédéric Joliot in Paris, confirmed the break-up and observed that extra neutrons are also emitted in the fission process. Thus a nuclear chain reaction might be feasible in which nuclear energy could be released on a large scale. With war threatening, military authorities in Europe and America began to take notice. The German Uranium Project was the first to emerge. Then there were similar ones in France, Britain, and the United States; finally, the US–Britain Manhattan Project led to the construction of nuclear bombs before the end of the war [1]. Later on, Soviet and Japanese physicists also became involved with nuclear energy.

Many of the nuclear and cosmic-ray scientists in America and Britain who had been actively working on meson physics participated in war projects: notably, Robert Oppenheimer and his students, together with a number of prominent European immigrants, such as Bethe, Fermi, Rossi, Teller, Weisskopf and Wigner. Working on the atomic bomb project in America, they were joined by British experts, e.g., Chadwick, Peierls and John Cockcroft. Among the leaders of the German project were Heisenberg and von Weizsäcker. Clearly these activities, and such other projects as radar, were given a much higher priority than fundamental research on the nuclear forces.

On the other hand, some fundamental research continued in neutral countries like Ireland, Switzerland and Sweden, and also in some countries under German occupation, such as Holland and Denmark. Elementary particle research also continued, to some extent, in the Soviet Union until it was invaded in June 1941, and in the United States and Japan until the Pearl Harbor attack in December 1941. Perhaps surprisingly, Heisenberg was returning to the study of fundamental research on elementary particles at about the same time as the American and Soviet atomic bomb projects were being initiated.

In the autumn of 1939, just as the war was starting, Yukawa cut short his European visit and returned to Japan via the United States. In Berkeley he met Robert Oppenheimer, who was beginning to adopt a more friendly attitude towards meson theory. During the following two years the Berkeley group of Oppenheimer and his students were working very actively on various meson problems. One of the students, Shuichi Kusaka, left for Princeton in 1942, where he continued the work with Pauli [2]. In August 1941, Kusaka wrote to Yukawa in Kyoto about the American work [3]:

> Thank you very much for your letter and the reprints. I am sorry this letter has been delayed because I was taking a vacation in Vancouver. Regarding Dr Sakata's work on the pseudoscalar meson, Professor Oppenheimer noted a mistake in the calculation and Nelson has gone over the calculation [4]. He found that the lifetime of the meson is far too short if the constants of β-decay are used. As far as the scattering of mesons, Schwinger has treated the problem of a meson field strongly coupled to a spatially extended source and obtained isobaric states as suggested by Heitler and Bhabha. I presume you have seen the recent work in *Phys. Rev.* by Oppenheimer and Schwinger which tells some of their results.
>
> Pauli is visiting here at present, and will stay for about a month. There are also several of Oppenheimer's former students, Carlson, Serber, Lamb, Dancoff, Morrison and Keller here at present, but they will leave soon. Also Schwinger, Rarita, Christy and Cooper are no longer here, so we shall not have many people left. The only addition is Sachs who has taken Schwinger's place as Oppenheimer's assistant. The main topic of interest here now is the nuclear force due to mesons strongly coupled to extended sources. Carlson is making some calculations in shower theory assuming the primaries as protons which produce many (about eight) mesons in a single collision. I am working on commutation relations for field quantities of a particle with spin $\frac{3}{2}$ to see whether they commute outside the light-cone in the presence of an electromagnetic field.

Kusaka's letter essentially marks the termination of a lively scientific exchange between the schools of Yukawa and Oppenheimer, for a few months later Japan was at war with the United States and the Second World War was fully underway. In the present chapter we discuss the meson research which took place in the West, namely in the United States and in the West-European countries, including the UK.

In 1939 theorists in the United States had begun to consider the problems of nuclear forces and cosmic rays which had been pioneered by Japanese and European physicists. One of their major concerns was the theory of particles of higher spins and its possible application to the experimental data (section 2). Another way suggested to bring about a *rapprochement* of theory with experiment was to use mixtures of fields of different space-time character (different spins) or to revive a theory involving the exchange of pairs of fermions (section 3). From Europe there arrived still another new suggestion, the strong-coupling theory of Gregor Wentzel, which was especially studied in Princeton by Pauli and his associates, and also taken up in Japan, especially by Tomonaga, who generalized the theory to intermediate-coupling (section 4). The war conditions in Britain absorbed the services of most British physicists in war projects, few were able to work on fundamental problems of physics. However, in Dublin, Walter Heitler and Erwin Schrödinger found a refuge and continued their research. At the end of the war, in October 1945, Pauli summarized the results of this wartime research on elementary particles in a letter to his friend Hendrik Casimir (section 5).

11.2 Field theories with higher spins (1939–41)

Among the important physicists migrating to the United States shortly before the outbreak of the European war, we can count Enrico Fermi and Bruno Rossi. On 2 January 1939 Fermi arrived, just two weeks before Niels Bohr brought news of nuclear fission to New York. Rossi visited Chicago in June 1939, in time to participate in the Chicago cosmic-ray symposium that we discussed in chapter 10. Both Fermi and Rossi gave up their homes and academic positions as a result of the anti-Jewish laws, which had been put into effect in Italy in 1938. Likewise, Pauli left his professorship in Zürich in August 1940 to spend the war years at the Institute for Advanced Study in Princeton. Being of Jewish origin and acquiring German citizenship as a result of the Austrian *Anschluss* in 1937, he was afraid of the consequences for him if Germany were to invade Switzerland [5]. Pauli became important in the American meson physics effort, started earlier by other émigrés, Lothar Nordheim and Hans Bethe (see chapter 9).

One of the concerns of this period was to formulate the quantum theory of particles of various spins. Aside from the recognition that accepted invariance principles, especially relativistic covariance, strongly limited the forms that theories of free particles could take (as shown, e.g., in Dirac (1936), Fierz (1939) and Pauli (1940, 1941)), there was the widely accepted possibility that there might be other, as yet unknown, fundamental particles. Thus, in April 1940 Pauli wrote from Zürich to Bhabha in India [6]:

> Quite independent from the theory of the scattering of charged mesons by protons I believe in the existence of much more particles than known until now, particularly in particles with arbitrary values of the spin and of the charge...Did you read Bethe's paper...about the meson-theory of nuclear forces? He finds some difficulties to understand the sign of the quadrupol-moment [*sic*] of the deuteron with reasonable assumptions (what he calls 'neutral theory' does not seem to me reasonable). Besides Fierz has some empirical indications...in agreement with the necessity of what Bethe calls 'the necessity of cutting off'. *So I think the present meson-theory of nuclear forces is completely wrong and has to be modified by the introduction of particles with more general values of charge and spin in the intermediate [states].* Of course it is the opinion of myself and also of Heitler [that these] particles could under favourable circumstances also be [obtained as] *real* particles in the cosmic rays.

Two additional ways of treating fundamental particles should be mentioned, because they were used in meson (and mesotron) calculations:

the Kemmer–Duffin and Rarita–Schwinger methods. Relativistic particles of spin $\frac{1}{2}$ were usually treated by Dirac's first-order wave equation and particles of spin 0 and 1 by, respectively, the Pauli–Weisskopf theory (using the Klein–Gordon equation) and the Proca theory (using generalized Maxwell equations). Kemmer (1939) sought to formulate the spin-0 and spin-1 theories in a form more closely resembling the Dirac equation, introducing generalized Dirac matrices that had been proposed independently by Duffin (1938). The new matrices obey generalized Dirac anticommutation relations and have three inequivalent irreducible representations, of dimensions 10, 5 and 1. Of these, the one-dimensional one is trivial, the 5×5 matrices of the five-dimensional representation give the equivalent of the spin-0 Pauli-Weisskopf theory and the 10×10 matrices of the ten-dimensional representation give the equivalent of the spin-1 Proca theory. Thus no new physical results are obtained, but a uniformly consistent treatment is gained of the quantum-mechanical operators, matrix elements, etc., for the three case of spins 0, $\frac{1}{2}$ and 1. Kemmer's method was further developed and applied to the vector meson theory in Japan (Taketani and Sakata 1940; Sakata and Taketani 1940).

Rarita and Schwinger (1941b) showed a simplified way of treating free particles of half-integer spin (greater than $\frac{1}{2}$), by using a hybrid spinor–tensor wavefunction $\Psi_{a\alpha\beta\gamma}\ldots$, having one Dirac spin-$\frac{1}{2}$ index a and k tensor indices α, β, γ, etc. (Each index takes on four values.) They found that upon application of the Dirac equation (operator $\mathbf{D}_a$) and the "usual supplementary conditions of the integral spin theory", namely:

$$\mathbf{D}_a \Psi_{a\alpha\beta\gamma} \cdots = 0, \qquad \Psi_{a\alpha\alpha\gamma} \cdots = 0, \qquad \partial_\alpha \Psi_{a\alpha\beta\gamma} \cdots = 0, \qquad (11.1)$$

the wavefunction describes a particle of spin $(k+\frac{1}{2})$, having $2(k+\frac{1}{2})+1$ degrees of freedom. The Rarita-Schwinger formalism is a consistent theory of free particles and it has been useful for applications in lowest order of perturbation theory, but it suffers from the usual difficulties of higher-spin theories with electromagnetic interaction [7].

A brief review of the meson situation as it appeared during the war years will help to explain why so many new approaches were attempted during this period. Kemmer (1938b) considered four theories (see chapter 7): two of spin 0, (a) scalar and (b) pseudoscalar, and two of spin 1, (c) vector and (d) pseudovector. The scalar theory, however, gives the wrong sign for the n–p interaction in the ${}^3S_1$ ground state and, in fact, gives no spin dependence of the nuclear force. To get spin-dependence in a spin-0 theory requires a meson–nucleon interaction

$$H' = g\Psi^* \boldsymbol{\sigma} \cdot \text{grad } \Psi\phi \qquad (11.2)$$

Since $\boldsymbol{\sigma} \cdot \text{grad}$ is a pseudoscalar operator, it follows that the meson field ϕ is also a pseudoscalar for H' to have even parity, as required by the strong interactions. In the three cases (b), (c) and (d), Kemmer showed that the n–p interaction has spin-dependent terms involving $\boldsymbol{\sigma}_N \cdot \boldsymbol{\sigma}_P$ and

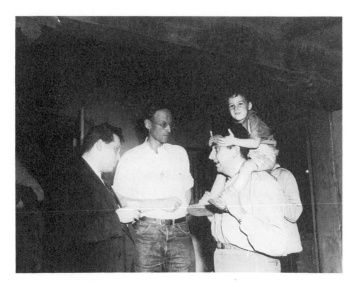

J. Schwinger (1918–1994) (left) and E. Teller (born 1908) (right) in the 1940s.
Photograph from Los Alamos Scientific Laboratory and reproduced by permission of
AIP Emilio Segrè Visual Archives.

$(\boldsymbol{\sigma}_N \cdot \mathrm{grad})(\boldsymbol{\sigma}_P \cdot \mathrm{grad})$. Case (d) provides repulsive forces both in the ^{1}S and in the ^{3}S state of the two-nucleon system, in contradiction to experiment. Case (c) has two independent coupling constants, which can be chosen to adjust the proportions of these terms. That was the case selected by Kemmer.

However, by 1940 it had become clear that no Yukawa-type meson theory could accommodate *all* that had been anticipated for a theory of nuclear forces and the cosmic-ray meson or mesotron — that is, for nuclear forces: the states of the deuteron, nuclear magnetic moments, β-decay and charge-independence; for the mesotron: nuclear and electromagnetic interaction cross-sections, and a mesotron decay lifetime predictable from nuclear β-decay rates. For example, β-decay and mesotron decay rates were not quantitatively compatible; hence a return to the four-fermion theory was considered for β-decay, and the assignment of an independent coupling constant for mesotron decay. This, of course, meant giving up a unified description of the two decay phenomena.

Other difficulties were also apparent in the interactions of mesotrons with nuclei and electrons. The observed nuclear cross-sections appeared much too small when compared with the known nuclear force strength and the electromagnetic cross-sections (as seen, e.g., in showers and bursts) were measurably much smaller than predicted for charged mesons of spin 1 [8]. In addition, the charge-independence of nuclear forces required a neutral meson, but no neutral mesotron (neutretto) had been observed. To compound all these problems, unacceptable high-energy divergences were reflected in unacceptable divergences at small distances, even at moderate energies (terms in the internuclear potential $\sim 1/r^3$).

Possible explanations were offered. First, the meson and mesotron were simply different particles. Thus Bethe, using spin-1 mesons (with cut-offs) for nuclear forces alone, considered three cases: charged (no neutral), neutral (no charged) and symmetric meson theories, and concluded that only the purely neutral theory could give the right sign and magnitude of the nucleon magnetic moments and of the deuteron quadrupole moments [9]. A second approach involved mixed vector and pseudoscalar fields, with the mix adjusted to cancel the objectionable $1/r^3$ singularity in the nuclear potential.

11.3 Mixture and pair theories (1939–44)

A theory of mixed vector and pseudoscalar fields was proposed by Christian Møller and Léon Rosenfeld (Møller and Rosenfeld 1939a, b). The cancellation of the terms that were singular at small distances worked, however, only in the static case. The divergences would reappear in higher field-theoretical approximations. In addition, if the mesons of the two types had equal masses, then the deuteron's quadrupole moment vanished. However, Schwinger (1942) derived the desirable result that, if the vector meson were chosen to have a mass greater than the pseudoscalar, then the deuteron quadrupole moment had the right sign, while the vector meson might decay rapidly enough (into a photon and the pseudoscalar meson) to be unobserved in the cosmic rays.

During the first half of the decade of the 1940s, after his move to Princeton, Pauli became one of the principal players in the game of meson theory. His fundamental work with Markus Fierz on the quantum theory of free fields — yielding the spin-statistics connection — was highly regarded by the meson theorists, but he himself at first remained skeptical about their latest inventions. In his usual warm but brusque style, he wrote a postcard from Zürich to Møller and Rosenfeld in Copenhagen, as follows [10]:

Dear authors of numerous *Nature*-notes! [11].
In contrast to uranium fission I follow the meson theory with a certain interest — however, also with skepticism, so long as no one comes to root out the evil [*Übel*] of infinities and divergences. — Still, there seems to be a certain measure of good sense in the notion of understanding the quadrupole moment of the deuteron as a relativistic effect and saying it should be zero in non-relativistic approximation. On the other hand, your patent mixture [of fields of spin 0 and 1], which makes certain singularities in the proton–neutron interaction zero, seems to me quite naive [*recht kindlich*] — since the theory still diverges in higher approximations. Thus I am more sympathetic to Heisenberg's, "If you're going to cut, then go all the way". [*Wenn schon abschneiden, dann gleich gründlich*].

Pauli then asked the authors for a manuscript or proof copy of Møller and Rosenfeld (1940), in order to read their ideas in greater detail, and closed with: "How much more beautiful would have been the Solvay Conference than what is taking place instead!" [12]

In spite of his initial reluctance, most of Pauli's papers in the first half of the 1940s were on the meson theory, with Oppenheimer's encouragement [13]. As we shall discuss below, Pauli confined his work to the strong-coupling theory. In several of these papers (Pauli and Kusaka 1943, Pauli 1943a, b), he dealt with the "patent mixture" of Møller and Rosenfeld, although his skepticism regarding the appearance of infinities in higher order was confirmed by other authors (Ferretti 1943; Hu 1945).

Pauli was little concerned with the newer pair theories, which were essentially modifications of the Fermi-field theory (see chapter 3). One of the alternative versions mentioned at the end of chapter 3 was the electron-pair theory of Gamow and Teller (1937), which was further developed in the hands of Critchfield *et al.* (1939). Hoping to achieve the same results for nuclear forces as Bethe had with his neutral meson theory, but without losing the Yukawa connection to the charged cosmic-ray mesons (or mesotrons), Robert E. Marshak revived the pair theory in 1940, but *his* pair consisted of oppositely charged "heavy electrons", i.e., heavy particles of spin-$\frac{1}{2}$ which he wanted to identify with the mesotrons. He also adopted Bethe's "single-force hypothesis" and chose to use a tensor interaction between the nucleon and the heavy electron, having the form

$$H' = g(\Psi^* \boldsymbol{\sigma}_n \Psi) \cdot (\psi^* \boldsymbol{\sigma}_e \psi), \qquad (11.3)$$

where the first bracketed quantities refer to the nucleon, the second to the heavy electron. For the force between two nucleons he found: "The potential function between two nuclear particles behaves at large distances, r, as $e^{-2kr}/r^{2.5}$, $k \approx \mu c/\hbar$, so that the range is effectively one-half the single meson range." He also found that, at small distance, "the potential goes as $1/r^5$ so that one has to cut off in the same way as in the original electron–neutrino theory" [14]. Except for the calculated meson scattering cross-section being much too *large* in comparison with the observed one, the Japanese theorist Tatsuoki Miyazima concluded on the basis of his study of the nuclear forces with heavy-electron pairs that "the possibility . . . of the mesotron of spin-$\frac{1}{2}$ is not totally excluded" [15].

In the early 1940s the pair theories attracted widespread interest. Although mainly concerned with his strong-coupling theory, Gregor Wentzel wrote two papers on the pair theory of nuclear forces (Wentzel 1941b and 1942) [16]. In these papers he was mainly concerned with the qualitative behaviour of the forces and thus, arguing that the spin of the exchanged particles was of minor importance in such an investigation, he used spinless particles as an approximation. Because the large coupling constant of meson theories makes the usual weak coupling perturbation of doubtful validity, Wentzel applied a non-perturbative method to calculate the dependence of the force on the separation of two nucleons [17]. Comparing with the second-order perturbation calculation, he concluded that the latter gave the correct distance-dependence, but "too high a value for the absolute strengths of the forces" [18]. Stressing the importance of the cut-off distance that must be introduced because of the $1/r^5$ singularity in the two-nucleon potential, Wentzel argued that the mass of the field particle plays only a minor role — it could even be as small as the electron mass without making any qualitative difference: "In contrast to Yukawa's meson theory, there is in the meson pair theory *no relation between the nuclear radius and the mass of the field particle* ... and one has from this standpoint no basis to give preference to a 'meson-pair theory' over an electron-pair theory." [19]

Turning back to the American efforts, we can summarize with what Pauli wrote to Wentzel from Princeton at the end of 1944 [20]:

> On pair theory: [Joseph Maria] Jauch [21] has posted me his result and it seems to me right, that the perturbation calculation of Marshak and Weisskopf is now generally accepted to be senseless, but there is one, as I believe, correct calculation of the scattering of spin-$\frac{1}{2}$ mesotrons by J.W. Weinberg [1941]. Oppenheimer has recently been interested in the possibility of bringing nuclear forces and scattering theory into agreement, with pair theory and more strongly spin-dependent nuclear forces, and came to a completely negative result.

Oppenheimer's student Joseph W. Weinberg based his paper on the suggestion of Critchfield and Teller (1938), replacing their electron pair field by a spin-$\frac{1}{2}$ meson pair field. As in their model, Weinberg's calculation treats the nucleus as an extended source of mesotrons, differing from Marshak and Weisskopf, who treated the nucleon as a point source. Weinberg concluded that a "rigorous" formula for the scattering cross-section, with its "constants" adjusted to fit the nuclear force's range and strength, gave about 10^{-24} cm^2 for low-energy scattering, "as contrasted with the cross-section of about 10^{-28} cm^2 observed for cosmic ray mesotrons" [22].

We note here, finally, that the attractiveness of pair theory, in spite of its evident failings, persisted as late as 1950, i.e., even long after the distinction between pion and muon had become clear. Namely, Wentzel published in that year a paper proposing that a pion might be a pair of strongly bound muons (Wentzel 1950). The binding energy was to arise from the possibility of a muon pair transforming into a proton–antineutron pair and back [23].

11.4 Strong- and intermediate-coupling theories (1939–44)

From 1939 to 1945 some new approaches were applied in nuclear forces and meson scattering to avoid the use of weak-coupling perturbation theory, which had the disqualifying difficulty (in addition to the usual field-theoretical divergences) that the expansion of probability amplitudes in ascending powers of the coupling constant becomes useless. For, even if the series were asymptotic (as is the case in QED), the terms of lower powers cannot give useful approximations (as they do in QED) because in meson theory the coupling constant is not small. Of course, the size of the coupling constant depends on what coupling is assumed. For pseudoscalar mesons there are two forms of coupling: direct or pseudoscalar, $G\psi^*\gamma^5\psi\phi$ and derivative or pseudovector, $(iF/\mu)\psi^*\gamma^5\gamma^\mu\psi\,\partial\phi/\partial x_\mu$. (We use the notation of Bethe and De Hoffmann (1955), who gave as a rough estimate $g^2 = G^2/(4\pi) \approx 10$ and $f^2 = F^2/(4\pi) \approx 0.1$.) An "equivalence theorem" shows that for many purposes the two forms of coupling give the same result.

As we have already discussed in chapter 9, section 5, the relatively strong coupling of meson fields to nucleons led to the suggestion that processes involving the production or exchange of many mesons would play a predominant role and therefore, by analogy with the situation in electrodynamics, a classical field method would be appropriate. This approach may be traced to Guido Beck's suggestion that the populous meson clouds surrounding the point-like cores of the proton and neutron would guarantee the charge-independence of nuclear forces *without* the necessity of a neutral meson (Beck 1938a). Heisenberg coupled his suggestion with the idea of a fundamental length advanced earlier (Heisenberg 1936, 1938a, b). He also emphasized that the inertia, or field reaction on the source, is far more important in the meson case than that is in electrodynamics, in which it is usually neglected, and suggested that this could lead to a saturation of the meson scattering cross-section at high energy (Heisenberg 1939b). At the same time, Bhabha considered a classical meson theory having similar goals (Bhabha 1939a, b, 1940a, 1941).

In two very influential papers, Wentzel (1940, 1941a) undertook to construct a new kind of perturbation expansion, in *falling* rather than *ascending* powers of the coupling constant [24]. Although he chose the artificial model of charged scalar theory to illustrate his methods, he argued that the results should at least qualitatively resemble those in a more realistic theory. In a later review article (Wentzel 1947) he listed the other papers based on his theory, as follows [25]:

- Serber and Dancoff (1943), charged scalar and neutral pseudoscalar theories;
- Pauli and Dancoff (1942), symmetrical pseudoscalar theory
- Pauli and Kusaka (1943), pseudoscalar, vector and mixture theories;
- Wentzel (1943a), symmetrical vector theory with two coupling parameters.

We now discuss all of the papers mentioned, as well as some other Japanese ones.

In Wentzel's simplified model of 1940, he placed N nucleons (which he calls "proton–neutrons") at some points of a cubic lattice and calculated the self-energy due to their meson fields. The nucleons are permanently fixed, equivalent to them having infinite mass. Wentzel pointed out that the situation is very different from the corresponding electromagnetic case, because of the non-commutability of the isospins of the nucleons. (This would also be the case for spins, but Wentzel's scalar field has no spin-

S.M. Dancoff (1913–1951). Photograph reproduced by permission of AIP Emilio
Segrè Visual Archives.

dependence.) That is, emitting a meson changes the charge of the nucleon,
whereas emitting a photon does not. The Hamiltonian of the problem reads

$$H = H^0 - G \sum_j [(\tau_1^j + i\tau_2^j)\psi_j + (\tau_1^j - i\tau_2^j)\psi_j^*], \qquad (11.4)$$

where the τ terms are the nucleon isospin operators, ψ is the charged scalar
meson field and j is the number of a lattice point. For a single nucleon
Wentzel found the approximate energy

$$E_1 = -G^2/\mu^2 + \mu(n + \tfrac{1}{2}) + [\mu^4/(4G^2)]m^2, \qquad (11.5)$$

where μ is the meson mass (in units such that $\hbar = c = 1$), n is the number of
mesons present and m is a half-integral charge quantum number. Strong
coupling has been assumed, i.e., $G^2 \gg \mu^3$.

Wentzel interpreted this result as follows [26]:

Here the term $-G^2/\mu^2$ represents a negative self-energy of the proton–neutron
and to that is added the energy of n mesons... [From $G^2 \gg \mu^3$] the sum is large
compared with the meson mass μ. [In lowest approximation], the eigenvalue
does not depend on the charge quantum number m, so that every state, even the
ground state $n = 0$, has the same energy for arbitrary integer charge.
Apparently this means that *through the binding of charged mesons to the*

proton–neutron, states could result which carry arbitrarily high integer charge and which all have the same or nearly the same mass [original emphasis]. In higher approximation, naturally, the *m*-degeneracy falls away... The states with the same *n* (e.g., *n* = 0) obtain a *quadratic dependence of the energy or mass on the charge number*: the proton and neutron ($m = \pm\frac{1}{2}$) have the smallest masses; the next adjacent isobars, namely doubly positive proton and negative proton ($m = \pm\frac{3}{2}$) have already a mass higher by $\mu^4/(2G^2)$, although this mass increase is small compared with the meson mass.

To obtain the above result, Wentzel imposed the additional condition on *l*, the lattice spacing, that $\mu l \gg 1$, remarking that: "Obviously such a strong 'cut-off' can lead to nothing of physical interest." However, he used that as the starting point for several steps of further generalization. One result for the one-nucleon problem was that there scattering of mesons occurred with a cross-section of order $4\pi/\mu^2$ [27]. Since μ here represents the inverse Compton wavelength of the meson, it is clear that this is a much larger cross-section than that observed for the mesotron. In a further step of generalization, Wentzel treated the case of nuclear forces among *N* particles, at least in the approximation g^0, i.e., the zero-point energy of the vacuum field, and obtained a result proportional to *N*. He could not let the lattice spacing *l* approach zero, because this led to infinite binding energy.

Concluding the paper, Wentzel sharply distinguished his isobar states, derived as a consequence of the Yukawa theory, from the "hypothesis recently published by Heitler [1940], according to which nuclear particles *a priori*, i.e., independent of the coupling with the meson field, are ascribed energetically higher states with the charge values −1 and +2". An assumption similar to Heitler's was made by Bhabha (1940c), who drew the consequence that although the scattering cross-section at low energies is large, it "does not deprive the meson of its penetrating power since it decreases for high energies as p^{-2} due to the p^6 term in the denominator which expresses the effect of radiation reaction" [28]. Bhabha's paper was based on classical meson theory (Bhabha 1940b).

The principal features exhibited by strong coupling and brought out by Heisenberg, Wentzel and Bhabha were incorporated in the work that followed, making up what Bethe and De Hoffmann (1955) later called the "old strong coupling theory". That is, the later papers took into account the field reaction on nucleon charge and spin, made the assumption of finite nuclear extension (rather than point-like, as in weak-coupling perturbation theory), used nucleon isobaric states and approximated with classical meson fields. The use of extended nucleons meant that they were treated non-relativistically (the mesons, however, always required relativistic treatment). Curiously, with one exception, Wentzel's new perturbation method of expansion in falling powers of the coupling constant was never mentioned.

The important exception was the work of Sin-itiro Tomonaga, who greatly extended and generalized Wentzel's method into a new and more practical intermediate coupling method. He proposed and developed this method in wartime Japan, in a series of papers, both alone (Tomonaga 1941, 1946a, b) and with Miyazima (Miyazima and Tomonaga 1942, 1943). However, these papers remained unknown outside Japan until after the war.

Tomonaga set forth his aim in his 1941 paper [29]:

> In the present work it will be shown how the behaviour of a meson emitted virtually from a proton or neutron can be investigated by the introduction of the Hartree approximation [i.e., the self-consistent field], not only in the limits of strong or weak coupling but also in cases lying between them. There results an approximation formula for the self-energy of the nuclear particle which agrees exactly for weak or strong coupling, respectively, with the results of the perturbation method or the method of Wentzel.

He began by writing the Schrödinger equation in configuration space for a nucleon in interaction with its meson field, and then expressing it as a series of integral equations (in momentum space) for Fock wave functions representing states of different numbers of positive and negative mesons. Like Wentzel, he used at first a simplified model of longitudinally polarized charged mesons, but said that it can easily be generalized to include transverse vector mesons or pseudoscalar mesons. He determined the self-energy by the Ritz variational method and so obtained his formula, which "agrees exactly" both with weak- and with strong-coupling limits, and is thus the interpolation formula sought (and which Wentzel had anticipated in his first strong-coupling paper).

In Miyazima and Tomonaga (1942), the variation method is improved by the introduction of more variational parameters; the authors applied it to calculate the multiple production of mesons by the "collision of a nuclear particle [nucleon] with a heavy particle [nucleus]". In 1943, they treated the problem of nuclear forces by the method of the classical meson field. Possible objections to this are first raised, but then resolved, as follows [30]:

> In this case, however, it is not clear how to deal with the spins of the nucleons. The magnitude of the spin-vector σ or τ is a given quantity and we can never regard it as large enough to be treated as a classical quantity. In spite of this circumstance, we look upon it, for the moment, as a unit vector in the spin and the isotopic spin space respectively so that a perfectly classical treatment of the problem becomes possible. Of course we have no right *a priori* to assert the validity of this procedure, but it can be shown that, in the limit of the large coupling constant, the classical result agrees well with both of the quantum-mechanical approximations which were introduced by Wentzel and by one of the authors recently. The classical approximation may, therefore, provide us with a comparatively powerful means of investigating the influence of the reaction of the mesotron field on the nucleons in the case of large coupling.

In the United States meanwhile, before war work fully occupied their attention [31], Oppenheimer and his associates at Berkeley applied the strong-coupling theory in an effort to understand the small scattering cross-sections of cosmic-ray mesons. Oppenheimer and Schwinger wrote [32]:

> We have, in part, generalized Heisenberg's treatment, and considered the classical problem of the coupling of neutral and charged, scalar and pseudo-scalar mesotrons to an extended, spatially fixed source. These problems are all rigorously soluble, for all values of the coupling constant and source size. In

addition, we have treated Wentzel's quantum problem of the charged scalar field, using an extended source instead of a lattice space, in the limit where the coupling constant is large... and have made the analogous calculation for the neutral pseudo-scalar in the corresponding limit.

The paper discussed the motivation and quoted some results, but the details of the calculation were left for a promised forthcoming paper by Schwinger, to which subsequent authors referred, although it was never actually published. After presenting their results for meson scattering from a nucleon (in the charged, neutral and symmetrical theories) the Berkeley authors concluded [33]:

It is thus clear that pseudo-scalar theories can give a scattering small enough to agree with that observed, but that scalar theories could, at most, do so with a choice of [coupling constant] far too small to account for nuclear forces. In fact, the experimental scattering results demand a value of a [nuclear size] of the order of the proton Compton wave length $\hbar/(Mc)$ or possibly slightly smaller. Indeed this length marks the extreme limit of the validity of the methods we are using, and of the classical localizability of the source.

A year later, Pauli and Dancoff drew conclusions similar to the above for meson scattering, i.e., a sufficiently small value of the scattering in pseudoscalar symmetric theory, by assuming a very small nuclear size. However, they also tried to calculate the nucleon magnetic moments, obtaining the result [34]

$$\pm[10g^2/(36\mu a) + \tfrac{1}{6}] \text{ proton magnetons,}$$

where again g denotes the pseudoscalar coupling constant, μ the inverse meson Compton wavelength and a the nucleon size. Although the assumption of strong coupling is consistent with the order of magnitude of the moments, the proton and neutron moments are far from being equal and opposite, as the result suggests. (These are the full moments and not the "anomalous", non-Dirac parts, which *are* nearly equal and opposite.)

Serber and Dancoff (1943) looked more carefully at the consequences of the Oppenheimer–Schwinger assumptions for the nuclear force problem, making use, as they acknowledged, of unpublished work by those authors and by Pauli. Their conclusion: "It is found impossible to obtain spin-dependent forces which, at the same time, extend to small separations and are of sufficient strength to account for the properties of the deuteron." [35]

Pauli and Kusaka (1943) then investigated the nuclear force problem using Schwinger's modification of the Møller–Rosenfeld "patent mixture" of pseudoscalar and vector-meson fields. They found a very good agreement with some properties of the deuteron, including the quadrupole moment, as in the weak coupling case, but not for its magnetic moment. Thus they reached the following negative conclusions [36]:

Our results show that the theory here considered suffers from two grave difficulties; it gives a magnetic moment for the deuteron a value only a few percent of the observed value, and it predicts instability of highly charged nuclei. These results seem to be properties of all strong coupling theories, and there does not seem to be any way of overcoming them. These difficulties are not present in the weak coupling theory, and thus it seems to be desirable to go back and reconsider the arguments which led us to take up the strong coupling theory in favor of the weak coupling theory. The main difficulties in the weak coupling theories are the divergences due to the treatment of the heavy particles as a point source, and the large scattering cross section of the meson. As already pointed out [Pauli 1943b], the first difficulty can be solved by using the λ-process developed by Wentzel and Dirac, and the second by using the theory of radiation damping developed by Heitler and Wilson. In addition, the weak coupling theory developed in this way has the advantage of relativistic invariance which the strong coupling theory does not have on account of the finite size of the source. *Thus there is no reason now to consider the strong coupling theory, and we should go back to the weak coupling theory.*

In spite of this pessimism, however, Pauli gave a series of six lectures in the autumn of 1944 at the Massachusetts Institute of Technology, organized by I.I. Rabi of the Radiation Laboratory, his main subject being strong coupling theory. These lectures were later published as a small book called *The Meson Theory of Nuclear Forces* (Pauli 1946). Its "Concluding Remarks", reprinted identically in the second edition (Pauli 1948), reiterated the unsatisfactory state of the subject:

> There are, at present, essentially two ways of approaching the problem of nuclear forces: (1) the non-relativistic theory with finite size nucleons; (2) the relativistic theory. Meson pair theories have not been considered here. The relativistic theory without "cutting-off" failed. With the use of the λ-limiting it also failed, because the magnetic moment of the proton becomes less than unity and the neutron moment becomes positive. In order to explain the quadrupole moment of the deuteron relativistic terms were required, which even in the mixed theory become infinite, of order r^{-3} [37]. It appears that, at the present time, the more restricted first approach—i.e., disregarding the demand of relativistic invariance—is more hopeful. In this theory, neither the weak nor the strong coupling approximation yields correct values for the magnetic moments of the nucleons, but intermediate coupling might give the correct results.

11.5 Meson physics in Britain and Ireland (1941–45)

Since 1937, meson physicists in Britain had played a central role, as we reported in chapter 8. The leaders in this were German immigrants (Kemmer, Heitler, Fröhlich, and Peierls) and the Indian physicist Bhabha. The native experts on nuclear and cosmic-ray physics, such as Blackett, on the whole, showed a skeptical reserve. At the outbreak of the European war, Bhabha returned to India. Some of the other meson theorists were interned as enemy aliens. Thus Heitler and Fröhlich spent time in three internment camps: in South Devon, in the Midlands and finally on the Isle of Man. "We determined not to do research on our special fields", Heitler recalled later, but he and Fröhlich gave lectures on quantum mechanics to physics students and attended the lectures of other scholars among the internees. "In autumn

W. Heitler (1904–1981) (left) and L. Rosenfeld (1904–1974) in 1934. Photograph reproduced by permission of AIP Emilio Segrè Visual Archives (Segrè Collection).

of 1940, Herbert and I were back in Bristol", Heitler continued: "During the winter we had 12 heavy air raids on Bristol, burning down the centre and other districts of the city. In the spring [of 1941] I received a call to the newly founded Dublin Institute for Advanced Studies. The formalities of moving took a lot of time, hence there was little opportunity for joint work." [38]

Subsequently, Fröhlich changed his field to solid state physics. Peierls, on the other hand, participated fully in the British nuclear programme, playing a major role in the early considerations of nuclear weapons. Thus meson theory in Britain was put on hold for the duration of the war.

On the other hand, the topic was taken up at the new institute in Dublin. Erwin Schrödinger, who had been dismissed from his chair in Graz in the spring of 1938, after the Austrian *Anschluß*, went first to Oxford and then to the University of Ghent, Belgium. In October 1939, he arrived at the Dublin Institute to assume the post of director of theoretical physics. He worked mainly on general relativity and unified field theory, but also showed some interest in meson theory.

In the autumn of 1941, he offered an associate professorship to Heitler, who accepted, remaining in Dublin until 1949, when he replaced Wentzel in Zürich. (Wentzel had by then accepted a call to the University of Chicago.) During the first part of his stay in Dublin, Heitler developed a theory referred to as "radiation damping" (Heitler 1941; Heitler and Peng 1942b). Pauli described it later in this way [39]:

Heitler has given a correspondence scheme... by which he can eliminate in a Lorentz invariant manner the divergences occurring in the treatment of scattering processes. This scheme consists [of] adding a new rule to the existing formalism of quantum mechanics. He hopes to obtain thus an approximate theory which would have the same relationship to a future quantum mechanics that Bohr's quantization of classical orbits had to quantum mechanics.

The Chinese student H.W. Peng (who later studied for his doctorate with Max Born in Edinburgh), worked with Heitler on the production of mesons in proton–proton collisions (Heitler and Peng 1943, 1945). With this and his other research, Heitler helped to establish the reputation of the Dublin Institute during and after the war.

11.6 Pauli's evaluation of the work on meson theory in the West during the Second World War (1945)

On 11 October 1945, Pauli wrote a letter from Princeton (in English) to Hendrick Casimir in Holland, as follows [40]:

Dear Casimir!
I was very glad to hear from you a bit more as I read your letter of September 12, which arrived today.
   It seems to me that the physicists in Holland are much overestimating the content *auf* American and English journals from 1940–45. They will soon realize that there is not much interesting in it with a few exceptions. For me it is much more interesting to read your Dutch *"Physica"* of [these] years than the American and English journals. It is different, however, with the neutral countries: Sweden, Ireland, Switzerland and also in Russian quite interesting work has been done (Alichanow's cosmic-ray expedition in the Caucasus, Landau's papers on liquid helium, Iwanenko and Sokolow on meson-theory) [41]. Here much work has been done on technical problems (microwaves, Radar) and on heavy nuclei. The latter is still kept secret (except the so-called "Smyth-report" which has been published as a book [Smyth 1945]). Fortunately I was not involved at all in this secret work and was sitting quietly and lonely in Princeton working on meson-theory (and also on Dirac's crazy quantum electrodynamic, in which I don't believe anymore [Dirac 1942]). Oppenheimer stimulated my interest in this field [in] 1941 and I have published a number of papers about it without solving definitely any problem. At present the most interesting facts are Amaldi's experiments (*Naturwissenschaften* 1942; now papers of him and pupils are in print in *Physical Review*) on neutron–proton scattering for 12–14 MeV. They find a pronounced forward scattering with which the present theories are difficult to reconcile (see Hulthén's papers in the Swedish-Arkiv) [42]. Wentzel in Zürich tries to explain this result with his "strong coupling theory" (according to which stable isobars, that means excited states of the proton with higher charges and spin, should exist). I worked in this direction too, but I have great doubt whether the theory is correct. Mimeographed notes of 6 lectures on meson-theory which I held one year ago at the M.I.T. in Cambridge, Mass. will be available soon and I shall send a copy to you or to Rosenfeld [Pauli 1946]. You will find many quotations in it, also of related papers of Heitler and of Heisenberg [Heitler and Peng 1942a, b;

Heisenberg 1943a, b]. The problem of the nature of the nuclear forces is still as it was in 1940 and I guess that the secret material on "term-zoology" of heavy nuclei (or whatever else it is) will be of no help in this respect when it will be published. It is just as unsatisfactory to invent arbitrarily a new field for every meson which is or will be discovered with an arbitrary interaction with heavy particles as to invent any arbitrary interaction between heavy particles themselves. And it is still more arbitrary to introduce a finite shape of the proton to make the theory convergent (as it is done in the strong coupling theory of Wentzel mentioned above). I think that experiments with high energy protons and neutrons could help quite a lot in finding out the laws which the nuclear forces obey.

For applications of known principles, which need much money and industrial facilities (as for instance the production of "plutonium" and the separation of the uranium isotopes), this country is ideal. But nothing really original was made in this country, neither in experimental nor in theoretical physics. For the work during the wartime the same holds for England, also; but I am interested how it will go on in both countries.

It will be very wise, therefore, if you will prepare yourself and the Dutch physicists for the great disappointment which they will have as soon as the American and British journals will arrive. But they will also have some encouragement: they will quickly find out that their own physics is less deteriorated than is the case for the American and English physicists and that it will *not* be so hard for them "to catch up again" than they think: a few weeks will be sufficient to you and others to learn everything of scientific interest which happened during these "lost years".

I am sending you today a package with reprints, please divide them among persons who are interested. There is a paper of Onsager included (I had several copies of it) which I think is a masterpiece of mathematical analysis [Onsager 1944]. It contains the rigorous solution of the Kramers–Wannier order–disorder problem for the two-dimensional model [Kramers and Wannier 1941](unfortunately the method cannot be generalized for three dimensional crystals).

I am greatly indebted to this Institute for the asylum it gave me during the years 1940 until now. Recently they even offered me a permanent professorship at the Institute (it is in some respect the succession of Einstein, who is now in the retiring age). On the other hand, I am still Professor in Zürich, where my position was kept open for me. Therefore I feel to have some obligation toward my colleagues in Zürich and I decided first to make a trip to Zürich as soon as possible and to consider the farther future later. My decision to leave Zürich in 1940 was greatly influenced by the circumstance that I was not a Swiss citizen. This same circumstance makes it difficult again for me to obtain a passport at present but nevertheless I hope to make the trip in a few months (may be already end of December or January). So we shall see on what side of the Atlantic we shall have our "chat on many subjects".

I know how bad the material situation in Europe is, and it is true that the material side of life is very well and undisturbed here. I cannot say the same about the spiritual situation. I wonder how the spiritual side of life will develop in Europe. Are people there very nationalist? For me, of course, it is not possible to consider myself as belonging to a single country (that would contradict the whole course of my life). I feel, however, that I am European. This concept, again, is not recognized in Europe, which makes the situation rather complicated for me.

The past years have been rather lonesome, particularly '42 and '43. Last year I saw Uhlenbeck regularly at the M.I.T. during my course there. I am often thinking on Kramers' old statement about me, that "my heart is better than my

mind". How is he? What is he working about? Rosenfeld wrote me that Kramers was for a short time in Switzerland.
With best regards to you and your wife from both of us
Yours W. Pauli

P.S. In case that Dutch physicists should still think that the American and British journals of the war years are still interesting after the arrival of the journals — please let me know.

Notes to text

[1] For the Manhattan project, see: Smyth (1945), Groves (1962), Smith (1965), Rhodes (1986). For the German Uranium Project, see: Irving (1967), Rechenberg (1988b), Walker (1989) and Powers (1993).

[2] Kusaka (1915–1947) came from Osaka with his parents to Canada at the age of five. He studied physics at the University of Vancouver, at MIT and at Berkeley. As an enemy alien, under threat of deportation, he assumed a teaching position at Smith College; later, however, he was allowed to do military research at the Aberdeen Proving Ground in Maryland. He was appointed to an assistant professorship at Princeton after the war, but drowned in the sea at Beach Haven, New York in August 1947.

[3] Kusaka to Yukawa, 30 August 1941 [YHAL].

[4] Sakata (1941a,b) and Nelson (1941).

[5] As Pauli wrote to Frank Aydelotte on 29 May 1940: "Actually I suppose I am after German law 75 per cent Jewish. This would mean that in the case of a German occupation I would be really menaced and treated as a Jew." (Quoted in Meyenn (1993, p XXVIII).) This article gives a good concise account of the emigrations. See also Jackmann and Borden (1983) and Stuewer (1984).

[6] Pauli to Bhabha, 12 April 1940. [WPSC3], p 29. Original emphasis.

[7] Examples of applications of Rarita-Schwinger formalism are Kusaka (1941) and Brown and Telegdi (1958).

[8] The predictions were given credence, in spite of the knowledge that the theory was not suitable for high energies.

[9] "These results are very regrettable since only the symmetric theory gives a natural explanation of the β-decay and of the extra magnetic moments of neutron and proton". (Bethe 1940b, abstract, p 390).

[10] Pauli to Møller and Rosenfeld, 25 October 1939. [WPSC3], p 822.

[11] Pauli meant Møller and Rosenfeld (1939a, b).

[12] That is, the recent outbreak of the European war.

[13] In 1948, Pauli wrote to Oppenheimer that "...in 1941 you pushed me in a direction that was then entirely new for me (meson theory)". Pauli to Oppenheimer, 6 January 1948. [WPSC3], p 493. A similar remark is made in Pauli to Casimir, 11 October 1945, [WPSC3], pp 320–3, which we quote in full below.

[14] Marshak (1940, abstract).

[15] Miyazima (1941, p 173).

[16] For another discussion of pair theories, see Mukherji (1974, pp 58–64).

[17] The method, determining the field energy as the exact eigenvalue of a quadratic form, is credited in Wentzel (1942) to Wigner *et al.* (1939).

[18] Wentzel (1941b, abstract).

[19] Wentzel (1942, p 126) (original emphasis).

[20] Pauli to Wentzel, 30 December 1941. In this letter he also says: "Schwinger wants to revive the Rosenfeld–Møller in a modified form and with stronger

coupling [Schwinger 1942], but no one believes him".

[21] Joseph Maria Jauch, Pauli's assistant in Zürich.
[22] Weinberg (1941, abstract). Weinberg's footnote 6 explains that the smaller prediction arose from the small coupling constant obtained by "cutting off" the highly divergent nuclear interaction at a small distance. This result is reminiscent of the attempt to let the small Fermi β-decay coupling constant account for the strong nuclear force in the original Fermi-field theory (see chapter 3), by taking advantage of the divergent small-distance-dependence of the force. In Miyazima (1941), in contrast, it is claimed that the small-scattering cross-section of Marshak and Weisskopf is due to their omitting a factor of $(2\pi)^{-6}$ in calculating the two-nucleon potential, which resulted in the coupling G being too small by $(2\pi)^6$.
[23] Wentzel said that this model showed "a certain (limited) resemblance to the model proposed by Fermi and Yang" (Wentzel 1950, p 711). The model referred to is that of Fermi and Yang (1949), a predecessor of the quark model.
[24] Realizing that this also might not be a good approximation, Wentzel said: "It seems worthwhile to consider also the opposite case [to weak], that of strong coupling, all the more so that it could yield the possibility of interpolating into the intermediate region of medium coupling strength." (Wentzel 1940, p 270)
[25] Wentzel (1947, p 7, footnote 24).
[26] Wentzel (1940, p 276).
[27] Wentzel (1940, p 290).
[28] Bhabha (1940c, p 101). Bhabha claimed, however, to have derived his isobaric states on the basis of classical meson theory (Bhabha 1940b).
[29] Tomonaga (1941, p 247, abstract).
[30] Miyazima and Tomonaga (1943, p 278).
[31] Oppenheimer became director of the Los Alamos Laboratory at the beginning of 1943. Schwinger worked on the Manhattan Project in Chicago at the beginning of 1943 and then moved later in the year to the Radiation Laboratory at MIT.
[32] Oppenheimer and Schwinger (1941, p 151).
[33] Oppenheimer and Schwinger (1941, p 151).
[34] Pauli and Dancoff (1942, p 87).
[35] Serber and Dancoff (1943, abstract). See also Dancoff and Serber (1942) and Dancoff (1939b).
[36] Pauli and Kusaka (1943, p 415). [Our emphasis added.]
[37] That was shown by Pauli's associate Ning Hu (1945).
[38] Heitler 1973. p 424.
[39] Pauli (1946, chapter 7, second edition, p 41)
[40] Pauli to Casimir, 11 October 1945.[WPSC3], pp 320–2. We quote this letter in full.
[41] Alichanow and Alichanian (1945); Landau (1941, 1944) and Iwanenko and Sokolow (1940, 1942).
[42] Amaldi *et al.* (1942); Ageno *et al.* (1947); Hulthén (1943a, b, 1944a, b). The results of Amaldi *et al.* were later shown to be erroneous.

Chapter 12

Meson Physics During the War (East)

12.1 The Soviet Union and Germany (1940–43)

In his letter to Casimir in October 1945, summarizing the progress in fundamental physics during the war, Pauli referred to "quite interesting work" done in the Soviet Union. Although Stalin and Hitler had signed a "non-aggression pact" just before the outbreak of the war, and the Soviet Union had moved into the Baltic states (Estonia, Latvia and Lithuania) and half of Poland, within the Soviet's prewar boundaries there was an uneasy peace, which ended with the German invasion of June 1941. During this troubled interlude, however, scientists engaged in their normal scientific pursuits.

For example, a nuclear physics conference, sponsored by the Physico-Mathematical Section of the Academy of Sciences of the USSR, was held from 20–24 November 1940 in Moscow (Lifschitz 1941). Reports at this meeting covered a wide range of topics, but were mainly concerned with two major themes: first, cosmic rays, mesons and nuclear forces, and second, fission of heavy nuclei. In the first category, Igor Tamm talked on "The theory of the mesotron and nuclear forces"; Lev Landau on "The radius of elementary particles"; Isaak Pomeranchuk on "The production of meson pairs by positron annihilation", also on "The scattering of mesotrons by mesotrons"; Dmitri Iwanenko and A. Sokolov on "Scattering of mesotrons by neutrons and protons according to Proca's theory"; Landau and J. Smorodinski on "Radiation effects of particles with a spin 1"; V. Veksler and N. Dobrotin on "Secondary mesotrons"; and Landau on "The cascade theory of showers". This work, on the same level as Western work on similar topics, often contained original ideas.

On the second major theme, fission, experts delivered speeches such as I. Kurchatov, "The fission of heavy nuclei"; G. Flerov and K.Petrzhak, "Spontaneous fission of uranium"; V. Berestetski and A. Migdal, "The mechanism of the fission of heavy nuclei"; etc. Although at this time these subjects were being investigated in secret nuclear projects, this Moscow conference still echoed the old pre-war tradition of open scientific research. Only after a year had passed , with the Soviet Union fighting for its survival, did the situation change drastically. As Evgeni Feinberg recalled [1]:

L. Landau (1908–1968) (left) and R.E. Marshak (1916–1992) in 1956. Photograph reproduced by permission of AIP Emilio Segrè Visual Archives (Marshak Collection).

The Soviet nuclear physicist G.N. Flerov suspected at the end of 1941 that there existed a uranium project in the USA, because all or nearly all specialists of nuclear physics completely ceased publishing in American scientific journals. Flerov promptly informed the presidential board of the Academy of Sciences about his observation, and his intervention played an important role in re-assuming our research work in this most difficult period of war for our country.

The secret nuclear energy and bomb project in the Soviet Union came into being a full year later (in 1943), after Hitler's armies had been beaten back. Feinberg made another interesting observation about the nuclear projects:

In June 1943 Heisenberg, as an editor, signed the preface to a book collecting scientific essays on cosmic rays. This collective book, dedicated in honour of Sommerfeld's seventy-fifth birthday, treated problems unrelated to uranium and represented very valuable scientific material. Of the 15 chapters, 12 were written by leading people in the "Uranium Project": Heisenberg contributed five, Weizsäcker two, Flügge two, and Wirtz, Bagge and Bopp each one. In the same year, Heisenberg published two papers which laid the foundation for a totally new direction in the topic of fundamental quantum theory of fields and particles. They exhibited *not the slightest reference to practical application*, not to speak of reactors or bombs.

We shall summarize the German research on cosmic rays and Heisenberg's new approach to which Feinberg referred in section 2, the Japanese wartime work in section 3; Section 4 deals with Italian physics and Section 5 with the transition from war to peace.

12.2 Cosmic-ray physics in Germany and the theory of the S-matrix (1941–44)

Unlike Enrico Fermi, who had worked on the slow neutron bombardment of nuclei since 1934, Heisenberg had not been involved in the physics using neutrons when he was called at the end of September 1939 to join the secret Uranium Project of the Army Weapons Bureau. Still, nearly simultaneously with Fermi working in the United States, Heisenberg developed a theory of the nuclear chain reactor and began to prepare experiments with the aim of large-scale production of energy. In the spring of 1942, Heisenberg and his Leipzig colleague Robert Döpel even had a slight lead in the race, but it was Fermi who by 2 December 1942 achieved the first chain reaction in a pile made of natural uranium, with graphite as moderator. The German uranium-heavy water reactor, on the other hand, was not even critical by the end of the war in the spring of 1945, due to lack of adequate materials.

On 1 July 1942 — the German Uranium Project having passed from military to civilian supervision — Heisenberg took over the direction of the Kaiser Wilhelm-Institut für Physik (KWI) in Berlin; the nominal director, the Dutch physicist Peter Debye, who had taken a leave of absence at the Chemistry Department of Cornell University in January 1940, did not return. In Berlin, a strong group of nuclear physicists were working on experiments related to the chain reactor: besides Weizsäcker, there were Karl Wirtz, Fritz Bopp, and Horst Korsching. Since autumn 1940, Heisenberg had been a consultant of this Berlin group, and in summer 1941 he started a colloquium at the Institute dealing with current topics, such as proteins and other macro-molecules [2]. The colloquium on cosmic rays, on the other hand, began in the winter term 1940-41 and continued in the years that followed. Speakers came from other institutions in Berlin and also from outside [3,4].

The cosmic-ray lectures were published as a book, edited by Heisenberg and dedicated to Sommerfeld on his 75th birthday (Heisenberg 1943a). The preface declared:

> Research on cosmic radiation has been hit especially badly by the misfortunes of our time. On the one hand, it must take its place in most laboratories behind that of other fields. On the other hand, information about results obtained in foreign countries is hindered by the lack of transfer of news. Finally, no detailed reviews have been published in Germany since the beginning of the war, because the physicist engaged at the war front lacks time for extended research of this type. Owing to the fundamental importance of this branch of physics, however, it seems justified to collect a series of colloquium talks giving an overview of the present status of cosmic-ray studies and to publishing them as a book.

Heisenberg emphasized that the talks mainly reviewed the existing literature and that the American work was available only up to summer 1941. Only a little original research was included: mentioned in particular by the editor were Gerhard Molière's investigation of large air showers, Flügge's on the neutron distribution in the atmosphere, and Heisenberg's own simplified theory of cascade showers, which he had presented earlier in 1939 in lectures in Leipzig and in Lafayette, Indiana. He considered this book to be a useful preparation for later serious study of the topic at the KWI für Physik and to keep up interest in the field in Germany and abroad [5].

In the summer of 1942, Heisenberg began to write down some original ideas on a fundamental aspect of elementary particle physics, ideas he had entertained for some years. However, it was a visit from his Italian friend Gian Carlo Wick in June 1942 that provided him with an opportunity to discuss his theory and encouraged him to proceed to publication. In September of the same year he submitted a first paper on it, dedicated to Hans Geiger on his sixtieth birthday (Heisenberg 1943c) [6].

The new ideas arose during unsuccessful attempts during the 1930s to arrive at a satisfactory description of elementary particle interactions. All the schemes then considered — whether quantum electrodynamics, the Fermi-field theory, or Yukawa's meson theory — exhibited fundamental difficulties, indicated by the appearance of infinities in predictions that could be removed only by artificial cut-off procedures [7]. In considering radical means to remove these difficulties, Heisenberg had made (in 1938 and 1939) two proposals, which he regarded as connected: first, the existence of a fundamental length l_0, such that the known relativistic quantum field theory broke down at distances smaller than l_0; second, the presence of nonlinear interactions between elementary particles (see chapter 10). In 1942, he suggested a third approach, which he again did not regard as inconsistent with the two preceding proposals. "The present paper", he stated in the introduction, "attempts to isolate from the conceptual scheme of the quantum theory of wavefields those concepts that probably will not be hit by future alteration and which may therefore represent a constituent of the future theory." [8]

To select the features that would probably be retained in a future theory, Heisenberg used a method that had served him previously with great success in his invention of quantum mechanics in 1925. Now he asked, in the new situation of 1942, what are the *observable quantities* in the present theory of elementary particles? As the answer he found: (i) the discrete energy eigenvalues of closed stationary systems, and (ii) the asymptotic behaviour of wavefunctions in scattering, emission and absorption processes. Mathematically, all these quantities should be described by a "scattering" or "characteristic" matrix, the unitary S-matrix, which he wrote as

$$\mathbf{S} = \exp(i\eta),$$

introducing a Hermitean matrix η that would replace the defective Hamiltonian function in the quantum theory of wavefields.

Obviously, the task was then to calculate the relativistic η matrix. In October 1942, Heisenberg submitted a second paper discussing the "observable quantities", giving several examples of theories (Heisenberg 1943d):

(i)   a δ-function-like interaction, leading only to scattering, with no bound state of the particles;
(ii) a distance-dependent interaction resulting in finite cross-sections for the scattering of particles with arbitrarily high energies;
(iii) an interaction implying the creation of new particles.

The last two examples could be not be adequately described by the then available quantum theory and they possessed no analogues in classical theory, which might serve as the correspondence limits of a future quantum theory. Thus the S-matrix appeared to provide for the moment the only tool that could be used to treat the processes which had been observed in the high-energy cosmic rays (including the "explosive showers" which had so fascinated Heisenberg).

On a lecturing tour in Switzerland in November 1942, Heisenberg propagated his S-matrix theory, so that Wentzel in Zürich and André Mercier in Bern got to know of it very soon. Ernst Stueckelberg regretted very much that his doctor did not allow him to leave Geneva to hear Heisenberg and that he "therefore missed the very interesting talk" [9]. However, he studied the published papers very carefully and related the S-matrix to his own new quantum theoretical formulation of a point electron with finite self-energy [10].

On another lecture tour abroad to Holland in October 1943, Heisenberg received further approval for his theory when he met Hendrik Kramers in Leiden, Ralph Kronig in Delft and Léon Rosenfeld in Utrecht. Kramers suggested that the S-matrix should be considered as an *analytical function* of the incoming and outgoing momentum variables k_i' and k_i'', as Heisenberg explained in a third paper, adding [11]

> The zeros of the S-matrix for imaginary k_i' provide the positions of the stationary states. For the eigenvalues of η this result means that the poles of η lying on the imaginary k-axis determine the position of the stationary states.

Heisenberg had intended to publish a more detailed paper together with his old friend Kramers, but he confined his letter to correcting and improving various parts of his preliminary manuscript. "I can only rejoice", Kramers replied, "that my remark on the analytic character of **S** has borne fruit in your consideration, and if you write this in a manuscript, my role in this is also described." [12]

Although Kramers found the means of communication too slow for effective collaboration, Christian Møller of Copenhagen, especially after Heisenberg visited in January and April 1944 [13], began to work actively on S-matrix theory, and exchanged letters with its inventor in Berlin [14]. After the war, he published a series of papers containing his results (Møller 1945, 1946, 1947) and became one of the leading experts in the field in the years to come [15].

12.3 Japanese meson physics after Pearl Harbor (1941–45)

On 7 December 1941 at dawn, Japanese planes bombed the American naval base at Pearl Harbor, thus hurtling Japan and the United States into the Second World War. By the same act, Japanese science was placed behind a wall of isolation, which was partly removed with the surrender of the Japanese state in August 1945. Full communication and freedom of research for nuclear and elementary particle physicists was restored only with the signing of the peace treaty in 1952 [16]. The Japanese physicists, especially the younger ones, had been inevitably drawn into war-related activities during the war, but they still found it possible to carry out pure physics research — which was not generally the case in the other warring countries. A good example of this may be seen in the work of Tomonaga and Miyazima on intermediate coupling meson theory, as discussed above [17].

Another example is that regular semi-annual meetings were held by Riken (the Japanese acronym for the Institute of Physical and Chemical Research, based in Tokyo). After the general meetings (and at other times), an informal group who called themselves Meiso Kai or the Meson Club, held one- or two-day meetings. These were summarized in Yukawa's laboratory diary, as described by Satio Hayakawa [18]:

> The first two meetings, those on 12 June and 13 December, 1941, were called "theory meetings". The meetings held on 24 April and 13 June, 1942, have the name "illusion meetings", which is a pun on the Japanese word for meson. Those on 12 December, 1942, on 19 June, and on 26 and 27 September, 1943 were named "meson meeting", "informal meeting on mesons" and "discussion meeting on mesons", respectively. The last one was a two-day meeting and its proceedings were published in mimeographed form. The final one, on 18 and 19 November, 1944, was sponsored by the Science Research Council and was called the "elementary particle theory meeting". Each of the first six meetings was attended by about 20 physicists, whereas about 50 attended each of the last two, which were of a more formal character.

By 1941, when it had become clear that there were various problems confronting the meson theory regarding its role both in strong and weak nuclear forces and in cosmic-ray phenomena, including the mesotron decay, Yukawa sent a note to a German journal (received January 1942) about these difficulties and their possible resolutions, entitled (in English translation) "Remarks on the nature of the mesotrons". In the introduction, he pointed out that there were various versions of meson theory that had been proposed, and thus [19]:

> The following questions can be posed to choose among all the theoretical possibilities:
> 1. properties of the nuclear forces,
> 2. anomalous magnetic dipole moments of the nucleons,
> 3. the electric quadrupole moment of the deuteron,
> 4. regularities in the β-decay,
> 5. the mesotron decay,
> 6. interactions of comic-ray mesotrons with the nucleons,

7. interactions of these with photons,
8. the role of neutral mesotrons.

Evidently, items 1-4 relate to the nuclear forces, while items 5-7 are closely connected to the cosmic rays, including decay of the free mesotrons (since those particles of short lifetime were to be found only in the cosmic rays). Item 8, on neutral mesotrons, concerned the question: why are neutral mesons (mesotrons), which are required by the charge-independence of nuclear forces, not present in the cosmic rays? Yukawa's reply was that calculation showed that their lifetime for spontaneous decay is much shorter than that of the charged mesons, and so they might well escape observation [20].

We now turn our attention to items 5-7 and to the suggestions for dealing with them that were made at the meetings of the Meson Club. One solution for the puzzling discrepancies, offered at the meeting of June 1942 and discussed at all subsequent meetings, was that the nuclear-force meson and the cosmic-ray meson were different particles, but related to one another. (Another solution, of course, was Tomonaga's intermediate coupling scheme, discussed earlier [21].) The evidence from cosmic rays was thought to rule out spins higher than 0 or $\frac{1}{2}$ for the cosmic-ray meson, because it was believed that high-energy charged particles with higher spin would produce large cascade showers (bursts) — and this effect was not observed (Christy and Kusaka 1941a,b; Oppenheimer 1941) [22]. The nuclear-force (Yukawa) meson was taken to be a boson of either spin 0 or spin 1, although by this time (in 1942) the pseudoscalar meson of spin 0 was preferred because it gave a better fit to the properties of nuclear forces.

In his diary No. IV of the year 1941-42, Yukawa constructed a table indicating how well the various meson theories fitted the experimental data. We reproduce it here in table 2 [23].

A little later a fundamentally new idea arose in the discussions in Japan, which changed the above picture considerably. Yasutaka Tanikawa of Nagoya University was the first to suggest that two different mesons might explain the anomalies, apparently influenced by the two-meson mixture that Møller and Rosenfeld (1939a) had proposed, mainly to cancel the divergence of the nuclear force at small distances, and not in order to explain cosmic-ray phenomena. Tanikawa discussed his suggestion with Sakata in the course of an outing and picnic of the Yukawa group at Nara on 6 May 1942 [24]. He developed his theory with both mesons as bosons (as in the mixture theory), whereas Sakata, working with Takesi Inoue, preferred the cosmic-ray meson to be a fermion with spin $\frac{1}{2}$, which turned out later to be an astute choice.

Sakata and Inoue (1942, 1946), using the notation Y for the Yukawa meson and N, P and E, for neutron, proton and electron, respectively, wrote of the cosmic-ray meson [25]:

In our theory, it is assumed that the meson is a Fermi particle with spin $\hbar/2$ and furthermore ... the following interactions are introduced:

$$m^{\pm} \rightarrow n + Y^{\pm}, \text{ etc.,}$$

Table 2

Data	Theory			
	Vector	Pseudoscalar	Mixed	Spinor
Nuclear forces	Good?	Good	Good	Good
Deuteron	Good?	Good	Good	Good
Quadrupole meson scattering	Very good (longitudinal)	? (Heitler–Ma)	? (Heitler–Ma)	?
β-decay	Gamow–Teller?	Gamow–Teller?	Good	?
Meson-decay	Good	Good	Good	?
Burst	Good?	Good	?	Good
Authors	Bethe and Kobayasi	Oppenheimer	Møller and Rosenfeld	Marshak

($m^{\pm}$: *negatively and positively* charged meson, n: neutral meson which is assumed in the following discussions to have a negligible mass and consequently may be considered equivalent to the neutrino).

They also considered the possible existence of a neutral Yukawa meson Y^0, as demanded by charge-independent nuclear forces, and the possibility that $m^{\pm}$ and n can interact with it. (All these interactions of a fermion and a boson can be considered as "Yukawa interactions", characterized by dimensionless strengths of the type $g^2/(\hbar c)$.) For the Y–nucleon interaction Sakata and Inoue assumed a strength $\approx 10^{-1}$; for the coupling strength of Y to m, they assumed about 10^{-2}. As in the original Yukawa theory, the authors took the decay Y to e + n to have a strength of about 10^{-15}, saying that with these choices "it is possible to account for the phenomena in atomic nuclei and cosmic rays consistently, without aiming to touch the inherent difficulties of field theory".

Next, Sakata and Inoue considered the masses of the mesons, noting that the value deduced from nuclear forces was larger than that of the cosmic-ray meson (they estimated by a factor of about two). Together with the interactions that they assumed, this meant that Y would decay spontaneously into m, its rate depending, of course, on the assumed mass ratio. For the masses approaching equality, Y would decay predominately into e +ν. "Consequently, the lifetime of the Yukawa particle does not become greater than 10^{-8} s." [26]

The authors further assumed that $m^{\pm}$ would decay into $e^{\pm} + \nu +$ n, via intermediate states containing Y. For the latter, they considered both vector and pseudoscalar cases, and, instead of predicting the lifetime of $m^{\pm}$, they used the measured cosmic-ray lifetime of $\approx 10^{-6}$ s, together with nuclear β-decay lifetimes, to fix the nuclear interaction constants. From these determinations, they concluded that in either case they could make the predicted scattering cross-sections of the cosmic-ray meson consistent with its small observed upper limit.

The "two-meson theory hiking" on 6 May 1942 in Nara. From left to right: Y. Tanikawa, ?, S. Sakata, ?, M. Kobayashi, ?, and H. Yukawa. Photograph reproduced by permission of Yukawa Hall Archival Library.

The alternative version put forward by Tanikawa (1943, 1947) did not lead to so detailed a discussion, but it was clear that with suitable choice of coupling constants, there was sufficient freedom to explain the properties of the distinct nuclear and cosmic-ray mesons in a broad qualitative sense. The specific example treated in the third two-meson theory (Marshak and Bethe 1947) did not involve a Yukawa meson at all: the authors chose spin $\frac{1}{2}$ for the strongly interacting meson, based upon an earlier pair theory of nuclear forces (Marshak 1940) [27].

Normal scientific communication between the warring parties was, of course, impossible during the war, and that was often true also for those countries with a common alignment–e.g., the Western allies and the Soviet Union after June 1941. Between Japan and Germany, only the vulnerable transport by way of submarines was possible. (Although only a part of the following material is directly relevant to the problem of fundamental nuclear forces, we include it here for its interest and because it is not generally accessible.) Between April 1942 and July 1944, an effort was made to establish systematic contact between Germany and Japan by submarine, mainly to exchange items of war matériel. However, in five attempts, only one Japanese ship succeeded, voyaging from Kure in Japan to the French port of Brest and back (1 June–25 December 1943). Others only made it one way, either from Germany to Japan or from Japan to France [28]. It appears that in at least two cases, the submarine connection was used by physicists. We know that by the end of 1943, Yukawa and others in Japan had available a mimeographed copy of Heisenberg's second S-matrix paper [29]. Furthermore, Heisenberg sent a letter dated 18 January 1944, to Yoshio Nishina with Captain M. Nomaguchi of the Japanese Navy, a brother of the physicist Kanetake Ariyama, who had been a visitor in Heisenberg's Leipzig institute during the 1930s [30].

By whatever path Heisenberg's papers (1943c, d) arrived in Japan, they were distributed there among meson physicists, who studied them carefully. Tomonaga, especially, used the concept of the "characteristic matrix" in his work on ultra-short-wave circuits in 1944 [31]. In those days, also, Tomonaga and Yukawa were considering more general relativistic field theories and were thus open to new ideas, as we gather from entries in Yukawa's diaries [32]. Under the date 6 September 1944, in Diary VIII, Yukawa began to write down considerations, including formulae, on an entirely new topic for him: nuclear fission.

These notes included scattering cross-sections and references to experiments by Enrico Fermi and Herbert Anderson (involving 0.2 tons of U_3O_8) and by Siegfried Flügge in Germany on the cross-section of uranium for neutrons. Also there are neutron transport equations and their solutions, and finally an estimate for a uranium pile (3.5 tons of U_3O_8 and other values as well). After a few more days had passed, the uranium problem had practically disappeared from the diary. Had the desire to get a hand on nuclear energy finally reached the Far East?

Relatively little is known abroad about the Japanese nuclear-energy programme [33]. It started in April 1940 with an army official seeking the advice of Ryokichi Sagane of Tokyo Imperial University and continued through consultation with the navy, with the modest conclusion of a final meeting of March 1943 that "it would probably be difficult even for the United States to realize the application of atomic power during the war" [34]. Still Nishina began a uranium project (the Army-funded NT-project) at Riken, led by Tadashi Takeuchi, a cosmic-ray physicist. It ran from March 1943 until the end of the war, but it lacked a method to separate the uranium isotopes and suffered also from lack of uranium (after futile searches had been made in Japan, and in Korea and the other occupied countries). A second programme (the navy-funded F-project) was also begun in 1943. Professor Bunsaku Arakatsu of Kyoto Imperial University at a leisurely pace conducted this attempt to separate the uranium isotopes by an ultra-centrifuge method. His team involved 19 scientists, mostly junior people, and included Yukawa and Minoru Kobayashi.

The F-project produced several theoretical papers and "a stable sample of purely metallic uranium for the first time in Japan; it was about the size of a postage stamp, three centimetres on each side and about one millimetre thick" [35]. At the first and last general meeting of the Kyoto group, on 21

July 1945, "Yukawa gave a report surveying 'International Research on Nuclear Power', based on information from neutral countries, in which he reiterated the belief that no nation was capable of harnessing the atom to military uses in the immediate future" [36]. Two weeks later the Enola Gay dropped its uranium bomb on Hiroshima.

12.4 Nuclear and cosmic-ray physics in wartime Italy (1942–44)

In the year 1938, the potent Italian nuclear physics community underwent dramatic change, as recalled by one of Fermi's collaborators, Emilio Segrè [37]:

> The Rome group was in effect dissolved by uncontrollable forces. I had gone to the United States in July 1938 for a visit and, when I was dismissed from my Palermo post, settled in Berkeley. Fermi left Italy in December of 1938. In the summer of 1939 Rasetti emigrated to Canada, and Amaldi came to the United States to look for a job, although his wife was reluctant to leave Italy. He applied for a passport for his family, but without waiting for it he left Italy alone to explore American possibilities. Before he could find a suitable job, however, Germany invaded Poland, a passport for the family was refused, and he returned to Italy in October 1939, after having spent some time in Berkeley and with Fermi in Ann Arbor [Michigan] and Leonia, New Jersey.

If one adds that another member of Fermi's Rome group, Bruno Pontecorvo, who had in 1936 been awarded a grant to visit Frédéric Joliot in Paris, remained in France until the German occupation forced him to emigrate, first to the USA and then (in 1943) to Canada, one can speak of a nearly complete exodus of the whole Fermi school.

The reason for the flight at this time can be found in Italian political history. Since 1936, Fascist Italy had moved closer to Nazi Germany, partly as a result of the Italian invasion of Ethiopia; the two dictatorial powers had formed the "Axis" in September 1937. A year later, in October 1938, Italy adopted anti-Jewish laws: the professional activities of the Italian Jews were severely limited, and they were excluded from the civil service and the armed forces. Since Pontecorvo, Rasetti, Rossi, and Segrè were Jewish, as was Fermi's wife, they all left their homeland.

Of the senior people, only Gian Carlo Wick and Eduardo Amaldi remained in Rome—the latter mainly on his wife's account [38]. The experimental physicist Amaldi had the task of replacing the ingenious Fermi in taking over the nuclear physics programme, together with Professor Gilberto Bernardini of the University of Bologna, who carried out his cosmic-ray experiments in Rome [39]. In Segrè's opinion, Amaldi "succeeded admirably" in spite of wartime difficulties [40]. As the war drew to an end, physicists in Rome performed a remarkable series of experiments having important unexpected consequences.

The first experiment of this series was done in 1940 by Bernardini and a recent Rome graduate, Marcello Conversi (Bernardini and Conversi 1940).

Their aim was to study the deflection of cosmic-ray particles in magnetized iron and in this way to determine the relative number of particles of a given sign of the electric charge. Two iron blocks were arranged vertically, magnetized in opposite horizontal directions. This arrangement, which focused particles of one sign and defocused those of the opposite sign, was a "magnetic lens". It had been used earlier for the same purpose by Bruno Rossi (Rossi 1931). However, the new work used two lenses, arranged vertically with counters above, between and below each lens. The first lens selected and the second analysed the selected distribution. In this way a large separation (about 35%) was achieved [41].

We shall discuss the remaining experiments of this series in chapter 13 and we shall now consider the relationships of the Italian physicists with their colleagues abroad. At first they maintained contact with their friends who had emigrated and published papers in the *Physical Review*. However, after Pearl Harbor, Italy, like Germany, declared war on the USA, after which Amaldi and his students and associates published either in Italian or German journals. They also maintained some relationships with nuclear and cosmic-ray physicists in Germany. We have already mentioned the contacts between Wick and Heisenberg. In June 1942, Wick wrote to Heisenberg, "I have been invited to give a seminar talk in Munich. At this opportunity I want to visit you in Leipzig (or Berlin)...it would really be a great pleasure to spend a few days with you in your institute." Upon receiving this letter, Heisenberg "rejoiced" and invited him to give a talk on some topic concerning cosmic rays in Leipzig or Berlin [42].

Heisenberg and Wick continued to correspond over the next two years, and Heisenberg sent Wick proof sheets of his first two S-matrix articles, on which Wick commented in detail [43]. They also discussed Italian cosmic-ray experiments by G. Cocconi and collaborators and similar work by J. Juilfs in Berlin [44], as well as recent notes of Wick on neutron diffusion (which, of course, interested the Berlin group of the Uranium Project) [45]. Wick explained in detail the neutron–proton scattering experiments of Amaldi, which seemed to contradict theoretical calculations of Karl-Heinz Höcker, a student of von Weizsäcker. In autumn 1943, on a postcard to Heisenberg he wrote [46]:

> One cannot write much in these hard times...But in any case I would like to know whether you, your wife and children are well. At this moment I can look from my window on a wonderful sunset, and the city lies quiet and peaceful as ever; one would like to ask whether this is all just a dream. The future now is quite uncertain for me and for the other people in the institute.

Indeed, it was a quite uncertain period for Rome and the rest of Italy. The Allied invasion of Sicily in June 1943 led to the bombing of Rome, the incarceration of Mussolini, his replacement in the Fascist government by Marshall Badoglio and the signing of the Italian armistice on 8 September. However, the control of all of Italy north of Naples passed into the hands of German troops, who liberated Mussolini, while at the same time the new

Italian government declared war on Germany in October. Rome remained under German occupation until 4 June 1944, when the Allies took the capital. In the meantime, Rome was open to air raids, as were the German cities [47].

"You might ask whether I can really work now", wrote Wick to Heisenberg in April 1944, and went on:

> Well, this is not always the case; on some days we have too many bomb alarms and sometimes the situation is too tense. Furthermore, we hear so much sad news. But one must try to muddle through, as long as one succeeds in finding food. The other evils that can occur nearly depend only on chance and one cannot resist them.

This last letter closed with "God bless your family" and "cordial greetings from Amaldi".

12.5 From war to peace: the years 1945–47

Among the scientists of the former Axis powers, after losing the war, the Italian physicists enjoyed the smoothest transition into peacetime work. Their research in nuclear physics was not directly restricted; they were able to continue their wartime research and to transmit their results personally and in print to their colleagues in Western Europe and the United States, conditions of which German and Japanese physicists could only dream [48]. The latter were restricted in their research by orders of the Allied occupation authorities and much of their instrumentation was destroyed.

On 9 August 1945, the day the plutonium bomb was dropped on Nagasaki, three pressure gauges were also dropped by parachute; to each was attached a copy of a letter addressed to Professor Riyukichi Sagane in Tokyo. Written by Luis Alvarez, Philip Morrison and Robert Serber (but not signed), it stated that it was sent by "three of your former colleagues, during your stay in the United States". In particular, the letter read [49]:

> We are sending this as a personal message to urge that you use your influence as a reputable nuclear physicist, to convince the Japanese General Staff of the terrible consequences which will be suffered by your people if you continue in this war... Within the space of three weeks, we have proof-fired one bomb in the American desert, exploded one in Hiroshima, and fired the third this morning. We implore you to confirm these facts to your leaders and to do your utmost to stop the destruction and waste of life which can only result in the total annihilation of all your cities if continued. As scientists we deplore the use to which a beautiful discovery has been put, but we can assure you that unless Japan surrenders at once, this rain of atomic bombs will increase many fold in fury.

Sagane saw the letter only after the war was over.

Under the Allied (essentially American) occupation, which lasted from the Japanese surrender in August 1945 until the peace treaty, effective on 28 March 1952, experimental research in nuclear and related fields was prohibited. On 23 November 1945 all four cyclotrons in Japan were dismantled and thrown into the sea. The destruction of the cyclotron from Arakatsu's Kyoto laboratory was witnessed by Walter Michels, an American naval officer and scientist, who wrote [50]:

> I have just taken part in an act of vandalism which is paralleled only by those perpetrated during the 15th and 16th centuries, or possibly, by the book burnings which accompanied the establishment of the Nazi party in Germany...My shame today is not only in the deed just done, but also in the fact that my profession, which presented the world with atomic energy, has not been able to control that blind unreasoning fear which has come into our national thinking with the start of the atomic age.

Thus the Japanese nuclear physicists entered the post-war era with most of their cities destroyed and their equipment dismantled or in disarray. "After the war ended," recalled Satio Hayakawa, "Tomonaga was still in bad health and depressed by the poor economic conditions. He told me that it would be almost impossible to continue physics research, and that he would work in biology instead." [51] In spite of the hostile conditions, however, Tomonaga and a group of about a dozen brilliant young theorists under his direction carried out an ambitious research program in quantum field theory, extending his super-many-time formalism, which he had developed during the war, both to renormalized quantum electrodynamics and to meson theory (as we will later discuss).

In July 1946, the first issue of an important new English-language journal was published at Kyoto University with Yukawa as editor. It proposed to publish "mainly those papers from the fields of quantum mechanics, statistical mechanics and the theory of elementary particles". There was a paper shortage, so that initially articles were to be restricted to twenty pages, and the publication was to be monthly. This journal, The Progress of Theoretical Physics, soon acquired an excellent international reputation.

One of the first tasks of the new journal was to publish in English results developed during the war and available only in the Japanese language, including conference talks, papers of Tomonaga opening new paths in quantum field theory, and the two-meson theories of Tanikawa and of Sakata and Inoue [52]. The next two volumes, published in 1947 and 1948, also included material from the wartime research (including that of the Meson Club), as well as new theoretical research.

Experimental research was still in a precarious state. Takehito Takabayasi wrote [53]:

> Nearly two years after the end of World War II may be viewed as a period of preparation for the resumption of research in the theory of elementary particles, both in Japan and in other countries. Since epoch-making experiments had yet to be reported, it was natural that long-standing problems like the difficulty of quantum field theory came under attack. By that time various methods had been proposed outside Japan: cut-off methods, subtraction methods, the indefinite metric, the theory of the S-matrix. Japan was still somewhat isolated but Tomonaga and Sakata now turned to the divergence problem, as well as Yukawa, but each of them chose a different approach.

From left to right: S. Tomonaga (1906–1979), H. Yukawa (1907–1989) and S. Sakata (1911–1970) in the 1950s. Photograph from University of Tsukuba, Tomonaga Memorial Room and reproduced by permission of AIP Emilio Segrè Visual Archives.

The Sakata group analysed the electrodynamics of Bopp (1940), which originated from the Born-Infeld theory. They constructed a kind of field mixture theory, in order to cancel the divergence of electron self-energy. That was the outgrowth of an idea of Stueckelberg (1939b). The Tomonaga group, applying the super-many-time formalism, were investigating the problem of field reaction. Starting from a critical analysis of the subtraction methods advanced by Dirac and Heitler, and through the detailed examination of the C-meson theory [of Sakata], they were proceeding toward the "self-consistent subtraction method". Yukawa, on the other hand, moved in the direction of giving shape to the idea of *maru* [circle], and intended to eliminate the divergences by generalizing the field theory in such a way as to attribute finite extensions to elementary particles.

In working through these ideas, and we may also include Born's reciprocity principle of 1938 (considered by Yukawa), the Japanese theoreticians soon succeeded in reaching the front line of research. When Yukawa accepted an invitation to go to Princeton's Institute for Advanced Study in October 1948, he opened a new era of international collaboration. Tomonaga went there the following year, and many students followed. Some, like Yoichiro Nambu, Toichiro Kinoshita and Susumu Okubo, stayed, prospered, and contributed to the scientific reputation of the United States (see Kaneseki (1974)).

Post-war conditions in Germany resembled, of course, the situation in Japan — cities destroyed and shortage of food. In addition, the most Eastern part of the country was given to Poland and its population expelled as refugees to the West, while the rest of the country was divided into four occupation zones: American, British, French and Soviet. Insofar as the physicists were concerned, as in Japan they were excluded from nuclear physics and related fields.

Through the emigration of the 1930s, the number of nuclear and cosmic-ray physicists had been much reduced, and during the war many of those remaining had been involved in the Uranium Project [54]. Beginning at the end of April 1945, the American ALSOS Mission seized nine of the leading participants in that project and interned them, as well as Max von Laue, in a country house in England, Farm Hall, from July 1945 until January 1946 [55]. After being questioned there, they were released, brought into the British occupation zone and allowed to resume scientific work under the restriction that they were to avoid any applied nuclear physics. Although the Western occupation authorities gradually softened this aspect of Allied Regulation No. 25 after 1950, the restrictions remained in principle until much later and experimental research suffered from this as well as from a shortage of funds.

In his first letter to Niels Bohr after being released, Heisenberg reported on the conditions in Göttingen where he could resume work (together with other members of his former Berlin institute) [56]:

> The plan of the British and American authorities is that I should rebuild the Kaiser Wilhelm Institute for Physics in the AVA (Aerodynamische Versuchsanstalt). We shall not be allowed to return to Berlin; the institute building there has been occupied by American troops, after having lost earlier all its apparatus and equipment through the Russian troops. Luckily the larger part of the apparatus had been brought during the war to Hechingen in Württemberg; there we had established a small improvised institute, where I liked to work and we also carried out experiments on the use of atomic energy for machines. It is a pity that we are not permitted also to return there. In spite of the friendly assistance of the British authorities we shall have some difficulties in Göttingen, since the French authorities will certainly wish to keep the institute in Hechingen.

Heisenberg then described his theoretical work of the previous months (on the theory of turbulence, with von Weizsäcker, on superconductivity and on S-matrix theory, with Richard Becker) and inquired: "Has Møller achieved any progress in these questions?" A little later, he queried Møller himself, who promptly replied: "I have written two papers on the subject–as soon as possible I shall send you reprints (at the moment it is not allowed to send reprints)." [57] Møller further stated that he had lectured in England in February 1946 on Heisenberg's theory and "his own modest contributions" to it, and "they were very much interested in this matter in England and at the conference in Cambridge at the end of July the problem of how to determine the S-matrix will be one of the points of discussion".

This positive news pleased Heisenberg greatly. Møller, who had worked out the detailed formalism of the S-matrix in his two papers, was the appropriate spokesman for this approach at the first international conference after the war, held in Cambridge, England, on 22–27 July 1946 [58]. German scientists were not invited. Besides Møller's talk, related topics were addressed by Pauli, Heitler and Stueckelberg; Max Born wrote to Heisenberg that he had "tried at the Cambridge Conference...to connect these things with my own, rather nebulous ideas" [59].

The S-matrix was at that time one of the most discussed new schemes in elementary particle physics. Pauli and his associates, Shi-Tsun Ma and Res Jost, soon criticized certain features of the mathematical scheme and Pauli wrote to Heisenberg expressing his view that "the S-matrix is not...a primary fundamental concept" but rather "has the character of something complicated and derived" [60]. However, Dirac's student Richard John Eden extended Heisenberg's theory in his Cambridge Ph.D. thesis and subsequent papers [61].

Despite having been much debated for several years, by the Eighth Solvay Conference in fall 1948 the S-matrix was no longer considered so important, as Pauli informed Heisenberg (who was not asked to attend): "I was at Brussels, where I also met Teller, and where Bohr and Oppenheimer talked much about the elementary particles in the discussion." [62] Instead, the newly discovered pi-meson and other experimental results, and the divergence difficulties of quantum field theory stood in the foreground of the reports and discussions [63]. Interestingly, the recent progress of renormalized QED in the USA and Japan did not attract much attention either. Although it was to become the prototype for the description of all elementary particle interactions in the future, the great pioneers, such as Dirac, Heisenberg and Yukawa, did not value renormalization highly, but regarded it as a mere technical device and preferred to consider their own ideas.

Notes to text

[1] Feinberg (1989, p 39).
[2] Heisenberg and Otto Hahn of the KWI für Chemie were officially appointed in July 1941 by the Kaiser Wilhelm Society to advise on the scientific programme at the KWI für Physik. They also consulted the other directors (e.g., for biochemistry) in planning the colloquium. In the summer term, Heisenberg scheduled a series of lectures on "radiation biology and physics of proteins", in which besides Heisenberg, Weizsäcker and Wirtz of Physics, Nikolaj Timoféef of Physical Chemistry and Adolf Butenandt of Biochemistry participated.
[3] According to the programme in The Werner-Heisenberg Archiv in Munich, the following topics were announced:
   1. New work on the decay time and mass of mesons (Heisenberg).
   2. Constitution of cosmic rays at sea level (Bopp).
   3. Secondary processes and the spin of the meson (Weizsäcker).
   4. Comparison of different forms of meson theory and consequences for meson production (Heisenberg).
   5. The origin of the penetrating component (Flügge).
[4] Missing was Heisenberg's Leipzig assistant, Hans Euler. In difficulty with the Nazi regime, he refused to join the Uranium Project, instead volunteering for

the air force. In summer 1941, while on a reconnaissance flight in the Caucuses, he disappeared.

[5] Cosmic-ray studies did constitute much of the early post-war programme of the institute. Also, the book was later translated (in 1946) and became the first post-war book on the subject in English.

[6] Discussions with Wick are referred to in footnote 1 on p 533 of Heisenberg's paper. For similar ideas in non-relativistic nuclear physics see Wheeler (1937).

[7] For a review of Heisenberg's and Pauli's struggles with these infinities, see Rechenberg (1993b).

[8] Heisenberg (1943c, p 513). Heisenberg was unaware that a "scattering matrix" had previously been introduced by John Wheeler (see note 6).

[9] Stueckelberg to Heisenberg, 13 February 1943.

[10] Stueckelberg (1942, 1943, 1944); see also Stueckelberg (1945, 1946) and Stueckelberg and Rivier (1946).

[11] Heisenberg (1944, p 95). See also, for a review, Heisenberg (1945).

[12] Kramers to Heisenberg, 12 April 1944.

[13] Heisenberg went first to save the Bohr Institute from occupation by the Germans, after Bohr had escaped in September 1943.

[14] Møller to Heisenberg, 28 December 1943 and 10 July 1944.

[15] For a historical account of the early S-matrix, see Rechenberg (1989) and Grythe (1982), and also Oehme (1989).

[16] "On 30 January 1947, the Far Eastern Committee, which was the organization for policy making of the allied powers decided that 'all research in Japan of either a fundamental or applied nature in the field of atomic energy should be prohibited'." (For the source of this quotation and other details, including the post-war destruction of the Japanese cyclotrons, see Konuma 1989, especially p 536.)

[17] Chapter 11, section 4.

[18] Hayakawa (1983, p 98).

[19] Yukawa (1942). Yukawa's note belonged to a set of papers by several Japanese authors, submitted to the *Zeitschrift für Physik*, dedicated to Heisenberg's fortieth birthday on 5 December 1941, but they appeared delayed (without the dedication) in the issue of Volume **119**, dated 9 July 1942. See Yukawa (1941) for an English review.

[20] This was shown for vector mesons decaying into three photons by Sakata and Tanikawa (1940), who obtained a lifetime of about 10^{-16} s. For spin-0 mesons, for which two-photon decay is allowed, the lifetime is even shorter.

[21] Chapter 11, Section 3.

[22] Yukawa (1942) pointed out that the calculated burst frequency should be reliable, because it is independent of a high-frequency cut-off. However, the strong increase in interaction with energy that is responsible for the QED divergences of vector mesons should render that theory suspect for large interaction energies. That point seems to have been generally ignored at the time.

[23] [YHAL] holds 18 diaries of Yukawa, beginning with I, dated 1938–39 (April 1938 to November 1939).

[24] A documented history is given in Brown *et al.* (1991) as follows: Appendix B, S. Nakamura, "On a history of the two-meson theory by Japanese workers", p 46; Appendix C, R. Kawabe, "Two-meson theory in Japan during WW II", pp 47–9; and Appendix D, Record of "meson meetings", based on document [YHAL] EDT 20.

[25] Sakata and Inoue (1946, p 144).

[26] Sakata and Inoue (1946, p 146).

[27] The Marshak-Bethe two-meson theory was proposed five years after the Japanese theories, and as Hans Bethe wrote in a letter to V. Mukherji: "The

paper by Marshak and myself... got the spin assignments for the two mesons wrong; in fact, this was against my better judgment at the time, but Marshak (who had originally made the suggestion) wanted it this way, and I did not want to fight over it." (Mukherji 1974, p 85, footnote 173c)

[28] For example, Submarine 29, a German gift to Japan, made the trip in August 1943. A Japanese Class I submarine arrived on 11 March 1944 at the French port of Lorient, left there on 16 April and was sunk on 26 July close to Hong Kong by an American torpedo. (We thank Rokuo Kawabe for this information.)

[29] Letter of Kawabe to Rechenberg 1985.

[30] Copies of the letters are contained in [WHA]: Heisenberg to Nomaguchi and to Nishina, both dated 18 January 1944. Captain Noguchi left the U-boat before it was sunk and returned safely to Japan, but the letters went down with the ship.

[31] Tomonaga's paper, "A general theory of ultra-short wave circuits", completed during 1944, was submitted in August 1946 and published in *J. Phys. Soc. Japan* 2 (1947), pp 158–71 and *ibid.* 3, pp 93–105. The paper refers to Breit (1940), but not to Heisenberg (1943c,d). See also Schwinger (1983b) for Tomonaga's wartime work and its similarity to Schwinger's.

[32] From Diary VI onwards, one finds references going beyond standard meson theory to, e.g., Max Born's reciprocity idea, and the Born–Infeld nonlinear electrodynamics, as well as to Heisenberg's S-matrix theory.

[33] See, however, Dower (1978) and Weiner (1978).

[34] Dower (1978, p 47).

[35] Dower (1978, p 50).

[36] Dower (1978, p 54, footnote 46).

[37] Segrè (1970, p 98).

[38] See quotation above, note 37. Also see L. Fermi (1954, pp 128–9).

[39] Bernardini had been part of Rossi's group in Florence, together with Giuseppe Occhialini, who joined the Russian-born Italian cosmic-ray physicist Gleb Wataghin in São Paolo, Brazil, where he remained until the war's end.

[40] Segrè (1970, p 98).

[41] Measurements made between 1941 and 1943 using this technique were summarized in the *Physical Review* (Bernardini et al. 1945).

[42] Wick to Heisenberg (undated); Heisenberg to Wick, 19 June 1942. [WHA] After visiting Munich, Leipzig and Berlin, Wick visited Erich Regener in Stuttgart. (See telegram of Regener to Wick in Berlin-Dahlem, 6 July 1942 [WHA].)

[43] Heisenberg to Wick, 10 February 1942; Wick to Heisenberg, 15 March 1943.

[44] Cocconi et al. (1943) and Juilfs (1943) tried to obtain the mesotron lifetime from various assumptions. Heisenberg to Wick, note 43.

[45] Wick to Heisenberg, 22 February 1943; Heisenberg to Wick, 8 March 1943.

[46] Wick to Heisenberg, 27 October 1943.

[47] In November, Heisenberg's former house and institute in Leipzig were bombed, as Heisenberg wrote in his Christmas letter to Wick, who replied on 10 April 1944: "Even here in Rome many bombs have been dropped, some even in the vicinity of my present apartment. I wanted to write to you that I got married; this was in December, but I didn't think that you would receive my letter."

[48] However, the Italians found that other circumstances prevented them from continuing to do nuclear physics. Thus in an interview with Charles Weiner, Amaldi said of fission research: "Well, at a certain moment after the bomb exploded we discussed at length what to do. We felt that in the United States and other countries there were certainly piles. We could not have a pile. We could not, probably, compete with the others in neutron physics. And then we decided that the only thing to do was to concentrate our effort in cosmic rays." Later in the interview, Amaldi described his meeting with Fermi in the United States at a conference in Princeton in 1946: "At a certain moment I found, when

we started talking about neutrons, he was talking completely freely up to a certain point, and then it was clear he did not want to give more information–not because he did not want, but he could not about fission. I found that extremely unpleasant...So I did not want to work in a field where the people were not able to talk freely." (Interview by C. Weiner, 9-10 April 1969, in [AHQP]).

[49] Weiner (1978, p 359). See also Alvarez (1987, pp 144–5).
[50] Reproduced in Weiner (1975, p 360). Many of Michels' American colleagues had similar thoughts and some (including Ernest Lawrence, who had helped with the installation of the first Japanese cyclotron) tried, with little success, to collect funds to replace the cyclotrons.
[51] Hayakawa (1991, p 158). For the post-war Tomonaga group, see also Hayakawa (1983).
[52] *Progress of Theoretical Physics* 1 (1946).
[53] Takabayasi (1991, p 273).
[54] See, e.g., in Heisenberg (1992): H. Rechenberg, "Aspekte der Physik im Dritten Reich", pp 7–20; "Die moderne theoretische Physik in Deutschland, Vorbemerkungen", pp 23–7; and " 'Deutsche' und 'jüdische' Physik, Vorbemerkungen", pp 71–7.
[55] For ALSOS, see Goudsmit (1947). Also see Frank (1993) and commentary in Rechenberg (1994). The restrictions placed on research in post-war Germany have been discussed by Cassidy (1994).
[56] Heisenberg to Bohr, 25 April 1946.
[57] Heisenberg to Møller, 1 June 1946; Møller to Heisenberg, 7 July 1946. (The post was still censored.)
[58] *International Conference* (1947).
[59] Born to Heisenberg, 25 April 1946.
[60] Pauli to Heisenberg, 9 September 1946.
[61] Eden (1948, 1949). For a historical account, see Rechenberg (1989), especially pp 563–70 and literature cited there, for a more detailed historical account.
[62] Pauli to Heisenberg, 20 October 1948.
[63] Institut International de Physique (E. Solvay) (1950).

Chapter 13

The Meson Paradox is Resolved—
and a Clear View of the
Nuclear Forces Emerges

13.1 Introduction

In his path-breaking paper of November 1934 Yukawa introduced "heavy quanta" of the nuclear forces with different constants g and g', denoting respectively their couplings to proton–neutron and electron–neutrino. From the binding energies of nuclei and from the rates of β-decay, he deduced that these interactions differed in strength by at least eight orders of magnitude. Yukawa's keen proposal, after the discovery of the cosmic-ray meson in 1937, occupied many physicists for more than a decade and led to many contradictory conclusions. Only in 1947 did it become clear that there were two "mesons", one being that proposed by Yukawa, having strong, weak and electromagnetic interactions, now known as the π-meson or pion; the other being the muon, a product of pion decay, without strong interaction properties. (Neither particle plays the dominant role in β-decay.)

The discovery of the pion and the muon resolved puzzles which had plagued physicists for more than a decade, while temporarily putting an end to unified theories of the two nuclear forces, whether Yukawa's or that of the Fermi-field. It was experiment that gave the solution, once again demonstrating the primacy of experiment over theory in the dramatic story of the nuclear forces until 1947. Still, we must warn against viewing historical events in such an oversimplified fashion, since without the heavily theory-laden interpretation of the empirical results of nuclear and cosmic-ray physics, especially the latter, the identification of the two mesons could not have been achieved.

Of course, particle accelerators were very important in the 1930s for the study of nuclear forces: in the establishment of the charge-independence symmetry of the strong nuclear force and in probing the properties of light nuclei, to give but two examples out of many [1]. However, the accelerators of the 1930s did not reach energies able to distinguish between different theories of meson exchange or to investigate the properties of the mesons themselves. Higher energy accelerators, capable of producing mesons, only began to appear in the late 1940s.

Until then, high-energy nuclear physicists (as they would be called later) had to rely on the only available natural source of fast particles, the cosmic rays. Ingenuity of a high order was needed to extract information from this source, extremely heterogeneous in particle type, energy and direction, and moreover, passing through a complex medium, the atmosphere, with which it interacted to present an ever changing composition with increasing depth. In such a situation, discoveries were bound to be determined, as well as limited, by instrumentation. That remained as true in the 1940s as in earlier times.

The main instruments in use to study high-energy interactions were cloud chambers and counter arrays, often used in combination with each other and with magnets, and then of increasing importance throughout the 1940s, nuclear photographic emulsions [2]. All of these techniques produced seminal discoveries, especially in the year 1947. We shall discuss them and their implications in the sections that follow.

Before doing so, let us show with an example how experiment and theory played together, not always successfully, in obtaining and interpreting a cosmic-ray result. On 13 December 1944, Louis Leprince-Ringuet and Michel Lhéritier submitted a note to *Comptes Rendus* concerning a cloud chamber event which they wrote indicated "the probable existence of a particle of mass $(990 \pm 12\%)$ m_e in the cosmic radiation" [3]. Two years later, Hans Bethe and Frederick De Hoffmann discussed theoretically the analysis of particle tracks in a cloud chamber (magnetic curvature yielding the momentum of a particle) and claimed [4]:

> It is shown that multiple scattering by the atoms of a gas in a cloud chamber can cause large apparent curvatures of tracks. This effect makes the measurement of curvature in a magnetic field meaningless if the energy of the particle is small and the magnetic field low. An analysis shows that *all* published meson tracks are compatible with a unique mass of about 200 electron masses.

Although they were not able to associate the French event with a mass as low as $200m_e$, they showed it to be compatible with a primary proton and concluded "that an entirely new particle could only be established by much more than one event and by measurements of much smaller error" [5]. However, notwithstanding this criticism, a new particle of the mass observed by Leprince-Ringuet and Lhéritier was soon afterwards confirmed by the Manchester cloud chamber group in England.

In section 2 we consider the work, begun in wartime in Manchester, using counter arrays and absorbers, and afterwards incorporating a large cloud chamber. Experiments undertaken to study the "penetrating particles" in penetrating showers shed light on the production of mesons and, by implication, on the nature of the primary cosmic rays. However, they also revealed completely unexpected new phenomena, the so-called strange particles.

In section 3 we shall return to the central problem of mesons and nuclear forces. We shall describe experiments in Rome, also begun in wartime, using counters and magnets, which showed conclusively that the cosmic-ray meson had no strong interaction with nuclei. We then discuss the theoretical interpretations of the Rome experiments, in part carried out in North America.

In section 4, we discuss the improved nuclear emulsion technique and its application by the Bristol group to the study of cosmic rays at mountain altitude. These experiments clearly solved the meson paradox, as we will show in section 5, by revealing the presence of *two* particles of different intermediate masses, one decaying into the other, as envisaged earlier by Sakata and Inoue.

Section 6 will deal with the artificial production of mesons in high-energy accelerators, beginning in 1948. The final step in our main history will be the discovery of the neutral pion in 1950. With this section, therefore, we will reach our main goal, dealing directly with the development of the concept of nuclear forces. That which remains will take the form of an epilogue, the final chapter, which will give a brief overview of the further development of these forces within the framework of elementary particle physics, until we reach the current Standard Model.

13.2 Cloud chambers and counter arrays: penetrating showers and strange particles (1940–47)

In earlier chapters we have stressed certain difficulties with meson theory when confronted with the cosmic rays. These included the mass discrepancy and, even more seriously, the lifetime discrepancy and the high penetrability of the cosmic-ray mesons. All of these obstacles appeared to rule out the possibility that the cosmic-ray meson was the heavy quantum of the strong nuclear force. We have discussed at some length the generally unsuccessful attempts to cure those theoretical ills.

Yet another problem arose from the large production of mesons, which were found to be copious at sea level, constituting about 75% of the charged particles there. The puzzle could be connected with uncertainties concerning: first, the nature of the primary cosmic rays that brought the energy necessary to create the mesons into the atmosphere; and secondly, the actual production mechanism [6]. Inquiries into both of these issues were posed in sharpened form by the consideration of *penetrating showers*, first observed in the cloud chamber by Fussel (1937). According to a well-known treatise on cosmic rays: "The penetrating shower differs from the electronic shower in the high penetration of secondary particles and their small angular divergence when they emerge from a lead plate in a cloud chamber." [7]

These showers were rediscovered and thoroughly investigated in Brazil (Wataghin *et al.* 1940a, b; Souza Santos *et al.* 1941). The experimenters were using multiple-coincidence counter circuit arrays with massive amounts of lead absorber, suitably arranged to screen out the electronic component. The authors interpreted their results as showing the simultaneous production of several penetrating particles. The same conclusion was reached by a British group, using an arrangement of trays of counters in coincidence separated by a large thickness of lead. Janóssy and Ingelby in Manchester argued [8]:

> It can be calculated that a cascade shower which penetrates 50 cm of lead must have an initial energy of at least 1019 eV. If electrons and photons of this energy were sufficiently frequent to account for our observations, they would have to transfer an enormous amount of energy into ionization on their way through a lead absorber. This ionization can be shown to be larger than the total ionization produced by the whole cosmic-ray beam, and would therefore be incompatible with the known rate of ionization. If, therefore, the Bethe–Heitler theory is valid for very high energies, the showers cannot be cascades.

Heisenberg cited this observation as favouring the existence of the explosive showers he had predicted [9].

After the war, P.M.S. Blackett turned over his large magnet cloud chamber to his colleagues at Manchester to continue the study of penetrating showers that had been carried out earlier by the Janóssy group. This chamber was modernized and deepened by a team under the direction of Clifford C. Butler and in the years 1946–48 many thousands of photographs were taken, controlled by counter sets in various arrangements [10]. A 3 cm lead plate was placed inside the chamber to help in the identification of penetrating particles. Although the large amount of data did not lead to any firm conclusion about the nature of the primaries (while favouring their nucleonic nature [11]), two of the photographs showed truly startling events.

On 15 October 1946, an event was seen containing a pair of tracks forming a downward pointing V, of opening angle 66.6°, with its vertex just below the 3 cm lead plate. The physicists observed a second "forked track" on 23 May 1947; this one, appearing above the lead plate, showed an incoming track that branched off at an angle of 161.1°. George Rochester and Butler concluded, with help from Blackett and several of the Manchester theorists, that both events were decays of previously unknown particles, the first one a neutral particle labelled V^0; the second, a charged particle of unknown sign, labelled $V^\pm$ (Rochester and Butler 1947).

Although the V-particles aroused great interest, one could hardly make very much of just two examples [12]. According to Rochester [13]:

> After the early discoveries that promised so much, there followed several frustrating years, a period of strain for Butler and myself, when no further examples of the V-particles were found. Indeed the next V-particles to be found by any Manchester group were found in the summer of 1950. [See Butler (1985).]

Some colleagues did respond favourably to the Manchester sightings. Thus Walter Heitler wrote from Dublin: "Your evidence is quite convincing and clear." [14] Bruno Rossi agreed from MIT "that it would be quite difficult to explain the two pictures otherwise" [15].

INSTITIUID ARD-LEIGHINN BHAILE ATHA CLIATH
(DUBLIN INSTITUTE FOR ADVANCED STUDIES)

DIRECTOR :
   PROFESSOR WALTER HEITLER

Telephone 64746

SCHOOL OF THEORETICAL PHYSICS
64-65 MERRION SQUARE
DUBLIN

23rd Nov, 47.

Dear Rochester,

Many thanks for your manuscript and your beautiful photographs. Your evidence is quite convincing and clear and I have therefore no comment to make. And as to the theoretical (side) I do not wish to make any comments yet. I feel it somewhat futile to speculate at the present stage about the genealogy of all the many particles that have been discovered now (most of which have only been observed once.). Evidently, the field is for the experiment now to find out all the particles that exist and the modes of decay they undergo.

I am very much looking forward to discuss all these things with you when you are here.

With best wishes

Yours sincerely,

W. Heitler

Letter of Heitler to Rochester on the new particles. Reproduced by permission of George Rochester Collection.

Upon hearing of the V-particles at Caltech, Carl Anderson switched the programme of his cloud chamber group to search for similar events at Pasadena and also on White Mountain in California: the result achieved after two years was a welcome confirmation of the Manchester observations. To quote Rochester again [16]:

The sequel was an exciting letter from Anderson to Blackett dated 28 November 1949, which included the following paragraph: "Rochester and

Butler will be glad to hear that we have about 30 cases of forked tracks similar to those they described in their article in *Nature* about two years ago, and so far as we can see now their interpretation of these events as caused by new unstable particles seems to be borne out by our experiments."

After several additional years it became evident that the first two V-particles were probably of the type later known as K-mesons. During the 1950s the number of "strange particles" continued to multiply. They were called strange because they appeared to be produced copiously in hadronic collisions, but decayed with a lifetime more characteristic of the weak interaction. The solution to the puzzle was eventually shown to lie in their possession of a non-zero value of a new additive quantum number (called "strangeness") which is conserved in strong and electromagnetic interactions, so that their strong production makes pairs of strange particles of total strangeness zero [17]. Because this subject does not strictly belong to the period with which this book deals, although it came out of the experimental programme of meson physics, we turn now to more relevant experiments.

13.3 Experiments in Rome on the capture of mesons by nuclei and their interpretation (1943–47)

The scene now shifts to Rome, where a series of experiments begun during the war led to a quite unexpected outcome [18]. Marcello Conversi and Oreste Piccioni, two recent graduates of the University of Rome, set out to demonstrate the existence of meson decay and to measure the mean life of slow mesons (Conversi and Piccioni 1944a, 1946). The key technical development here was the invention of a new kind of delayed coincidence circuit, using what is now called a univibrator, that produces a pulse of controllable duration and has a very fast rise time. The starting and ending triggers arise from the stopping of a meson and its subsequent decay (Conversi and Piccioni 1943). The researchers had prepared the apparatus for the measurement at the University of Rome, but before the experiment could be performed the situation took a dramatic turn, requiring a shift of venue, as Conversi related [19]:

On July 19, 1943, when the electronic system was nearly completed, Rome was bombed for the first time by the American Air Force. Nearly 80 bombs fell within the perimeter of the University campus, and one of them fell just outside the window of our laboratory, a few minutes after I had left it, fortunately having moved the electronic system far away from the window ... we decided to move as soon as possible to some place near the Vatican City, presumably more protected from the air raids. So towards the end of July 1943, with the help of a few friends including Professor Eduardo Amaldi, we transported all our equipment to a semi-underground class room of the "Liceo Ginnasio Statale Virgilio". The class-room was soon transformed into a laboratory with the essential equipment. We completed the installation of the counters and the electronics and started to test the apparatus. But we were soon interrupted by the Italian armistice (September 9, 1943) and the consequent occupation of Rome by the German troups. Work was resumed about one month later under very difficult and sometimes dramatic conditions. Late in 1943 we finally started to run our first experiment.

Conversi and Piccioni demonstrated the meson's exponential decay, and obtained a mean life of 2.3 μs ($\pm 6.5\%$) for the meson at rest in iron, in agreement with the measurement that had already been made in the United States (Rossi and Nereson 1942), unknown to the Rome physicists. Next, they set out to test the earlier theoretical prediction (Tomonaga and Araki 1940) that stopped negative mesons should be captured, whereas positive mesons should decay. This experiment, using a thin iron absorber, obtained the result that the fraction of stopping mesons that decayed was 0.49 ± 0.07, very close to the theoretically expected value of 0.55 (Conversi and Piccioni 1944b).

Then came the idea of combining the fast electronic coincidence circuit and the magnetic lenses that Bernardini and Conversi had used in 1940 [20], in order to test the Tomonaga–Araki theory for positive and negative mesons separately, again using a thin iron absorber [21]. In this, Conversi and Piccioni were joined by another young investigator, Ettore Pancini, who was first in the Italian army and then, after the armistice, a leader in the Italian Resistance fighting for the liberation of Italy. Pancini was wounded in the fighting and, during a convalescent leave in 1943, he expressed interest in the meson work. Immediately after the end of the war, in September 1945, he became a scientific collaborator of Conversi and Piccioni.

The findings confirmed, as expected, the theoretical results of Tomonaga–Araki; that is, essentially all positive mesons decayed and all negative mesons were captured (Conversi *et al.* 1945). Next, in order to search for γ-rays resulting from capture of the negative mesons (and perhaps also "for the sake of experimental thoroughness"), a graphite absorber was used, but now the number of decays of negative mesons turned out to be quite comparable to the number of positive decays (Conversi *et al.* 1947a, b)! To quote Conversi again [22]:

> This result was entirely unexpected and we thought at first that something was wrong with our set-up. But all tests we did and repeated gave the confirmation that the apparatus worked properly. There was no instrumental explanation of the simple and clear fact that delayed coincidences were counted when we concentrated negative mesotrons on the graphite absorber, whereas they disappeared when the graphite was replaced with iron. This appeared astonishing to us because, to the best of our understanding, the predictions of Tomonaga and Araki clearly indicated that negative mesons slowed-down *in any material* should undergo nuclear capture in a time much shorter than 2.2 μs.

It had become apparent to the physicists in Rome that cosmic-ray mesons observed at sea level did not possess the strong interaction with nuclei that the meson theory of nuclear forces required of its heavy quantum. They thus

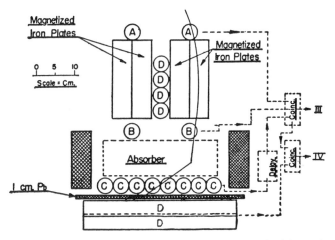

FIG. 1. Disposition of counters, absorber, and magnetized iron plates. All counters "D" are connected in parallel.

TABLE I. Results of measurements on β-decay rates for positive and negative mesons.

Sign	Absorber	III	IV	Hours	M/100 hours
(a) +	5 cm Fe	213	106	155.00'	67 ±6.5
(b) −	5 cm Fe	172	158	206.00'	3
(c) −	none	71	69	107.45'	−1
(d) +	4 cm C	170	101	179.20'	36 ±4.5
(e) −	4 cm C+5 cm Fe	218	146	243.00'	27 ±3.5
(f) −	6.2 cm Fe	128	120	240.00'	0

The decisive Rome experiment: equipment (top) and result (bottom). From Conversi *et al.* (1947a).

confirmed in a very direct manner that which had already been implied by experiments on meson penetration deep underground [23]. Instead of pushing this point further, they were content to present their experimental results, without any further interpretation. Quantitative interpretations of the weakness of interaction that they had demonstrated would come from abroad. Even before publication by the Rome group, Amaldi attended a conference in Princeton and, travelling through the USA, reported the news from Rome to his friends, including Enrico Fermi.

Fermi and his collaborators, Edward Teller and Victor Weisskopf, were the first to take notice of the importance of the Rome experiments (Fermi *et al.* 1947, received 7 February). Their paper went beyond the analysis presented by Tomonaga and Araki by a quantitative estimate of the time for a negative meson, slowed to about 2000 eV by ionization loss, to reach its lowest orbit around the nucleus. The authors reached this conclusion [24]:

The mesotron reaches its lowest orbit around the nucleus in most solids in not more than 10^{-12} second. This orbit is 200 times smaller than the radius for the K-shell, which is for carbon about 10 times the nuclear radius and for iron about twice the nuclear radius. After reaching this orbit the mesotron can be found within the nucleus with a probability of 1/1000 in the case of carbon and a probability 1/10 in the case of iron.

Fermi *et al.* also considered the capture time for a meson already in its lowest orbit and concluded—although the result is somewhat dependent on the type of meson and the interaction assumed—that conventional meson theories roughly predict capture times of 10^{-18} and 10^{-20} s in carbon and iron, respectively. These times are, of course, entirely negligible compared with the cosmic-ray meson's natural decay lifetime of about 10^{-6} s. The Rome experiments demonstrated that, in carbon, in which the meson's wavefunction has its greatest probability density mainly outside the nucleus, the capture time takes longer than the decay time, rather than being essentially instantaneous. The discrepancy found was a factor of order 10^{12}, thus requiring "a very drastic change in the forms of mesotron interactions" [25].

Fermi and Teller (1947) confirmed these results in a more detailed paper. They determined the slowing that takes place below 2 keV in a degenerate electron gas, treating the cases of metals and insulators, and finally considered the slowing in a gas. For solids, the time of capture into a K orbit was found to be about 10^{-13} s; for normal air, about 10^{-9} s. Both of these times are much shorter than the meson decay time.

Finally, Bruno Pontecorvo, at Chalk River Laboratory in Ontario, Canada, noted that when "allowance is made for the difference in disintegration energy and the difference in the volumes of the K-shell and of the meson orbit", the probability of capture of a bound negative meson is comparable to the probability of the ordinary electron K-capture process. Pontecorvo then declared: "We assume that this is significant and wish to discuss the possibility of a fundamental analogy between β-processes and processes of emission and absorption of charged mesons." [26]

In order to develop this analogy, Pontecorvo considered that the strong nuclear forces might be due entirely to neutral mesons. Also, he drew this conclusion about the weak interactions [27]:

> An immediate consequence of the experiments of the Rome group is that the usual interpretation of the β-process as a "two-step" process ("probable" production of virtual meson and subsequent β-decay of the meson) completely loses its validity, since it would predict too long β-lifetimes; the meson is no longer the particle responsible for nuclear β-processes, which are to be described according to the original Fermi picture (without mesons).

Thus Pontecorvo saw no reason to assume that the cosmic-ray meson had integral spin or that it decayed with the emission of an electron and a neutrino. Instead, he proposed that it had spin $\frac{1}{2}$ and decayed into an electron and two neutrinos ("or [by] some other process"). By the last remark,

Pontecorvo indicated that the meson *might* decay into an electron and a photon, and suggested that that process should be sought. Finally, he remarked that a neutral meson theory of nuclear forces might not be needed. Instead, one might turn to the possibility of the meson-pair theory proposed by Robert Marshak (1940).

Although the arguments of Fermi, Teller, Weisskopf and Pontecorvo were later identified as the beginning of such concepts as the mesotron as a "lepton" and "electron–muon" universality, and even as the first proposal of a "universal Fermi interaction", to draw such conclusions would be quite unhistorical. It anachronistically introduces, as we shall expound further in our concluding chapter, present conceptions into a past when they were, if they existed at all, mere speculations. In this connection, we can usefully recall that the name given originally to the cosmic-ray meson by its discoverers, Anderson and Neddermeyer, was "heavy electron", although they were not by any means claiming anything about its leptonic status. In reading the earlier papers, including those on the Rome experiments, it becomes clear that an essential element was still needed to understand what was going on adequately. That missing piece of the puzzle was waiting, indeed, on the threshold of discovery, as we shall explain in the sections that follow.

13.4 The nuclear emulsion technique for observing particle tracks (1939–48)

As we mentioned in chapter 4, photographic plates, which had already been used since the early days of the study of radioactivity, were exposed to cosmic rays by Blau and Wambacher in 1937 [28]. They discovered that fast incoming particle tracks produced events in which several secondary tracks emerged from a point; these events, called "stars", were interpreted as the breaking up of atomic nuclei in the emulsion. Within two years, Cecil Frank Powell began to apply the photographic method quantitatively to low-energy nuclear physics at Bristol. Powell had built (with G.E.F. Fertel) a Cockcroft–Walton accelerator of 750 keV energy to study the interactions of neutrons produced by bombarding light elements with deuterons (Powell 1950, 1972a). As Powell related in his "Fragment of autobiography" [29]:

> The original intention was to study the scattering of fast neutrons by protons using a Wilson Chamber filled with hydrogen, the neutrons being generated in the disintegration of light elements such as lithium, beryllium and boron by the fast deuterons from the generator. But about this time W[alter] Heitler who had been in Bristol for some years pointed out that Blau and Wambacher had successfully used "half-tone" photographic emulsion to detect particles in the cosmic radiation and, since the method had the advantage of extreme simplicity, he thought we might begin by sending similar plates on to a mountain to see if we could simulate the Viennese results.

After describing the processes involved in normal photography, Powell continued [30]:

C.F. Powell (1903–1969) (left) and C. Møller (1904–1992) in 1946.

The action of fast charged particles is similar to that of light. The moving particle produces changes in some of the grains through which it passes so that, after processing, the track shows up as a line of developed grains, like beads on an invisible string, which can be seen and recognized when the plate is examined under the microscope at high magnification. So in the experiments on the cosmic radiation it was only necessary to place a small number of [glass] plates coated with a suitable emulsion at high altitudes on a mountain where radiation is more intense than at sea level. To keep out the light, the plates are wrapped in black paper, which the cosmic rays readily penetrate, and they are recovered after a few weeks, brought home to the laboratory, and processed.

The Bristol researchers also found the "stars" of the type observed by Blau and Wambacher, and published a note in *Nature* (Heitler *et al.* 1939). More importantly, they investigated whether this method could attain the precision of others that were in use, such as counter and/or cloud chamber systems. To measure the particles' energies, they used the lengths of stopping tracks (i.e., their "ranges") and compared the same proton energy distributions (from the bombardment of boron by deuterons), measured both by the emulsion method and by counters (Powell and Fertel 1939). Finding very similar distributions, they concluded that the two methods had comparable resolution and added: "It will be seen [from the plots] that the photographic method has resolving power considerably higher than that achieved with the expansion chamber." [31] The authors pointed out that it was also a much less laborious experiment.

Afterwards, during the war, Powell was assigned to continue training students and did not participate in war-related research. He and his collaborators again used photographic plates to study nuclear reactions and scattering at the Liverpool cyclotron. Still, Powell kept in mind a return to the study of cosmic rays when he next had the opportunity. He mentioned in his autobiography that over the next 20 years the photographic method was widely applied at accelerators to the study of nuclear reactions, and recalled [32]:

> But in addition, during the war years, F.C. Champion and I found time from other duties to pursue the question of the scattering of homogeneous groups of neutrons by protons. Contrary to other observers at that time we were able to show that the scattering is isotropic in the centre-of-mass system of the collision up to energies of almost 10 MeV [33]. And during this time, I was also still looking at plates exposed on the Jungfraujoch immediately before the war and the possibilities of the cosmic-ray work.

As the war drew to a close, G.P.S. Occhialini was transported from Brazil to England by the good offices of Blackett (who was then an important scientific adviser to the British government) and in 1945 he joined Powell at Bristol [34]. Occhialini, in contrast to Powell, had always been a cosmic-ray physicist. He at once realized the potential value of the photographic technique, but also the drawback of the emulsions then available. He and Powell tried to persuade Ilford Ltd, who were with Kodak the main suppliers of photographic plates, to improve the emulsion by reducing the size of the silver bromide grains and by increasing the ratio of silver bromide to gelatin. The half-tone emulsions did not produce clear tracks from high-energy particles, whose ionizing power was much less than that of the slower accelerator-produced protons. Ilford agreed, and provided the first important breakthrough with the Ilford B and C emulsions, which had eight times the silver bromide density and were much more sensitive than the half-tone plate (Powell *et al.* 1946; Waller 1988) [35].

To continue the story with Powell's account [36]:

> Occhialini immediately took a few small plates coated with the new emulsions, about 2 dozen each 2 cm × 1 cm in area, with emulsions about 50 microns thick, and exposed them at the French Observatory in the Pyrenees at the Pic du Midi at an altitude of 3000 m. When they were recovered and developed in Bristol it was immediately apparent that a whole new world had been revealed. The track of a slow proton was so packed with developed grains that it appeared almost like a solid rod of silver, and the tiny volume of emulsion appeared under the microscope to be crowded with disintegrations produced by fast cosmic ray particles with much greater energies than any which could be generated artificially at the time. It was as if, suddenly, we had broken into a walled orchard, where protected trees had flourished and all kinds of exotic fruits had ripened in great profusion.

We shall consider some of these remarkable new fruits in the next section, with just a remark here about the next significant development in producing nuclear emulsions. Although C2 represented an enormous improvement over the earlier emulsion, it still did not reveal the tracks of particles of the *highest* energy [37]. This required "electron-sensitive" emulsion, such as Kodak NT4 in 1948 and Ilford G5 the following year. The latter became the standard and soon assisted in many important discoveries [38].

13.5 The discovery of the pion (1947)

The charm of exploring the new world revealed by the improved photographic plates brought with it the problem of an embarrassment of riches: where to begin? At Bristol, technicians were employed to man the microscopes (although the "manners"—or "scanners" as they were named—were mainly female, part-time employees, who were inevitably called "girls"). They were instructed which of the "events" viewed in the microscope were important and should be brought to the attention of the physicists in charge. Soon the Bristol workers found many interesting and puzzling events [39].

However, an important discovery made with the new emulsions was first published not by the Bristol group but by Donald H. Perkins, a young researcher at Imperial College, London, who had had a Royal Air Force plane fly some B2 emulsions, 50 μm thick, for several hours at 30000 feet. A number of stars resulted and, as he reported [40]:

> One of these disintegrations was of particular interest, for whereas all stars previously observed had been initiated by radiation not producing ionizing tracks in the emulsion [i.e., either neutral or singly charged but too energetic], the one in question appears to be due to nuclear capture of a charged particle, presumably a slow meson.

The "slow meson" track was identified by its ionization and scattering, both of which increased from the place where the particle entered the top of the emulsion to its entry point into the star. An estimate of the energy release in the star showed the nucleus to be one of the light nuclei in gelatin, not silver or bromine, and Perkins estimated the mass of the "meson" to lie in the range $(120–200)m_e$.

Before Perkins' article appeared, Occhialini and Powell (1947) submitted a paper reporting six examples (out of 800 stars) of the type of event that Perkins had found. Their mass estimates, based upon range and grain counts, $(100–230)m_e$, were similar to that of Perkins. To the Bristol experimenters it seemed clear that they had directly observed the capture of the negative cosmic-ray mesons that had been inferred by Conversi *et al.* (1945).

At this point a new player appeared upon the scene, Cesar M.G. Lattes, whom Occhialini knew from São Paolo. Powell and Occhialini invited him (and Ugo Camerini, also from Brazil) to join them. Lattes arrived in Bristol during the winter of 1946 and began to familiarize himself with the new emulsion techniques. He soon found that the addition of boron to the plates (called "loading"), originally intended for the detection of neutrons, greatly diminished a difficulty with the technique that occurred when plates were exposed for a period of weeks or months, namely, that of tracks "fading". For this reason, some of the plates that Occhialini exposed at the Pic du Midi were boron-loaded. Lattes related what happened when the laboratory workers began to study these plates [41]:

> After a few days of scanning, a young lady, Marietta Kurz, found an unusual event: one stopping meson and, emerging from its end, a new meson of about 600 μm range, all contained in the emulsion. A few days later, a second "double meson" was found; unfortunately, in this case the secondary did not stop in the emulsion, but one could guess, by studying its ionization (grain counting), that its extrapolated range was also about 600μm.

The scientists duly reported these results in *Nature* of 24 May, 1947 (Lattes *et al*. 1947a) and interpreted the two events as the decays of a heavier meson into a lighter one. Referring to the Rome investigations on nuclear capture of negative mesons, they noted that those experiments demonstrated that "the nuclear forces are several orders of magnitude smaller than has been assumed hitherto. Since our observations indicate a new mode of decay of mesons, it is possible that they may contribute to the solution of these difficulties." [42]

The Bristol group, convinced of the potential importance of these results, wanted to have a richer sample of these intriguing "double mesons" and accordingly dispatched Lattes with some boron-loaded plates to South America, where he exposed them at a meteorological station on Mt Chacultaya near La Paz, Bolivia, at a height of 5500 m. Upon his return, after an exposure of one month, the scanning turned out to be very successful, revealing 40 examples of decaying mesons, in 11 of which the secondary meson had been brought to rest in emulsion, thus allowing the determination of its range. In the definitive two-part paper (Lattes *et al*. 1947b), the authors for the first time denoted the primary meson by the symbol π and the secondary by μ. They referred to the observed decay of the π as its μ-decay, and they established, by the constancy of the μ's range at about 600 μm, that it is a two-body decay. Lattes *et al*. also detected many examples of the type that the Bristol researchers and Perkins, at Imperial College, had reported earlier. Calling these stopping mesons that produce nuclear disintegrations σ-mesons, they wrote: "Although we have no evidence that they are all of the same type, it will be convenient to refer, provisionally, to all mesons producing nuclear disintegrations when moving at low velocities, as σ-mesons." [43]

In a section headed "Interpretation of the Observations", the authors finally proposed [44]:

> We assume that the π-mesons, and most of the σ-mesons, are, respectively, positive and negative particles of the same type, which are produced in

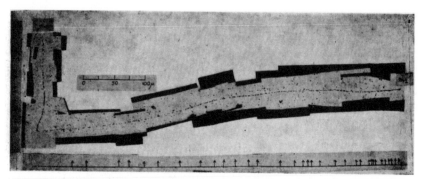

Fig. 1. Observation by Mrs. I. Roberts. Photomicrograph with Cooke × 45 'Fluorite' objective. Ilford 'Nuclear Research', boron-loaded C2 emulsion. m_1 is the primary and m_2 the secondary meson. The arrows, in this and the following photographs, indicate points where changes in direction greater than 2° occur, as observed under the microscope. All the photographs are completely unretouched

observe a single secondary particle. Of these latter events, the secondary particle is in four cases a hydrogen or heavier nucleus; in four other cases the identification is uncertain, and in the last two cases it is a second meson.

Fig. 1 is a reproduction of a mosaic of photomicrographs which shows that a particle, m_1, has come to the end of its range in the emulsion. The frequent points of scattering and the rapid change of grain-density towards the end of the range show that the track was produced by a meson. It will be seen from the figure that the track of a second particle, m_2, starts from the point where the first one ends, and that the second track also has all the characteristics of that of a particle of small mass. A similar event is shown in Fig. 2. In each case the chance that the observation corresponds to a chance juxtaposition of two tracks from unrelated events is less than 1 in 10⁹.

Grain-counts indicate that the masses of the primary particles in Figs. 1 and 2 are 350 ± 80 and $330 \pm 50 \, m_e$, respectively; and that of the secondary particle in Fig. 1, $330 \pm 50 \, m_e$, the limits of error corresponding only to the standard deviations associated with the finite numbers of grains in the different tracks. All these values are deduced from

calibration curves corresponding to an average value of the fading in the plate, and they will be too high if the track was produced late in the exposure, and too low if early. We may assume, however, that the two-component tracks in each event were produced in quick succession and were therefore subject to the same degree of fading. In these circumstances the measurements indicate that if there is a difference in mass between a primary and a secondary meson, it is unlikely that it is of magnitude greater than $100 \, m_e$. The evidence provided by Fig. 2 is not so complete because the secondary particle passes out of the emulsion, but the variation in the grain density in the track indicates that it was then near the end of its range. We conclude that the secondary mesons were ejected with nearly equal energy.

We have attempted to interpret these two events in terms of an interaction of the primary meson with a nucleus in the emulsion which leads to the ejection of a second meson of the same mass as the first. Any reaction of the type represented by the equations

$$A_z^N + \mu_{-1}^* \to B_{z-1}^N + \mu_{+1}^*, \text{ or } A_z^N + \mu_{+1}^* \to C_{z+1}^N + \mu_{-1}^*, \quad (1)$$

in which A represents any stable nucleus known to be present in the emulsion, involves an absorption

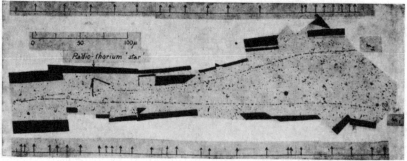

Fig. 2. Observation by Miss M. Kurz. Cooke × 45 'Fluorite' objective. Ilford 'Nuclear Research' emulsion, type C2, boron-loaded. The secondary meson, m_2, leaves the emulsion

Page with the first pictures of the meson decay (Lattes *et al.* 1947b). Reprinted with permission from Nature **160** 1947.

processes associated with explosive disintegration of nuclei, such as we have observed. The positive π-mesons suffer μ-decay and give rise to μ-mesons, for the fate of which our experiments give no evidence. On the other hand, the negative π-mesons, which appear in our experiments as σ-mesons, are captured by nuclei to produce disintegrations with the emission of heavy particles.

In addition, the authors concluded that "the greater part of the mesons observed at sea-level are μ-mesons formed by the decay in flight of π-mesons; and that positive and negative π-mesons are short-lived, with a mean life in the interval from 10^{-6} to 10^{-11} s" [45].

To observe photographically the muon decay, which yields an electron of high energy, required, of course, an electron-sensitive emulsion, first available from Kodak in 1948. Exposing such plates at the Pic du Midi, the Bristol group found nine complete π–μ decays (Brown *et al.* 1949): that is, a π-meson comes to rest and decays into a μ-meson and an invisible neutral particle (a neutrino); then, after a fixed range, the μ-meson comes to rest, emitting an electron. Since the emitted electron does not have a fixed range, it was assumed to be a three- (or more) body decay. Actually the invisible decay products consisted of two neutrinos [46]. The two-part paper in which these events are recorded contains many additional results, especially the confirmation of the reported existence of another new particle of mass about $1000m_e$. However, that harbinger of the particle explosion to follow goes beyond the scope of this chapter. (Note that Rosenfeld (1948), the most recent book on nuclear forces, appeared just after the above events.)

13.6 Artificial production of mesons and the discovery of the neutral pion (1948–50)

In his book *Inward Bound*, Abraham Pais stated [47]:

When in May 1947 the Powell group announced their discovery of pions in cosmic rays, they did not mention that these new particles had already since November 1946 been produced in the laboratory, to wit, by the α-particles accelerated to 380 MeV in Berkeley. The reason for this silence is simple: pions were being produced but no one knew that yet. So it remained until March 1948. There are at least four reasons why it took nearly a year and a half before pions were seen.

Pais explains why pions were not discovered in the United States. First, the Berkeley physicists had focused their efforts on readying the synchrocyclotron, to the neglect of developing detection techniques; secondly, they were unfamiliar with the new energy regime; thirdly, they succeeded in extracting the synchrocyclotron beam only well into 1948; and finally, a 380 MeV α-particle has a mean energy per nucleon of only 95 MeV, which is insufficient for producing a pion (of rest energy 140 MeV) in collision with a nucleon at rest.

The last objection, which appears to be most the serious one, can, however, be overcome with the help of the notion of "Fermi motion". That

C.M.S. Lattes (left) and E. Gardner in late 1940. Photograph reproduced by permission of AIP Emilio Segrè Visual Archives (Physics Today Collection).

is, the nucleons composing the α-particle are not at rest in the rest frame of the α-particle, but have a momentum distribution that is consistent with the Heisenberg uncertainty principle. The same is true, of course, for the nucleons in the target nucleus. As a result, the nucleon–nucleon collision energy in some fraction of the collisions would exceed the threshold energy for meson production. The objections regarding the unfamiliarity with detection techniques in the new energy regime were overcome when Cesar Lattes arrived in Berkeley, fresh from Bristol.

Lattes described his visit as follows [48]:

At the end of 1947, I left Bristol with a Rockefeller scholarship with the intention of trying to detect artificially produced pions at the 184-in. cyclotron that had started operation at Berkeley, California. The beam of α particles was only 380 MeV (95 MeV per nucleon), an energy insufficient for producing pions. I took my chance on the "favorable" collisions in which the internal momentum of a nucleon in the α and the momentum of the beam provided sufficient energy in the center-of-mass system. The results showed that mesons were indeed being produced.

The first publication, in collaboration with Eugene Gardner, used the fringing magnetic field of the cyclotron to focus negative mesons emerging from a thin carbon target roughly onto the edge of a stack of nuclear emulsions. The report stated of the tracks obtained upon developing the emulsions: "These show the same type of variation of grain density with residual range found in cosmic-ray meson tracks by Lattes, Occhialini, and Powell [1947b], and roughly two-thirds of them produce observable stars at the end of their range." [49] A rough mass value was inferred from the magnetic deflection, namely 313 ± 16 electron masses.

A second paper dealing with positive mesons (Burfening *et al.* 1949) appeared almost a year later. Again, the authors used their cyclotron's fringing field to focus the mesons, but this time they had to deal with a fairly large "background" of other positive particles, namely protons and α-particles. Nevertheless, they could still obtain meson tracks, again closely resembling those found by the Bristol group, with most of the "heavy" mesons coming to rest in the emulsion and decaying to "light" mesons with a range corresponding to a kinetic energy of 4 MeV. That is, Burfening and his collaborators were observing the typical process known as π–μ-decay.

Still one major disclosure regarding the pion was to be made, and this was done first at the accelerators and not in the cosmic rays, as practically all others had been. This was the discovery of the neutral pion. As we have reminded the reader at various places in this narrative, the charge-independence of nuclear forces (or isospin invariance), established experimentally in 1936, implied the use of either symmetric meson theory, as shown by Kemmer (1938c), or of purely neutral mesons, as considered later by Bethe and others. In either case, the existence of neutral mesons would be obligatory [50]. However, they were not directly observed in the cosmic rays until much later, although they might have been inferred as an interpretation of the origin of certain phenomena.

As late as 1948, a paper on the multiple production of mesons, which assumed that mesons are produced by the collision of primary protons in the upper atmosphere, argued as follows [51]:

> Most of the evidence of multiple production and of collision cross section comes from the typical cosmic ray, of energy say ten billion volts. It is possible so to interpret the very large air showers as to obtain evidence bearing on much higher energies. It seems to be true that these showers have their origin quite near the top of the atmosphere and that they involve an intimate mixture of meson and soft radiation. A natural, but by no means unique, interpretation of this is that the soft radiation arises primarily from the decay of neutral mesons, though little in our arguments would be altered were the decay process to involve transitions from heavier to lighter particles.

Two circumstances made it very difficult to observe the neutral pion in the cosmic rays: first, its predominant decay mode is into two neutrals (γ-rays); second, its lifetime is extremely short. Indeed, Sakata and Tanikawa had already in 1940 correctly estimated its lifetime to be about 10^{10} times shorter than that of the charged meson, and thus about 10^{-16} s. The distinction of detecting these particles through their decay was thus reserved for physicists at the accelerators. This was indeed the very first observation of a new particle at an accelerator! (The proton and neutron had been seen first in α-particle bombardment, and the positron and all the other new particles first in the cosmic rays.) The neutral pion accelerator experiments followed the production and observation of charged mesons at Berkeley, as described above.

The search was again carried out at the Berkeley 184-inch synchrocyclotron (SC), now running with a proton beam at a maximum energy of 340 MeV. The report, received 19 September 1949, stated [52]:

> Evidence is here presented for the production of high energy photons in the collision of protons of energies over 175 MeV with the cyclotron target. The yield and the spectrum seem to exclude a Bremsstrahlung origin, and a Doppler shift effect excludes a nuclear origin. The observations are consistent with a source associated with proton-nucleon impacts (as distinguished from proton-nucleus collisions).

The high-energy photons were allowed to materialize in thin sheets of tantalum, and the resulting electron pairs were detected by proportional counters in quadruple coincidence. The experimentalists used targets of different materials, and the yield of photons in all cases increased by a factor of about 100 as the proton energy was increased from 175 to 350 MeV. At the lower energies, the results were consistent with proton Bremsstrahlung, whereas the threshold for the process producing the large increase lay in the range 175–200 MeV. Taking into account the "Fermi energy" effect (as in the charged meson production case), the experimenters argued that [53]:

> the maximum energy available for inelastic processes produced by the 175 MeV incident protons is about 170 MeV. This energy is about twice the mean energy of the photons observed and suggests, perhaps, that the photons are produced in pairs, or that an intermediate particle with a rest mass equivalent to twice the photon energy is involved. It may also be noted that the threshold for producing π-mesons is also about 175 MeV, and that the yield vs. energy results are similar.

Bjorklund *et al.* concluded that although the neutral meson is "clearly not required at the present stage of the experiments", it is the only one of the (five) "hypotheses which seems to fit the experimental data" [54].

Following upon the work reported above, Jack Steinberger and his collaborators undertook a search for the neutral pion at another Berkeley accelerator, the electron synchrotron of Edwin McMillan [55]. In this case, the authors clearly named the neutral meson as the source of high-energy photons observed and showed, as the paper's title proclaimed: "Evidence for the production of neutral mesons by photons" (Steinberger *et al.* 1950). They based their confidence on the observation of coincidences between the two γ-rays produced in the decay, remarking [56]:

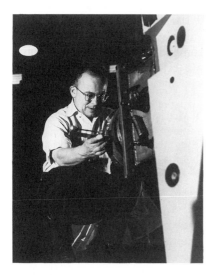

Left, W.K.H. Panofsky (born 1919) (photograph from Stanford University) and right, J. Steinberger (born 1921). Photographs reproduced by permission of AIP Emilio Segrè Visual Archives.

> The evidence [of the SC group] is already much in favor of the existence of a gamma-unstable neutral meson. However, until now, coincidences between the two gamma-rays have never been observed. We report here the detection of such coincidences, produced by the bombardment of various nuclei in the [330 MeV] x-ray beam of the Berkeley synchrotron. This must be regarded as strong additional evidence supporting the existence of the neutral meson.

After summarizing the properties of the detected photons, Steinberger *et al.* concluded [57]:

> It is clear from these properties that the gamma-rays are the decay products of neutral mesons. Since spin-$\frac{1}{2}$, and spin-1 mesons are forbidden to decay into two photons, the spin must be zero, excluding the possibility of very high intrinsic angular momenta. It seems reasonable, and it is in very good agreement with all observations, to assume that both charged and neutral mesons are of the same type. It then follows from the angular distribution of the x-ray produced π^+-mesons, and the high cross sections for making neutral mesons by x-rays, that the π-meson is a pseudoscalar.

Neutral pions were also produced and studied by the capture of negative pions in hydrogen and deuterium targets (Panofsky *et al.* 1951). Evidence for the pions being pseudoscalar particles was obtained later from the capture of negative pions in deuterium (Chinowsky and Steinberger 1954b).

Less than a month after the Berkeley synchrotron experimental report arrived at the *Physical Review*, the *Philosophical Magazine* received part of an extensive emulsion study of the high-energy cosmic rays (part 5) by members of the Bristol group, announcing their observation of neutral pions. We shall quote only their summary [58]:

> The spectrum of the γ-radiation in the atmosphere at 70000 ft has been determined by observations on the scattering of pairs of fast electrons recorded in photographic emulsions exposed in high-flying balloons. The detailed form of the spectrum is consistent with the assumption that the γ-rays originate by the decay of neutral mesons. It is found that the mass of the neutral mesons is (295 ± 20) m_e, and that they are created in nuclear explosions with an "energy spectrum" similar to that of the charged π-particles.

The authors also concluded that the mean life was less than 5×10^{-14} s and that: "The ratio of the number of neutral mesons to charged mesons, produced in nuclear explosions of great energy, is equal to 0.45 ± 0.10." [59]

With the discovery of the neutral pion, both at accelerators and in the cosmic rays, the story in which we have been engaged in this book has reached a certain closure. In the main part, we have traced the fundamental theories of nuclear forces from Heisenberg's first electron-exchange model to the completion of the ontology of Yukawa's meson theory. Of course, much more could be said of the details of fitting together the nuclear and particle (i.e., high-energy) aspects of these theories as the fifties approached; and of the less-than-neat overlap of this period with that of the "particle-explosion" which was to follow, presaged by the discovery of strange particles in 1947, which was also the year of the pion and of the first really successful quantum field theory, renormalized QED. In our final chapter, the "epilogue" which follows, we shall attempt to paint with broader brush-strokes a picture of the further development of the concept of nuclear forces within the emerging frame of the new elementary particle physics.

Notes to text

[1] See chapter 3 for details.
[2] For examples of such arrangements of counters, cloud chamber and absorber, see Bridge (1956). For cloud chamber pictures, see Rochester and Wilson (1952) and Gentner *et al.* (1954). For a photographic emulsion history and atlas, see Powell *et al.* (1959).
[3] Leprince-Ringuet and Lhéritier (1944), and for more detail (1946).
[4] Bethe and De Hoffmann (1946, p 821).
[5] *Ibid.*, p 828.
[6] The mesons, being unstable, obviously could not be the primaries themselves.
[7] Hayakawa (1969, p 16).
[8] Janóssy and Ingelby (1940, p 511). See also Janóssy (1942).
[9] Heisenberg (1943b, p 8). See also our chapter 3.

[10] This and the history of strange particles that follows is based largely on talks at several recent historical conferences: Rochester (1982, 1985, 1988, 1989, and Butler 1985).

[11] Rochester (1988, p 124). That the primary cosmic rays consist almost entirely of protons and heavier nuclei was shown conclusively in cloud chambers and in nuclear emulsions by Freier *et al.* (1948).

[12] Thus Heisenberg wrote to Blackett on 9 December 1947: "Many thanks for the copy of your paper on the new elementary particles, which we have studied here in Göttingen with greatest interest. The arguments in favour of the new particles seem to be very convincing, and from theoretical reasons I like the idea that there exist in nature still many elementary particles hitherto unknown with a rather short life time. On the other hand, of course, one would like to have still more pictures of these particles than just one or two." And after reporting that, visiting Bristol, C.F. Powell had told him of their recent discovery of two mesons, Heisenberg added: "As I wrote before, from theoretical reasons I like these new elementary particles and I feel that one finally will find an almost unlimited number of elementary particles, just in the same way as the hydrogen atom has got an unlimited number of stationary states." (Letter in the Blackett Archive, Royal Society, London.)

[13] Rochester (1989, pp 62–3).

[14] Heitler to Rochester, 23 November 1947 (Rochester Collection).

[15] Rossi to Rochester, 28 November 1947 (Rochester Collection).

[16] Rochester (1989, p 63).

[17] The new additive quantum number was introduced independently by Murray Gell-Mann and Kazuhiko Nishijima. See, e.g., Brown *et al.* (1989, pp 20–1).

[18] This account is based upon historical accounts by the participants: Conversi (1983, 1988) and Piccioni (1982, 1983); also, interviews in Rome (by LMB) with Conversi and Eduardo Amaldi in September 1979, with Amaldi (by Charles Weiner, [AHQP]) and original literature. See also chapter 12, section 4 for the first experiment of this series, carried out by Bernardini and Conversi.

[19] Conversi (1988, p 11).

[20] See chapter 12, section 4.

[21] According to Piccioni (1983), Gian Carlo Wick and Bruno Ferretti supported these experiments with detailed calculations.

[22] Conversi (1988, p 15).

[23] E.g., Nishina *et al.* (1941). Penetrating charged particles were observed under 1400 m of water equivalent. We recall that the depth of the atmosphere is only 10 m of water equivalent.

[24] Fermi *et al.* (1947, p 314).

[25] *Ibid.*, p 315.

[26] Pontecorvo (1947, p 246).

[27] *Ibid.*

[28] Blau and Wambacher (1937). For other historical references, see Perkins (1989, p 89).

[29] Powell (1972b, p 23).

[30] *Ibid.*

[31] Powell and Fertel (1939, p 115).

[32] Powell (1972b, pp 25–6).

[33] In particular, Amaldi *et al.* (1942) had reported a strong forward asymmetry in the scattering of 14 MeV neutrons on protons, which could not be explained by the theory. This result was confirmed to be wrong after the war by Amaldi *et al.* (1947).

[34] Occhialini, a colourful figure, had worked with Blackett earlier to construct the first counter-triggered cloud chamber, with which electron–positron pair production was discovered in 1932. As a refugee from fascist Italy, he had accompanied Gleb Wataghin to São Paolo, Brazil, where he did further work on cosmic-ray research. However, after Brazil entered the war he was classified as an "enemy alien" and lost his university position. He then worked as a mountain guide until he responded to Blackett's call.

[35] The letters A,B,C, etc. refer to decreasing grain size; numbers 0–5 are added to indicate increasing sensitivity. In addition to the measurement of ranges of stopping particles, which could be calibrated with particles of known energy from accelerators or radioactivities to determine their energies, a number of other techniques were applied that required a high density of small sensitive grains of uniform size. These included: grain counting, which yielded the rate of energy loss; multiple scattering, which measured momentum; and thickness of track or frequency of short secondary electron tracks (δ-rays), which were sensitive to charge and energy. Combining two or more of these measurements on a track determined the particle's mass. A "V" pointing at a star, but separated by a gap, could indicate a neutral particle and give a measure of its lifetime.

[36] Powell (1972b, p 26).

[37] The rate of ionization energy loss decreases with increasing energy, reaching a minimum at a kinetic energy of about twice the particle's rest energy (i.e., at about 2 GeV for protons and at about 1 MeV for electrons). It then increases slowly (logarithmically) for higher energy. Since the older plates could not reveal electron tracks at all (which also scatter greatly), whereas the newer plates could, the latter were called "electron-sensitive".

[38] See Lock (1988).

[39] Reminiscences of the exciting period in Bristol were presented at the 40th Anniversary International Conference, held in Bristol in 1987 (Foster and Fowler 1988), especially in the talks of P. Fowler (1988), C. Waller (1988) and D. Perkins (1988).

[40] Perkins (1947, p 126).

[41] Lattes (1983, p 308).

[42] Lattes et al. (1947a, p 696).

[43] Lattes et al. (1947b, p 486).

[44] *Ibid.*, p 490.

[45] *Ibid.*, p 492. For the mass determination see Lattes *et al.* (1948).

[46] The detailed electron spectrum from μ-decay was measured later electronically by Jack Steinberger, who concluded rather cautiously: "The experiment offers some evidence in favor of the hypothesis that the μ-meson disintegrates into 3 light particles". (Steinberger 1949, especially p 1136.)

[47] Pais (1986, p 479). For a contemporary account, see Serber (1950).

[48] Lattes (1983, pp 309–10).

[49] Gardner and Lattes (1948, p 270), also Gardner *et al.* (1950).

[50] Of course, after the discovery of the charged pion, purely neutral meson theories of nuclear forces were no longer of interest.

[51] Lewis *et al.* (1948, pp 128–9).

[52] Bjorklund *et al.* (1950, p 213). This careful statement is intended to rule out previous suggestions that the high-energy photons observed in cosmic rays were either *Bremsstrahlung* of primary or secondary protons or γ-rays from highly excited nuclei.

[53] Bjorklund *et al.* (1950, pp 216–17).

[54] *Ibid.*, p 218.

[55] A brief account is given in Steinberger (1989, p 309ff).

[56] Steinberger *et al.* (1950, p 802).

[57] *Ibid.*, p 805.

[58] Carlson *et al.* (1950, p 701).

[59] *Ibid.*

Epilogue

Chapter 14

The Strong Nuclear Forces after the Pion

14.1 Introduction

Our book began with a prologue, dealing with the nuclear forces before the discovery of the neutron in 1932. The main part of our narrative detailed the gradual emergence of a neutron-proton nucleus, with the forces carried between the nuclear particles by new heavy quanta called mesons. The earlier fundamental theories of the 1930s aspired to be unified relativistic quantum theories that would account for *all* the forces in the nucleus, other than electromagnetism. (The latter had its own troubles, which were perhaps even worse in the nuclear domain.) Yukawa's meson theory, which eventually became dominant, began as a unified field theory, but from about 1940 onwards most physicists agreed that the strong nuclear force should be treated separately from β-decay and meson decay.

We have described the different kinds of experimental information bearing on nuclear forces including: data from the systematics of light nuclei; results from scattering and nuclear reactions at the mega-electronvolt level, attainable from radioactive sources and the particle accelerators of the 1930s; and observations of the high-energy interactions of the cosmic rays. It became especially puzzling that, although the cosmic-ray mesons (the mesotrons) strongly resembled in some respects the mesons postulated by Yukawa, in other respects they could not have been more different. Especially baffling was their failure to exhibit any evidence of strong nuclear interaction.

Experiments with cyclotrons and other accelerators, and studies of nuclear systematics, showed that the strong nuclear forces were charge-independent, requiring that the meson theory should likewise be charge-independent. Kemmer accomplished this feat by coupling a meson field of isospin 1 to a nucleon field of spin $\frac{1}{2}$ (with the isospin invariant coupling having the form $\tau \cdot \phi$). This, however, implied the existence of *three* mesons, two charged and one neutral. Assuming that the cosmic-ray mesons were those of the Yukawa theory, where then was the neutral one?

With regard to the mechanical spin, only spin 0 and spin 1 were seriously considered (and briefly spin $\frac{1}{2}$), because field theories with higher spin led to even worse difficulties both with electromagnetic and with nuclear divergences than did the smaller spins. By the early 1940s, the pseudoscalar meson was widely preferred as giving a better account of the forces in light nuclei, the scalar being ruled out by, for example, the spin of the deuteron.

With the proof that the cosmic-ray meson had no strong interaction with the nucleus and the discovery in 1947 of the π–μ–e decay chain, physicists began to speak of a class of "weak interactions" of the β-decay type. Whether this was a universal class or not (which only became clear much later), those interactions have had, from that time until the present, a different destiny and history from the strong nuclear forces, although the weak and strong sectors of the Standard Model have a fascinating (but little understood) symmetry and complementarity. Henceforth, in this epilogue we shall consider mainly the strong nuclear force.

In section 2 we shall consider the consequences of taking isospin and charge-independence seriously in the phenomenological analysis of meson production by nuclear particles and by electrons, and also of meson scattering. In the early 1950s, mesons and meson beams were the most important products of the new large accelerators, and they gave rise to very surprising results: first "resonances", then "new particles", then the realization that the "resonances" could also be thought of as new particles.

In section 3, we report what happened when meson theories came to be viewed seriously as relativistic quantum field theories, having all the divergence difficulties of QED (and even more). The fact that theorists had found a way to handle the divergences of QED and obtain finite higher order correction in perturbation theory—namely the renormalization theory of QED—seemed to offer new possibilities of dealing with the divergences of meson theory. We shall briefly outline this QED revolution, and its application to a renormalized theory of mesons interacting with nucleons. We shall also describe attempts to deal with the problems arising from the large coupling of mesons to nucleons.

We then turn in section 4 to the emergence of many new particles, forming a rich hadron spectrum, and the attempts to understand that spectrum via group theory. That symmetry properties could be used to characterize the elementary particles and their interactions became fully recognized in the mid-1950s. Although field-theoretical methods could be applied to the weak interactions, for the strong interactions—whose large coupling constants rendered the usual perturbation expansion useless—new approaches were tried: dispersion theory, the analytic S-matrix and Regge poles. These methods were dominant from the mid-fifties to the late sixties (section 5).

Finally, in section 6 we turn to the current Standard Model of elementary particle interactions and compare it with some standard models of the past. Can we be certain that we have found the solution to the puzzles which began to appear in high-energy physics in the early 1930s?

14.2 Experiments with mesons produced at accelerators: isospin amplitudes

Unlike the situation in the war-torn countries of the East and West, physicists in America could return rapidly to their laboratories and studies [1]. Nuclear physicists were in an especially fortunate position, as they were widely regarded as the heroic saviours of the nation who had brought the war in the Pacific to a rapid successful conclusion by their "atomic bombs". For the physicists, this national recognition translated into glamour, power and funding, which attributes (not necessarily in that order) made the scientists very attractive to university officials as professors and as mentors of younger scientists [2]. In addition, they were able to apply the many useful techniques that they had developed and learned during the war to the problems of scientific research, from the electronics of radar to the accelerators used as mass spectrometers to separate uranium isotopes. There had also grown up an industrial infrastructure which could supply instruments (for a price now available) that would earlier have been constructed by hand in each laboratory in a laborious and time-consuming way [3].

In chapter 13 we have discussed the artificial production of charged mesons and the discovery of the neutral pion. These pregnant events occurred at the University of California at Berkeley, where by 1950 there were in operation three prototype accelerators: Ernest Lawrence's synchro-cyclotron, Edward McMillan's electron synchrotron and a 32 MeV proton linear accelerator built by Luis Alvarez and Wolfgang Panofsky [4]. Other electron synchrotrons with energies near 300 MeV were in operation at Purdue University, at Cornell University and at MIT. Cyclotrons in this energy range were operating in the early 1950s at the University of Rochester, at Columbia University and at the University of Chicago.

Most of these early accelerators had barely enough energy to produce pions, so that that they were restricted to a small range of meson dynamics. However, they yielded important information regarding the static properties of mesons. For example, a comparison of the two inverse reactions

$$p + p \to D + \pi^+, \tag{14.1}$$

$$\pi^+ + D \to p + p, \tag{14.2}$$

showed, using the principle of detailed balance, that the spin of the π^+ is zero, implying the same for π^-. Similarly, the study of the capture of π^- mesons in hydrogen and deuterium showed that the parity of π^- is odd (hence it is pseudoscalar) and also gave an accurate measure of the difference in mass between the charged and the neutral pion [5].

The first machine that produced mesons of sufficient energy to study their scattering (a classic method, since Rutherford's day, of probing the structure of projectile and target) was the Chicago cyclotron, built by a group of Fermi's collaborators, which in 1951 could accelerate protons to 450 MeV, with intense beams of well-defined energy. This made possible the study of

Rochester Conference 1952. From left to right: H. Yukawa (1907–1981), E.M. McMillan (1907–1991), C.D. Anderson (1905–1991) and E. Fermi (1901–1954). Photograph reproduced by permission of AIP Emilio Segrè Visual Archives (Marshak Collection).

production cross-sections of pions, charged and neutral, and the scattering of charged mesons; the trajectories of the charged pions were bent by the cyclotron's own magnet and formed into beams by channels cut into its external shielding. The first measurements on "transmission" of pion beams through a hydrogen target (measuring loss of pions by scattering or absorption from the beam) showed by its large value that pions definitely interact strongly with protons [6]. The dependence on energy for low energies showed p-wave behaviour, consistent with the pions being pseudoscalar.

Perhaps one of the more significant observations was that of a much higher absorption for π^+ than for π^- mesons, even though the positive pions should have exhibited only elastic scattering, whereas the negative ones could produce reactions such as charge exchange and radiative capture. At first this puzzled the Fermi group, although the situation had been anticipated by theoretical studies based upon the charge independence of nuclear forces.

The latter line of investigation was initiated by Heitler (1946), who pointed out that a system consisting of a meson (with isospin $I = 1$) and a nucleon (with $I = \frac{1}{2}$), combining as do angular momenta, would have a total isospin of either $\frac{3}{2}$ or $\frac{1}{2}$ [7]. The various sub-states would be either pure or mixed charge states. E.g., the state $(\frac{3}{2}, \frac{3}{2})$ would be pure $\pi^+ p$, while $(\frac{3}{2}, \frac{1}{2})$ would be $(\pi^+ p + \sqrt{2}\pi^- n)/\sqrt{3}$. As a result, a given mixture of $I = \frac{1}{2}$ and $I = \frac{3}{2}$ states produces a fixed ratio of scattering (including charge-exchange scattering) cross-sections. Heitler showed that his isospin amplitudes obeyed a number of relationships, some based upon charge symmetry, which he called trivial, and others holding only for charge-independence (*à la* Kemmer), which he called non-trivial.

Kenneth Watson took up these ideas in 1951 and applied them to meson production by nucleon collisions and by photons, as well as to scattering of mesons by nuclei (Watson 1951; Watson and Brueckner 1951). On the basis of preliminary results reported by the Chicago group on pion scattering in hydrogen, Keith Brueckner suggested, as a further application of isospin methods, that the scattering was dominated by a resonant state of spin $\frac{3}{2}$ and isotopic spin $\frac{3}{2}$ (the so-called 3–3 state). This was stimulated in part by the prediction of the existence of excited nucleon states, called nucleon isobars, by strong-coupling meson theories. Such a dominant resonant state would result in a cross-section ratio for scattering from protons of elastic π^+, elastic π^- and charge exchange of 9:1:2 [8]. The validity of charge-independence was well borne out as the Chicago group refined its methods and extended its results (Anderson *et al.* 1952, 1953) [9].

Isobars were also invoked to account for the production ratios of meson production by photons, either real or virtual, as provided by the new electron synchrotrons, for which weak-coupling theories gave a poor account (Brueckner and Case 1951). Experimentalists working at the new electron synchrotrons measured these photoproduction cross-sections and compared them with the predictions. Besides the "3–3 resonance" described above, they found two new pion–nucleon resonances (D_{13} and F_{15}, the "1" referring to isospin $\frac{1}{2}$, the "3" and "5" referring to total spin of $\frac{3}{2}$ and $\frac{5}{2}$ respectively, while D and F refer to the orbital angular momentum of the state) [10].

The emergence of nucleon resonances, in ever increasing number and type, began to be a characteristic feature of high-energy physics, and was intimately related to what became the "particle explosion" [11]. This became especially important in the 1960s, when a large number of strange particle resonances were discovered, especially at Berkeley and at Brookhaven National Laboratory on Long Island, New York [12], and later at CERN, near Geneva, Switzerland. These were the particles, together with the stable hadrons, the pion and the strange particles, which were eventually classified into the succeeding "flavour" groups, SU(3), SU(4) and SU(6), all being generalizations of the first "internal" symmetry group, the isospin group SU(2).

14.3 Renormalized QED and meson theories

In part B of this book, when we discussed the origin and development of Yukawa's meson theory, we emphasized that QED was the archetypal field theory whose concepts and methods were invoked to model the new theory of nuclear forces. In chapters 10 and 11, we pointed out that meson physics, in enlarging the class of physically interesting quantum field theories, stimulated an interest in the general properties of particles and fields, an example being the study by Fierz and Pauli of the general connection between spin and statistics. This kind of investigation reached its apex during the Second World War with the publications of Pauli and Wentzel [13].

However, QED, the exemplary theory, itself suffered from the disease of divergences, which was the more troubling because it entailed both conceptual and practical difficulties. The meson theories were even worse than QED with respect to divergences, and bore the further difficulty that the large coupling constant might render perturbation methods questionable, if not entirely useless, even without the divergence problem. The divergences of meson theories were troublesome not merely at high energies and for higher order correction terms, as in QED, but even for low-energy nuclear forces, as in the binding of the simplest nucleus, the deuteron, for which the nuclear force was unmanageably singular at small distances, unless arbitrary cut-offs were introduced.

In chapter 11 we considered Møller and Rosenfeld's meson-mixture theory, which they introduced mainly to provide a cancellation of the $1/r^3$ singularity in the vector-meson theory. Inspired in part by the work of these Copenhagen physicists, Sakata and Osamu Hara in Nagoya constructed another mixture theory, introducing a hypothetical new field, that of the C-meson (or *cohesive* meson), postulated to accompany virtual photon emission [14]. With its own cancelling divergence, the C-meson made the electromagnetic self-energy of a point particle (electron or idealized proton) finite, at least in the second-order perturbation approximation. Tomonaga and his collaborators then applied this approach to calculate the scattering of an electron by a fixed potential, again finding, in contrast to an earlier American calculation (Dancoff 1939a), that the higher-order correction due to the exchange of a virtual photon became finite in the C-meson theory (Ito *et al.* 1947). Finally, the Japanese authors were able to show that the result was finite, even without the C-meson, correcting an omission in Dancoff's earlier calculation (Koba and Tomonaga 1947). Using Tomonaga's earlier manifestly covariant formulation of QED (1946a, b), the Tokyo theorists were then able to produce finite renormalized QED results that were substantially equivalent to those that Julian Schwinger obtained independently in the United States at about the same time [15].

The Japanese did much of this work during the war (although its delayed publication in English took place after the war's end) and they based it largely upon theoretical constructions. In contrast, the American theorists (Bethe, Schwinger, Richard Feynman and Freeman Dyson from England, who was Bethe's student) produced a renormalized theory of QED after finishing their martial duties and they were greatly stimulated by experiments that were carried out with techniques that originated from wartime research. These experiments showed that the electron's magnetic moment (Kush and Foley 1948) and the fine structure of the hydrogen atom (Lamb and Retherford 1947) possessed features that were, from the standpoint of Dirac's electron theory, anomalous. Carried out at Columbia University in New York, they made use of new electronic techniques created for the

M. Gell-Mann (born 1929) (left) and R.P. Feynman (1918–1988). Photograph reproduced by permission of AIP Emilio Segrè Visual Archives (Marshak Collection).

development of military radar, as well as the molecular beam methods of Columbia's I.I. Rabi.

The new experimental results were reported at a conference held at Shelter Island, New York in June 1947 and created a sensation among the other participants. All of the theorists, with the exception of Hendrik Kramers, were Americans (some of them refugees) who were fresh from the triumphs of their wartime accomplishments [16], and they were determined to rise to this new challenge [17]. In particular, Bethe used a relativistic cut-off to perform a subtraction, suggested at Shelter Island by Kramers, to obtain an estimate of the QED correction that agreed well with the reported hydrogen spectrum anomaly. At a second conference, held at the end of March 1948, Schwinger and Feynman described new approaches to QED that provided consistent subtraction methods for treating any order of perturbation theory; later these were shown by Dyson to guarantee the finiteness in any order of perturbation theory, no matter how high. Some of the features of the Schwinger and Feynman approaches had been anticipated several years earlier by E.C.G. Stueckelberg in Geneva.

The QED perturbation expansion parameter, namely the dimensionless fine structure constant $\alpha = e^2/(\hbar c) \approx 1/137$, is small; as a result, the QED predictions are often astonishingly accurate. Even assuming that the same methods could be carried over to meson theory, the results would not compare with QED in accuracy, because the meson coupling to nucleons is much larger. Nevertheless, the meson theorists were heartened, at least at first, by the new possibilities opened up for quantum field theory. However, except for being able to use the simpler and more powerful calculational methods that were developed for renormalized QED, especially Feynman's diagrammatic methods, most of the hopes for a new renormalized meson theory were not to be realized.

For example, in the preface of one of the earliest books to use the new calculational methods, Robert Marshak pointed out that the new accelerators, as well as cosmic-ray experiments, were rapidly adding to the store of knowledge of mesons, thus presumably hastening the obsolescence of the book being presented and, he continued [18]:

> The theoretical situation in meson physics is even less encouraging for book writing: no genuine meson theory exists but only plausible conjectures which occasionally illuminate the complexities of the experimental material. Despite these formidable obstacles, the task has been undertaken for two reasons. First, many indisputable facts concerning π and μ mesons have been established and these seem worth recording... Secondly, by restricting ourselves to real meson processes and omitting consideration of all nuclear phenomena which involve mesons only as virtual transitions (e.g., *nuclear forces*), we have eliminated the most speculative and least satisfactory predictions of meson theory.

Even the two-volume work *Mesons and Fields*, published three years later, makes little use of renormalization techniques, except in considering QED problems in the first volume, which deals with fields [19]. It is almost entirely phenomenological in the second volume, in the preface to which the authors refer to Marshak's book as follows [20]:

> In particular Marshak deals with the production of π mesons at relativistic energies, especially with the multiple production... In his discussion of π mesons, Marshak uses weak coupling theory. Since we know the coupling constituent g^2 to be of order of magnitude ten, it is clear that an expansion in powers of g is not warranted. Hence the calculations on π mesons reported in Marshak's book should be used only to give qualitative ideas on the orders of magnitude to be expected; but even in this respect caution is indicated. On the other hand, Marshak's book contains the most complete collection of explicit formulae in the weak coupling approximation.

If we wish, therefore, to survey what became of fundamental theories of nuclear forces in the period after the discovery of the neutral pion, that is beyond 1950, we shall have to turn to the more phenomenological approaches discussed in Section 5 below.

14.4 The new particles and their symmetry properties

Even before the momentous discovery in 1947 of the pion in nuclear emulsion exposed to cosmic rays, the Manchester cloud chamber group of Rochester and Butler had viewed their first V-particle (on 15 October 1946); a second type of V-particle was observed in May 1947 (see chapter 13). These new mesons were only a sampling of many new particles which no one had "ordered" [21]. In 1951 cloud chambers also yielded up a new baryon called lambda (Λ), which decays weakly into a proton and a pion; in 1952 and 1953 the cosmic rays revealed the Xi-minus particle (Ξ^-) and the charged Sigma particles ($\Sigma^{\pm}$) in a cloud chamber and in nuclear emulsion, respectively. These objects were not only unexpected, but they also had a characteristic so unusual that they were christened "strange particles". Namely, although produced relatively copiously in nuclear collisions, they decayed with mean lives of 10^{-10} s or even 10^{-8} s, which on the nuclear time scale is absurdly long. (In contrast, the pion–nucleon resonances, also "particles", decay with mean lives of the order of 10^{-24} s [22].)

The reason suggested for this strange behaviour was that decay of the new particles by the strong interaction was forbidden by a selection rule, whose rare violation explained their long mean lives. On the other hand, they could be produced together with one or more similar particles without violating the selection rule. In 1953, Tadao Nakano and Kazuhiko Nishijima in Japan and, independently, Murray Gell-Mann in the USA, gave a definite formulation of the quantum number that was conserved in the "associated production" of strange particles and violated in their decay [23]. The theorists generalized a known relation for the electric charge (Qe) of a particle with isospin I and baryon number B, namely, $Q = I_3 + B/2$, as

$$Q = I_3 + B/2 + S/2 \tag{14.3}$$

where S is a new quantum number called "strangeness" which is non-zero for the strange particles [24]. The additive quantum number S can take on the values ± 1 and 0 (as well as other integer values); it is assumed to be conserved both in strong and in electromagnetic interactions, but it may be violated in weak interactions, such as those that occur in strange particle decay.

At the new accelerators operating in the giga-electronvolt range—the Brookhaven Cosmotron (1952) and the Berkeley Bevatron (1954) in the United States, the Dubna machine (1957) in the Soviet Union, the CERN proton synchrotron (PS) in Geneva (1959) and the alternating gradient synchrotron (AGS) at Brookhaven (1960)—the strange particles were produced and studied in associated production. The accelerators took over from the cosmic rays in the study of high-energy interactions.

In the mid-fifties, violations were discovered in the conservation of quantum numbers in the weak decays, beginning with space-reflection symmetry or parity P (Lee and Yang 1956). This observation solved the so-called "θ–τ puzzle" involving the strange mesons of mass about 500 MeV: all of them (in spite of their decay modes into states of opposite parity) could be made members of a single isospin multiplet, the K-mesons or kaons (Bridge 1956). The nuclear forces described as strong and weak could now be distinguished also by their symmetry properties: the strong interactions conserve parity, strangeness and isospin, whereas the weak interactions violate the corresponding conservation laws [25].

During the 1960s, many new "resonances" were produced at the new large particle accelerators operating in the giga-electronvolt range, both of the strange and of the non-strange variety. They generally satisfied the Gell-Mann–Nakano–Nishijima relation (14.3), which can be used to group particles into isospin families and to specify their charge states. For example, an isospin-$\frac{1}{2}$ family, with zero baryon number and strangeness of -1, has two members of charges 0 and -1. All of the new and old baryons and mesons (and resonances) were found to obey this rule. It does not apply to particles like the electron, muon and neutrino, which do not participate in strong interactions and are called *leptons*.

The large number of new particles suggested further classification, which was accomplished by Gell-Mann, and independently by Yuval Ne'eman in 1961. They carried this out in terms of the two-parameter symmetry group SU(3), which is a generalization of the one-parameter isospin group SU(2). In this scheme, strongly interacting particles (hadrons) are assembled into larger families of eight, ten, etc., members. However, so many hadronic families still resulted that physicists sought further simplification.

This took the form of the theory of quarks, originated by Gell-Mann (1964a) and, under the name "aces", by George Zweig (1964). In this theory, all hadrons are composed of sub-units (the quarks or aces), an SU(3) triplet of spin-$\frac{1}{2}$ particles of baryon number $B = \frac{1}{3}$. One of these particles (the strange quark) has strangeness -1 and isospin 0; the other two particles form an isospin doublet ($I = \frac{1}{2}$) of zero strangeness. As a consequence, according to equation (14.3), their charges are, respectively, $-\frac{1}{3}$, $+\frac{2}{3}$ and $-\frac{1}{3}$. In this picture, baryons are each composed of three quarks and mesons of one quark and one antiquark (the antiquark triplet having the opposite strangeness, charge and baryon number to the quark). The absence of evidence for free quarks is currently, in the so-called Standard Model, attributed to their being permanently confined in hadrons, although there is no accepted, clear explanation for why this is so.

In the next section we will deal with certain phenomenological approaches to particle production and scattering which also lead to the suggestion that the hadrons are composed of simpler units. Here we should like only to mention an earlier line of conjecture that gave rise to the same idea. In 1949 Fermi and his student Chen Ning Yang published a paper entitled "Are mesons elementary particles?" (Fermi and Yang 1949). In it they suggested that the ($I = 1$, $B = 0$) pions might be composed of a nucleon ($I = \frac{1}{2}$, $B = 1$) and an antinucleon ($I = \frac{1}{2}$, $B = -1$). This idea was later extended to include the concept of strangeness by Sakata (1956), who enlarged the nucleon doublet to a triplet by including the Λ-particle ($I = 0$, $B = 1$, $S = -1$). The Sakata model gave a good account of the families of mesons, but an inadequate account of the baryons.

14.5 Strong interactions without the pion field

The new high-energy and high-intensity particle accelerators of the 1950s permitted an intensive study of the strong interactions, which the cosmic rays did not afford. As we have frequently noted, strong interactions created serious problems for the quantum field-theoretical approach, because the (dimensionless) coupling constant was not small in comparison with unity, so that even in a renormalized relativistically covariant QFT, the perturbation method failed.

While some theorists tried to establish a sounder mathematical basis for renormalized QFT [26], others started from simplified models of pion–nucleon interaction (Chew 1954; Lee 1954; Chew *et al.* 1957) for relations that were independent of the principal deficiencies of QFT. Valuable relationships were obtained by extending an old quantum mechanical result of Hendrik Kramers and Ralph Kronig of the 1920s, the "dispersion relation" for the scattering of light or x-rays from atoms. The dispersion relations were derived from a causality assumption [27].

Marvin Goldberger and his collaborators relied on field-theoretical arguments to obtain the first dispersion relations for pion–nucleon scattering (Gell-Mann *et al.* 1954), but later investigators showed them to be valid on the basis of only a few assumptions (analyticity, causality and relativistic invariance). Dispersion relations predicted quantitatively high-energy pion–nucleon and (less successfully) kaon–nucleon interactions. This raised the question of whether, perhaps, further quantitative results might be obtained without any reference to the defective QFT. A major effort of particle theorists in the later 1950s and throughout the 1960s was devoted to this programme [28].

A result that strongly suggested a new approach to hadronic structure came from the electron scattering experiments of Robert Hofstadter and collaborators at the Stanford (California) accelerator Mark III, which measured the electromagnetic form factors of the nucleons, showing them to have structure and size of the order of a fermi (1 fermi = 10^{-15} m) [29]. Referring 30 years later to theoretical developments based on his work, Hofstadter recalled [30]:

> One of the most important was that of Yoichiro Nambu, who predicted that the proton and neutron form factors could be explained by the existence of a new heavy neutral meson (Nambu 1957)...It is now called the ω meson and was subsequently discovered by a Berkeley group in 1961 as a three-pion resonance decay produced by $\bar{p}p$ events at 1.61 GeV/c (Maglic *et al.* 1961). Nambu also developed a mass-spectral representation of the nucleon's form factor...Two years later the theoretical developments by Geoffrey Chew and associates (Chew *et al.* 1958) and P. Federbush and associates (Federbush *et al.* 1958) represent the basic works on the use of dispersion theory methods in developing the modern theory of the electromagnetic structure of the nucleon...In place of the often-tried and rather unsuccessful methods of perturbation theory, these newer methods showed that the nucleon form factors should be written simply as a *sum*, in terms of the masses of the intermediate states, particularly of those states with the lightest masses.

The Stanford experiments showed that the ω meson predicted by Nambu (1957) and a two-pion resonance, the ρ meson (Frazer and Fulco 1959), played most important roles in the electromagnetic structure of the nucleons.

G.F. Chew (born 1924). Photograph reproduced by permission of AIP Emilio Segrè Visual Archives.

At the same time, a new theory treating the hadrons as composite objects became available.

At the December 1960 Berkeley conference on "Strong Interactions", Geoffrey Chew presented "a unified dynamical approach to high- and low-energy interactions" (Chew 1961). He reported that a group of Berkeley theorists (including John Charap, Steve Frautschi, Marcel Froissart, Virendra Singh and B.M. Udgaonkar) were trying to replace the phenomenology used hitherto to treat scattering at high energy (optical and statistical models) and at low energy (S-matrix and dispersion-relation methods).

Taking the simplest problem, pion–pion scattering, as an example, he sketched it out as follows [31]:

> Frautschi and I are proposing (Chew and Frautschi 1960) the simplest extension of the original Mandelstam program [representing the scattering amplitudes by analytic functions of relativistic invariants of the problem] that we feel can conceivably accommodate the low-energy P resonances... We are encouraged, in fact, to propose a universal definition of strong interactions: They are always as strong as possible—consistent with the requirements of analyticity and unitarity.

Analysing the empirical data on scattering at low and high energies, Chew and his collaborators found a good fit if one took over the asymptotic (i.e., high-energy) representation which the Italian theorist Tullio Regge (1959, 1960) had shown to hold rigorously for potential scattering. Namely, the asymptotic behaviour of the amplitude for large invariant momentum transfer t is proportional to $s^{\alpha(t)}$, where s is the square of the relativistic energy, with Re(α) being positive for an attractive potential and increasing with the strength of the potential.

These ideas launched the "Regge-theory" of strong interactions, which dominated the description of hadronic scattering data throughout the 1960s. The partial-wave amplitudes are determined by the "Regge trajectories $\alpha(t)$" in the complex angular-momentum J-plane, with all hadrons lying on linear trajectories. That is, the members of each "family" of hadrons occupy as "Regge poles" the positions at real integer values of J, with a spacing of $\Delta J = 2$ [32].

By the late 1960s, this universal Regge theory of strong interactions had reached its highest state of sophistication: in addition to simple Regge trajectories, one had to consider daughter and parent trajectories, and conspiracies of trajectories. Besides Regge poles, there were Regge cuts to consider, greatly complicating the picture. Thus the whole approach began to lose predictive power, even before new data from Serpukhov exhibited rising total cross-sections, which contradicted the original basic assumptions. On the other hand, subtle and experimentally confirmed conclusions did seem to follow from the "duality" in the so-called "finite energy sum rules" which depended crucially on the Regge expansion of the high-energy amplitude [33]. However, the great days of the anti-QFT approach to hadron processes were over, as one might infer from a review article by Richard Eden (1971). One of the advocates of analytic S-matrix theory, Eden discussed only theorems that could be derived from quantum field theory.

Thus a promising scheme, respecting the fundamentally non-linear nature of elementary-particle physics, gradually faded out. On the one hand, not even its simplest, partly realistic, model could be solved exactly; on the other hand, the number of arbitrary parameters robbed it of predictive power. In that respect, a different non-QFT approach, also developed in the 1960s, seemed to offer a better hope, at least for a while. That approach was based upon "effective Lagrangians and field algebras with chiral symmetries" [34].

The idea of chiral symmetry arose in the successful description of weak interactions by the so-called $V–A$ theory. (V is the conserved vector current of a lepton or hadron pair; A is the corresponding partially conserved axial vector current [35].) Carrying this model of interacting currents over to the strong interactions, theorists in the 1960s constructed models of pion-nucleon interaction (e.g., the σ-model of Gell-Mann and Lévy (1961)) in which the axial vector current is conserved in the limit of vanishing pion mass ("soft-pion" limit). In a climate increasingly hostile to fundamental QFT and with the "current-algebra" approach favoured (Gell-Mann 1964b; Fubini 1965), soft-pion expressions for $\pi–\pi$ and $\pi–N$ scattering were derived (Weinberg 1967). Steven Weinberg then observed that the same results can be obtained from an effective Lagrangian, which is meant to be used only to lowest order in the coupling constant (now called "tree-approximation"). Weinberg's effective Lagrangian is obtained from the σ-model, according to Feza Gürsey, in the following way [36]:

1. The mass of the auxiliary σ-field is made to tend to infinity. The elimination of σ by this method results in a nonlinear partially chiral invariant Lagrangian that is identical with the old models.

2. By a unitary transformation all nonderivative couplings are changed to a derivative coupling form.... The new nucleon field has a complicated transformation property under the chiral group, but the π–N coupling is now explicitly P-wave coupling. The new Lagrangian can be more readily used as an effective Lagrangian and it leads exactly to the same predictions as current algebra for scattering lengths.

With the effective-Lagrangian method, Weinberg, Gürsey, Schwinger and others obtained interesting results in strong-interaction dynamics. It seemed an especially promising advance to replace the local chiral gauge group SU(2) × SU(2) by SU(6) × SU(6), although the construction of suitable effective Lagrangians encountered difficulties (Gürsey and Chang 1968). Therefore, in spite of initial successes, attention slipped away within a few years, not least because each different process seemed to require a different effective Lagrangian. This could hardly be a way to approach the goal of a fundamental theory of strong interactions.

14.6 Mesons and nuclear forces in the Standard Model

Abraham Pais conveys the following "main message" in the part of his book *Inward Bound* that is entitled "Essay on modern times: 1960–1983" [37]:

> Relativistic quantum theory is much healthier and much richer in new options than had been thought during the fifties and much of the sixties when, to be sure, quantum electrodynamics looked increasingly successful but the status of meson field theories remained highly problematical. As we now see it, the Yukawa-type interactions, unalterably important for low energy phenomena such as nuclear forces, are actually secondary manifestations of an underlying field theory, called quantum chromodynamics, not unlike the way Van der Waals forces are secondary consequences of electrodynamics. Furthermore, the Fermi interaction for weak processes, unalterably important for low energy phenomena such as β-decay, is also a secondary manifestation of an underlying field theory—whence the W and the Z.

That is, in the 1970s, nearly three decades after high-energy physicists resolved the meson–mesotron paradox and thus appeared to clear the road for a consistent meson theory, the very concept of fundamental nuclear forces acquired a new meaning through the convincing emergence of a new ontology. According to this structural concept, all hadrons, mesons and baryons are composed of fractionally-charged fermions, *quarks*, that interact through the exchange of massless vector bosons, *gluons*, which are the quanta of a new superstrong "colour" Yang–Mills gauge field (Yang and Mills 1954a, b). Quark–gluon and gluon–gluon interactions are the "fundamental" ones from which all other strong nuclear interactions emerge [38].

Quantum chromodynamics or QCD is thus a fundamental quantum field theory, which is renormalizable. It constitutes the strong interaction (or colour) sector of what has become known as the Standard Model (SM). The other sector of SM is electroweak theory, which successfully unifies quantum electrodynamics and the universal weak interaction. Although the QCD and electroweak sectors are subtly linked theoretically [39], they function nearly independently in practical application. We turn first to the strong nuclear forces, our main concern, and afterwards we make a few remarks on the electroweak sector.

Quantum chromodynamics, like QED, may be described as an unbroken renormalizable gauge theory [40], but it differs in that, instead of a single kind of photon (as in QED), there are eight colour gluons in QCD. These form a representation of the non-abelian colour group SU(3). (Non-abelian means that the result of successive gauge transformations, which mix the eight different gluons, depends upon the order in which the transformations are carried out, i.e., the gauge transformations are non-commuting operations.) The sources of QCD gluons are the quarks and the gluons themselves, i.e., particles that carry "colour charge", whereas electrically charged particles, such as electrons, are the sources of photons in QED. The self-interaction of the gluons is largely responsible for the behaviour of QCD being very different from that of QED, being the agent that is (probably) responsible for the so-called confinement of colour, i.e., the non-realization of free quarks and gluons, and for "asymptotic freedom", which is defined below.

Quantum chromodynamics was not the product of a single mind. As we have noted in section 14.4, Murray Gell-Mann and George Zweig introduced fractionally charged quarks independently in 1964 (Gell-Mann 1964a, Zweig 1964). Two years later, Yoichiro Nambu (1966) introduced the unbroken colour SU(3) gauge field, self-coupled and coupled to quarks [41]. (The terms *quark, gluon, colour* and *quantum chromodynamics* were invented by Gell-Mann.) In 1973, theorists discovered that QCD has a property called "asymptotic freedom", which means that the gluon–quark interaction strength approaches zero for large momentum transfer (Politzer 1973; Gross and Wilczek 1973). One consequence of this behaviour and the self-interaction of gluons is that the colour degrees of freedom are (probably) permanently confined. In other words, free quarks and free gluons do not occur in nature [42]. A further result of asymptotic freedom is that perturbation results are much more reliable at high energy, at which the coupling becomes weaker (unlike the case of QED and other weak coupling theories, for which the opposite is true).

Thus the history of QCD is as complex as that of QED, which did not become a fully successful and reasonably consistent theory until the introduction of full relativistic covariance and renormalization in the 1940s. Physicists and historians generally credit Paul Dirac, who first quantized the electromagnetic field in interaction with electrons, with the invention of QED. In the same sense, Yoichiro Nambu, who first formulated the theory of an SU(3) octet of massless vector gauge bosons interacting with quarks, must be considered to be the inventor of QCD, although he used integrally charged quarks. (Quark confinement, to avoid having real fractionally charged particles, would not be seriously considered until the seventies.) In 1973, eight years after Nambu's paper had been submitted, theorists made QCD the centre of their attention, without referring directly to Nambu's original work (Fritzsch *et al.* 1973).

What is now the status of the strong nuclear forces within the Standard Model? Nucleons and other baryons are composed of three "valence quarks", whose quantum numbers sum to the charges, isospins, spins, etc., of the particle in question. Similarly, pions and other mesons are each composed of a valence quark and a valence antiquark. The quarks and antiquarks all carry colour charge (red, green or blue, say), but the mesons and baryons are colourless, their wavefunctions being suitable linear combinations of the coloured quarks to form singlet (colourless) representations of the SU(3) colour group. In addition to the valence quarks, each physical particle contains significant quantities of virtual gluons and quark pairs of the various flavours. In the case of the nucleons, for example, this is equivalent to having "clouds" of pions and other kinds of mesons.

The interaction of two hadronic particles in this picture involves the exchange of gluons and of quark–antiquark pairs (i.e., mesons). Thus, for nuclear interactions, QCD is a generalization, rather than a rejection, of the meson theory of nuclear forces. Of course, it is a much more difficult theory to apply than meson theory, except for very-high-energy collisions, for which the coupling is weak and perturbation theory applies. Just as many chemical problems are very difficult to analyse by solving a many-body Schrödinger equation, even when it can be written down, so most of the problems for which meson theory was designed have not been successfully treated by QCD [43]. Thus much of "classical" meson physics, such as nuclear forces, meson lifetimes, meson–nucleon scattering and production, still seems to require less "fundamental" and more specific treatment than is available in QCD.

Before concluding, let us say a few words about the other part of the Standard Model, electroweak theory; both the "nuclear-force meson" (pion) and the "mesotron" (muon) play essential roles in its history [44]. One began by looking at renormalizable QFT descriptions of weak leptonic interactions, of which there were several versions, characterized by pairs consisting of a charged lepton and its corresponding neutrino [45]. At least these four steps were involved in passing from the earlier QFT of weak interaction to the new unified renormalizable electroweak theory:

(i) The fundamental four-fermion vertex was replaced by an interaction mediated by charged vector bosons $W^{\pm}$ coupled to fermion currents (either leptons or quarks).

(ii) The necessarily heavy $W^{\pm}$, together with a neutral heavy vector boson Z_0, form a triplet representation of a non-abelian gauge field of the Yang–Mills type.

(iii) The parity-violating weak currents and the parity-conserving electromagnetic currents can be united by combining the SU(2) gauge group of the massive vector bosons and the U(1) gauge group of the massless photon by the unified gauge group SU(2) × U(1).

(iv) The new symmetry group is spontaneously broken (by the Higgs mechanism) in such a way that the originally massless fundamental quartet-representation becomes a single massless photon and the three massive weak vector bosons.

These steps and others both of equal and of lesser importance (such as the discovery of three additional quarks, the "mixing angles" and the proof of renormalizability of the electroweak gauge theory) all took place between 1958 and 1971. They revealed the original and parity-violating four-fermion interaction to be a low-energy approximation to the theory [46].

Physicists, at the turn of the present century, began to unravel the intimate atomic structure of matter. The investigation of nuclei, nuclear forces and nuclear particles pushed the investigation of the micro-world immensely further. It has revealed much, but certainly not all, of the secrets of the micro-world. Niels Bohr liked to say that one should always be prepared to be surprised by nature. As we approach the new century, we look forward and anticipate that new fundamental features of matter will emerge, on the macroscopic as well as on the microscopic levels that we have been exploring in this work.

Notes to text

[1] For the post-war transitional period in some other countries see chapter 12, section 5.

[2] "Commenting in 1949 on the effects of World War II on physicists, [Lee] Dubridge noted that they had become more numerous, more affluent and much more famous." (Schweber 1994, p 144)

[3] It is beyond the scope of this book to expand upon this interesting post-war period, upon which there have been several historical studies. For overviews see, e.g., parts of Schweber (1994) and Brown *et al.* (1989).

[4] The circular machines with beams of relativistic energy were made possible by the discovery of the principle of phase stability, by Vladimir Veksler in the Soviet Union and independently by McMillan in the United States.

[5] See Panofsky *et al.* (1951), Marshak (1951) and Chinowsky and Steinberger (1954a, b).

[6] Reminiscences of this work of the Fermi–Anderson group have been given in Anderson (1955) and Anderson (1982).

[7] A meson–nucleon resonance had been anticipated earlier by Gregor Wentzel in his strong-coupling theory (Wentzel 1941a). See below.

[8] Brueckner (1952, p 109). The "charge exchange" reaction in this example is $\pi^0 + n = \pi^- + p$.

[9] These observations also led to a renewed interest in isospin invariance as it applied to low-energy nuclear reactions and the comparison of nuclear states. In this, as in many other aspects of this overview, we have insufficient space for details or original references. For the history of isospin and its applications see, e.g., Brown (1988).

[10] For a history of the photoproduced resonances, see Walker (1989).

[11] In the discussion following Anderson's review Eugene Wigner remarked: "I hope I am only repeating what has already been said and this is that the new 'particles' of high energy physics can be considered also to be resonances. There is no fundamental difference between an unstable 'particle' and a resonance". (Anderson 1982, p 161)

[12] For the history see Alvarez (1989), Steinberger (1989) and Maglic *et al.* (1961).

[13] Pauli (1940, 1941) and Wentzel (1943a, b).

[14] They were also influenced by the work of G. Mie, F. Bopp, H. Bhabha and M. Born. (See Hara (1991) for a historical study.)

[15] See Schwinger (1983a, b) for an historical account of Schwinger's work and its relationship to that of Tomonaga and his group.

[16] The experimentalists at the Shelter Island conference were Isadore I. Rabi and Willis E. Lamb (who was also a theorist) from Columbia University, and Bruno Rossi from MIT, who reported on new results in the cosmic rays, especially the experiments of the Rome group.

[17] The renormalization theory, its origins and its creators are much beyond the scope of this book. An excellent recent comprehensive account is Schweber (1994). See also Lamb (1983), Schwinger (1983a, b, 1993) and Brown (1993), as well as the Nobel Prize addresses of Lamb and Polykarp Kusch (1955), and of Feynman, Schwinger and Tomonaga (1965). Some of the most important papers have been collected in Schwinger (1958).

[18] Marshak (1952), p. v. (our emphasis)

[19] Volume I, *Fields*, Schweber *et al.* (1955); volume II, *Mesons*, Bethe and De Hoffmann (1955). Schweber (1961) is a much revised and extended version of volume I (written by Schweber). Although its treatment of renormalization is much more complete, including, e.g., the electrodynamics of scalar particles, only one eight-page section of this 900-page book deals with renormalization in meson theory.

[20] Bethe and De Hoffmann (1955, p xii).

[21] "Who ordered that?" was reportedly the response of I.I. Rabi in 1948 to the discovery of the muon, which apparently played no role in the structure of ordinary matter.

[22] Because of their relatively long mean lives, the strange particles leave observable tracks, or gaps in the case of neutral particles, in emulsion and in cloud chambers.

[23] In this broad overview, we shall not give extensive references, but refer the reader to other accounts such as Close *et al.* (1987) and Brown *et al.* (1989).

[24] The quantum number B is itself a generalization of the nucleon number which has the value 1 for the neutron and proton and 0 for mesons.

[25] For simplicity we have not included charge conjugation (C) in our list of symmetries that are conserved in strong and violated in weak interactions . This was pointed out by Reinhard Oehme (see Lee *et al.* (1957)). However, ordinary weak interactions (not super-weak ones) do conserve the product CP. (See the so-called TCP theorem, T being time-reversal, of Gerhart Lüders (1954).) For further history of weak interaction, see Cline and Riedasch (1986) and Maglic (1973).

[26] See Wightman (1989) and Lüders (1954), for example.

[27] For a historical treatment see Pickering (1989) and the literature quoted in this article. See also Pickering (1984) and Cushing (1986, 1990).

[28] Comparing the books on mesons and nuclear forces in the early 1950s with later ones, one sees how completely QFT disappeared from strong-interaction physics. See, e.g., Marshak (1952) and Bethe and De Hoffmann (1955), on the one hand, and Dean (1976), on the other.

[29] The qualitative existence of such structure was, of course, expected. See Hofstadter and McAllister (1955) and Yearian and Hofstadter (1958). For a

historical review see Hofstadter (1989).

[30]   Hofstadter (1989, p 137).

[31]   Chew (1961, p 467).

[32]   For a review of the status of Regge theory in the late 1960s, see Collins and Squires (1968). See also Cushing (1990).

[33]   See, e.g., Schmid (1969).

[34]   See the detailed review paper of Gasiorowicz and Geffen (1969).

[35]   Sudarshan and Marshak (1958) and Feynman and Gell-Mann (1958).

[36]   Gürsey (1968, pp 186–7).

[37]   Pais (1986, p 551).

[38]   Just as molecular binding and chemical reactions arise from the "fundamental" quantized structure of atoms and the electron–photon interaction, so analogously the "classic" nuclear forces are supposed to arise from the quark–gluon interactions.

[39]   In the Standard Model there are three generations, each consisting of one lepton pair and one quark pair (and their antiparticles). For example, the first generation has the electron and its neutrino and the "up" and "down" quarks. This quark–lepton symmetry is required for the cancellation of certain singular contributions to processes of higher-order perturbative rank, called "anomalies", that would otherwise prove fatal to the proof of renormalizability of either sector of the Standard Model.

[40]   The adjective "unbroken" distinguishes the colour Yang–Mills field from the electroweak field, which is also a non-abelian Yang–Mills gauge field, but is "spontaneously broken" (Brown and Cao 1991). A result of this kind of symmetry breaking is that some of the field components (which would otherwise be massless) acquire mass.

[41]   Although both Gell-Mann and Zweig introduced the quarks as the triplet representation of a broken non-gauge SU(3) group, this generalization of the isospin SU(2) group had nothing whatever to do with the unbroken gauge colour SU(3) group used by Nambu. (In 1964, only three quarks were needed, whereas the current Standard Model has three pairs of quarks. Each quark is considered a "flavour" and comes in three "colours.") Nambu's theory had nine quarks, with integer charges (Han and Nambu 1965).

[42]   A complete proof of these properties of QCD is still lacking, but lattice gauge models have been used to show the plausibility of confinement.

[43]   This observation is quite separate from other criticisms which have been made of the Standard Model, on the grounds that it is too incomplete to be the much desired "final" theory of elementary particles. For one thing, it does not unify the QCD and electroweak sectors (to say nothing of gravitation) within a single gauge group. Furthermore, it includes a theory neither of quarks nor of generations, but merely admits them *ad hoc*. Finally, it contains many adjustable parameters, including the quark masses and the various mixing angles of the electroweak sector. For this reason, other approaches are now very popular among theorists, especially the theories of "Higgs" fields and supersymmetric strings.

[44]   See Pais (1986, chapter 21), for the more recent history of electroweak theory. For the Standard Model in general, see Hoddeson *et al.* (1996), especially the editors' introduction.

[45]   In 1962 the neutrino associated with the muon was found to be different from that of β-decay. In 1974 a third-generation lepton, the charged τ-lepton was discovered, together with its partner neutrino.

[46]   For details see Hoddeson *et al.* (1996).

Bibliography

Archival sources include: Archive for the History of Quantum Physics [AHQP], New York, etc.; Yukawa Hall Archival Library [YHAL], Kyoto; Niels Bohr Archive, Copenhagen; Werner Heisenberg Archive [WHA], Munich; Dirac Archive, Churchill College [DACC], Cambridge, now Tallahassee Science Library, Tallahassee, FL; Robert Millikan Archive, Caltech, Pasadena, [RMA]; Blackett Archive, Royal Society, London; Powell Archive, Bristol; Karl von Meyenn, editor, *Wolfgang Pauli: Scientific Correspondence, Volume II: 1930–1939*, [WPSC2], (Berlin, 1985); Volume III: 1940–1949, [WPSC3], (Berlin, 1993). A useful bibliography to 1947 is given in R.T. Beyer, 1949 *Foundations of Nuclear Physics* (New York). See also Committee for YHAL 1982 and 1985, and Kuhn *et al.* 1967.

Abraham, M.
    1904: Die Grundhypothesen der Elektronentheorie *Phys. Z.* **5** pp 576–9
Ageno, M., Amaldi, E., Bocciarelli, D. and Trabachi, G.C.
    1947: On the scattering of fast neutrons by protons and deuterons *Phys. Rev.* **71** pp 20–31
Alichanow, A.I and Alichanian, A.I.
    1945: The composition of the soft component of the cosmic rays at an altitude of 3250 m above sea level *J. Phys. (Moscow)* **9** pp 73–86
Alvarez, L.W.
    1938: The capture of orbital electrons by nuclei *Phys. Rev.* **54** pp 486–97
    1987: *Alvarez, Adventures of a Physicist* (New York)
    1989: The hydrogen bubble chamber and the strange resonances, in Brown *et al.* (1989, pp 299–306)
Amaldi, E.
    1966: Ettore Majorana, man and scientist, in Zichichi (1966, pp 10–77)
    1987: The Fermi-Dirac statistics and the statistics of nuclei, in *Symmetries in Physics* ed M. Doncel, A. Hermann, L. Michel and A. Pais (Bellaterra) pp 251–78
Amaldi, E., Bocciarelli, D., Cacciapuoti, B.N. and Trabacchi, G.C.
    1947: The elastic scattering of fast neutrons by medium and heavy nuclei, in International Conference (1947, Vol. I, pp 97–113)
Amaldi, E., Bocciarelli, D., Ferretti, B., and Trabacchi, G.C.
    1942: Streuung von 14-MV-Neutronen an Protonen *Naturwiss.* **30** pp 582–4
Anderson, C.D.
    1932: The apparent existence of easily deflectable positives *Science* **76** pp 238–9
    1933: The positive electron *Phys. Rev.* **43** pp 491–4
    1961: Early work on the positron and muon *Am. J. Phys.* **29** pp 825–30
Anderson, C.D. and Anderson, H.L.
    1983: Unraveling the particle content of cosmic rays, in Brown and Hoddeson (1983, pp 131–54)
Anderson, C.D. and Neddermeyer, S.H.
    1936: Cloud chamber observations of cosmic rays at 4300 meters elevation and near sea-level *Phys. Rev.* **50** pp 263–71
    1938: Mesotron (intermediate particle) as name for the new particle of intermediate mass *Nature* **142** p 878
Anderson, H.L.
    1955: Meson experiments with Enrico Fermi *Rev. Mod. Phys.* **27** pp 269–72

1982: Early history of physics with accelerators, in Colloque International (1982, pp 101–62)
Anderson, H.L., Fermi, E., Long, E.A. and Nagle, D.E.
1952: Total cross sections of positive pions in hydrogen *Phys. Rev.* **85** p 936
Anderson, H.L., Fermi, E., Martin, R. and Nagle, D.E.
1953: Angular distribution of pions scattered by hydrogen *Phys. Rev.* **91** pp 155–68
Arley, N. and Heitler, W.
1938: Neutral particles in cosmic radiation *Nature* **142** pp 158–59
Aston, F.W.
1922: *Isotopes* (London)
1924: Mass spectra. Part II. Accelerated anode rays *Phil. Mag.* **47** pp 385–400
Auger, P.V.
1938: Sur les nouvelles particules lourds du rayonnement cosmique *Comptes rendus (Paris)* **206** pp 346–49
1946: *Die Kosmische Strahlung* (Bern)
1983: Some aspects of French physics in the 1930s, in Brown and Hoddeson (1983, pp 173–76)
Auger, P., Ehrenfest, P. Jr, Fréon, A. and Fournier, A.
1937: Sur la distribution angulaire des rayons corpusculaires cosmique durs *Comptes rendus (Paris)* **204** pp 257–59
Auger, P., Le-Prince Ringuet, L. and Ehrenfest, P. Jr
1936: Analyse du rayonnement cosmique à l'altitude de 3500 mètres *J. Phys. Rad.* **1** pp 58–64

Bagge, E.R.
1985: When muons and pions were born, in Sekido and Elliot (1985, pp 161–64)
Bainbridge, K.T. and Jordan, E.B.
1936: Mass spectrum analysis *Phys. Rev.* **50** pp 282–96
Barnóthy, J. and Forró, M.
1937: Messung der Ultrastrahlung in Bergwerken mit Koinzidenzmethoden *Z. Phys.* **104** pp 744–61
1939: Cosmic ray particles at great depth *Phys. Rev.* **55** pp 870–2
Bartholomew, J.
1989: *The Formation of Science in Japan* (New Haven)
Barut, A.O.
1982: Bemerkungen über C.F. von Weizsäckers Theorie der Kernkräfte, in *Physik, Philosophie und Politik* ed. Meyer-Abich (Munich) pp 335–41
Beck, G.
1933a: Conservation laws and β-emission *Nature* **132** 967
1933b: Energiesatz und Reversibilität der Elementarprozesse *Z. Phys.* **84** pp 811–3
1935: Report on theoretical considerations on the radioactive β-decay, in International Conference (1935, pp 31–42)
1938a: Structure of heavy elementary particles *Nature* **141** p 609
1938b: Solution exacte de la théorie quantique des champs et interaction de deux particules *J. Phys. Radium* **10** pp 200–201
Beck, G. and Sitte, K.
1933: Zur Theorie des β-Zerfalls *Z. Phys.* **89** pp 105–19
1934a: β-Emission of positive electrons *Nature* **133** p 722
1934b: Bemerkung zur Arbeit von E. Fermi: 'Versuch einer Theorie der β-Strahlen I' *Z. Phys.* **89** pp 259–60
Becquerel, H.
1896: Sur quelques propriétés nouvelles des radiations invisible émises per divers corps phosphorescents *Comptes rendus (Paris)* **122** pp 559–64
Belinfante, F.J.

1939a: A new form of the barytron equation and some related questions *Nature* **143** p 201

1939b: Undor calculus and charge-conjugation *Physica* **6** pp 849–69

1939c: The undor equation for the meson field *Physica* **6** pp 870–86

1939d: On the spin angular momentum for mesons *Physica* **6** pp 887–98

Bernardini, G.

1983: The intriguing history of the μ meson, in Brown and Hoddeson (1983, pp 155–72)

Bernardini, G., Cacciapuoti, B.N., Ferretti, B., Piccioni, O., and Wick, G.C.

1940. The genetic relation between the electronic and mesotronic components of cosmic rays near and above sea level *Phys. Rev.* **58** pp 1017–26

Bernardini, G. and Conversi, M.

1940: Sulla deflessione dei corpuscoli cosmici in un nucleo di ferro magnetizzato *La Ricerca Scientifica* **9** pp 1–11

Bernardini, G., Conversi, M., Pancini, E., Scrocco, E. and Wick, G.C.

1945: Researches on the magnetic deflection of the hard component of cosmic rays *Phys. Rev.* **68** pp 109–20

Bethe, H.A.

1932: Bremsformel für Elektronen relativistischer Geschwindigkeit *Z. Phys.* **76** pp 293–99

1933: Quantenmechanik der Ein- und Zwei-Elektronen-Probleme, in H. Geiger and K. Scheel (eds): *Handbuch der Physik*, vol. 23, part I (Berlin) pp 273–560

1937: Nuclear physics. B. Nuclear dynamics, theoretical *Rev. Mod. Phys.* **9** pp 69–244

1938: The barytron theory of nuclear forces *Phys. Rev.* **53** p 938

1939: The meson theory of nuclear forces *Phys. Rev.* **55** pp 1261–3

1940a: The meson theory of nuclear forces, I. General theory *Phys. Rev.* **57** pp 260–72

1940b: The meson theory of nuclear forces, II. Theory of the deuteron *Phys. Rev.* **57** pp 390–413

Bethe, H.A. and Bacher, R.F.

1936: Nuclear physics. A. Stationary states of nuclei *Rev. Mod. Phys.* **8** 81–229

Bethe, H.A. and De Hoffmann, F.

1946: Multiple scattering and the mass of the meson *Phys. Rev.* **70** pp 821–31

1955: *Mesons and Fields, Volume II, Mesons* (Evanston, IL)

Bethe, H.A. and Heitler, W.

1934: On the stopping power of fast particles and on the creation of positive electrons *Proc. Roy. Soc. (London)* A **146** pp 83–112

Bethe, H.A. and Marshak, R.E.

1938: Electron pair theory of heavy particle interaction *Phys. Rev.* **53** p 677

Bethe, H.A. and Morrison, P.

1956: *Elementary Nuclear Theory* 2nd edn (New York)

Bethe, H.A. and Nordheim, L.W.

1940: On the theory of meson decay *Phys. Rev.* **57** pp 998–1006

Bhabha, H.J.

1937: Negative protons in cosmic radiations *Nature* **139** pp 415–16

1938a: Nuclear forces, heavy electrons, and the β-decay *Nature* **141** pp 117–18

1938b: On the penetrating component of cosmic radiation *Proc. Roy. Soc. (London)* A **164** 257–93

1938c: On the theory of heavy electrons and nuclear forces *Proc. Roy. Soc. (London)* A **166** 501–27

1939a: The fundamental length introduced by the theory of the mesotron (meson) *Nature* **143** p 276–77

1939b: Classical theory of mesons *Proc. Roy. Soc. (London)* A **172** pp 384–408

1940a: Classical theory of spinning particles *Proc. Indian Acad. Sci.* A **11** pp 247–67

1940b: On elementary heavy particles with any integral charge *Proc. Indian Acad. Sci.* A **11** pp 347–68

1940c: Protons of double charge and the scattering of mesons *Phys. Rev.* **59** pp 100–101

1941: General classical theory of spinning particles in a meson field *Proc. Roy. Soc. (London)* A **178** pp 314–50

Bhabha, H.J., Carmichael, H. and Chou, C.N.

1939: Production of bursts and the spin of the meson *Proc. Indian Acad. Sci.* A **10** pp 221–23

Bhabha, H.J. and Heitler, W.

1937: The passage of fast electrons and the theory of cosmic showers *Proc. Roy. Soc. (London)* A **159** 432–58

Bjorklund, R., Crandall, W.E., Moyer, B.J. and York, H.F.

1950: High energy photons from proton-nucleus collisions *Phys. Rev.* **77** pp 213–18

Blackett, P.M.S.

1922: On the analysis of α-ray photographs *Proc. Roy. Soc.(London)* A **102** pp 294–308

1925: The ejection of protons from nitrogen, photographed by the Wilson method *Proc. Roy. Soc. (London)* A **107** pp 349–60

1935: The absorption of cosmic rays, in International Conference (1935 Vol. 1, pp 199–205)

1936: The measurement of the energy of cosmic rays. I.–The electromagnet and cloud chamber *Proc. Roy. Soc. (London)* A **154** pp 564–73

1938a: High altitude cosmic radiation *Nature* **142** pp 692–93

1938b: On the instability of the barytron and the temperature effect of cosmic rays *Phys. Rev.* **54** pp 973–74

1938c: Further evidence of the radioactive decay of mesotrons *Nature* **142** p 992

Blackett, P.M.S. and Brode, R.B.

1936: The measurement of the energy of cosmic rays, II–the curvature measurements and the energy spectrum *Proc. Roy. Soc. (London)* A **154** pp 573–87

Blackett, P.M.S. and Champion, F.C.

1931: The scattering of slow alpha particles by helium *Proc. Roy. Soc. (London)* A **130** pp 380–88

Blackett, P.M.S. and Occhialini, G.P.S.

1933: Some photographs of the tracks of penetrating radiation *Proc. Roy. Soc. (London)* A **139** pp 699–27

Blackett, P.M.S. and Wilson, J.G.

1937: The energy loss of cosmic ray particles in metal plates *Proc. Roy. Soc. (London)* A **160** pp 304–22

Blanpied, W.A.

1986: Pioneer scientists in pre-independence India *Physics Today* **39** No 3 pp 36–44

Blatt, J.M. and Weisskopf, V.

1952: *Theoretical Nuclear Physics* (New York)

Blau, M. and Wambacher, H.

1937: Disintegration process by cosmic rays with the simultaneous emission of several heavy particles *Nature* **140** 585

Bloch, F.

1933: Zur Bremsung rasch bewegter Teilchen beim Durchgang durch Materie *Ann. Phys. (Leipzig)* **16** pp 285–320

Bloch, F. and Nordsieck, A.

1937: Note on the radiation field of the electron *Phys. Rev.* **52** pp 54–59

Bøggild, J.K.
   1937: On the secondary effects of cosmic radiation *Ph.D. Thesis* (in Danish, Copenhagen)
Bohr, N.
   1913: On the constitution of atoms and molecules I, II, III *Phil. Mag.* **26** pp 1–25, 476–502 and 857–875
   1932: Chemistry and the quantum theory of atomic constitution *J. Chem. Soc. (London)* pp 349–384
   1936: Conservation laws in quantum theory *Nature* **138** pp 25–26
   1961: The Rutherford Memorial Lecture 1958 *Proc. Phys. Soc. (London)* **78** pp 1083–115
   1963: *On the Constitution of Atoms and Molecules: Papers of 1913* reprinted (Copenhagen and New York)
   1981: *Collected Works* Vol. 2, ed. U. Hoyer (Amsterdam)
   1986: *Collected Works Vol. 9: Nuclear Physics (1929–1952)* ed. R. Peierls (Amsterdam)
Booth, F. and Wilson, A.H.
   1940: Radiative processes involving fast mesons *Proc. Roy. Soc. (London)* A **175** pp 483–518
Born, M.
   1933: On the quantum theory of the electromagnetic field *Proc. Roy. Soc. (London)* A **143** pp 410–37
Born, M. and Infeld, L.
   1934: Foundations of the new field theory *Proc. Roy. Soc. (London)* A **144** pp 425–51
Born, M. and Nagendra Nath, N.S.
   1936: The neutrino theory of light *Proc. Indian Acad. Sci.* **3** pp 318–37 and 611–20
Bothe, W. and Becker, H.
   1930: Künstliche Erregung von Kern-γ-Strahlen *Z. Phys.* **66** pp 289–306
Bowen, I.S., Millikan, R.A. and Neher, M.V.
   1938a: The influence of the earth's magnetic field on the cosmic ray intensities up to the top of the atmosphere *Phys. Rev.* **52** pp 80–8
   1938b: The secondary nature of cosmic ray effects in the lower amosphere *Phys. Rev.* **53** p 214
   1938c: New evidence as to the nature of the uncoming cosmic rays, etc. *Phys. Rev.* **53** pp 217–23
   1938d: New light on the nature and origin of the incoming cosmic rays *Phys. Rev.* **53** pp 855–61
Bragg, W.H. and Kleeman, R.
   1905: On the α-particles of radium, and their loss of range in passing through various atoms and molecules *Phil. Mag.* **10** pp 318–41
Bramley, A.
   1937: The hard component of cosmic radiation *Phys. Rev.* **52** p 248
Breit, G.
   1940: Scattering matrix of radioactive states *Phys. Rev.* **58** pp 1068–74
Breit, G., Condon, E.U. and Present, R.D.
   1936: Theory of scattering of protons by protons *Phys. Rev.* **50** pp 825–45
Breit, G. and Feenberg, E.
   1936: The possibility of the same form of specific interaction for all nuclear particles *Phys. Rev.* **50** pp 850–56
Bridge, H.S.
   1956: Experimental results on charged K-mesons and hyperons, in *Progress in Cosmic Ray Physics* (Amsterdam) pp 143–252
Bridge, H.S., Hazen, W.E., Rossi, B. and Williams, R.W.

1948: A study of cosmic-ray bursts *Phys. Rev.* **74** pp 1083–102
Brillouin, L.
1939: Individuality of elementary particles, in *International Institute of Intellectual Cooperation* (1939, pp 119–171)
Brink, D.M.
1965: *Nuclear Forces* (Oxford)
Brode, R.B. and Starr, M.A.
1938: Nuclear disintegrations produced by cosmic rays *Phys. Rev.* **53** pp 3–5
Broglie, L. de
1932a: Sur une analogie entre l'électron de Dirac et l'onde électromagnétique *Comptes rendus (Paris)* **195** pp 536–537
1932b: Remarque sur le moment magnétique et le moment de rotation de l'électron *Comptes rendus (Paris)* **196** pp 577–88
1933: Sur la densité de l'énergie dans la théorie de la lumière *Comptes rendus (Paris)* **197** pp 1377–9
1934a: Sur la nature du photon *Comptes rendus (Paris)* **198** pp 135–7
1934b: Remarques sur la théorie de la lumière (Mémoir de l'Académie Royale des Sciences de Liége, 3e série, tome XIX)
1934c: L'équation d'ondes du photon *Comptes rendus (Paris)* **199** pp 445–448
Broglie, L. de, and Winter, J.
1934: Sur le spin du photon *Comptes rendus (Paris)* **199** pp 813–16
Bromberg, J.
1971: The impact of the neutron: Bohr and Heisenberg *Historical Studies in the Physical Sciences* **3** pp 307–41
Brown, L.M.
1978: The idea of the neutrino *Physics Today* **31** No 9 pp 23–28
1981: Yukawa's prediction of the meson *Centaurus* **25** pp 71–132
1985: How Yukawa arrived at the meson theory *Prog. Theor. Phys.*, Supp. No. **85** pp 13–19
1986: Hideki Yukawa and the meson theory *Physics Today* **39** No 5 pp 1–8
1988: Remarks on the history of isospin, in Festi-Val–Festschrift for Val Telegdi (Amsterdam) pp 39–47
1989: Yukawa in the 1930s: a gentle revolutionary *Historia Scientiarum* **36** pp 1–21
1990: Yukawa, Hideki Sakata, Shoichi *and* Tomonaga, Sin-itiro *Dictionary of Scientific Biography*, Supp. **II** pp 999–1005, 766–70 and 927–32
1993: *Renormalization: From Lorentz to Landau (and Beyond)* (New York)
1996: Nuclear forces, mesons, and isospin symmetry, in Hoddeson *et al.* (1996, pp 357–419)
Brown, L.M. and Cao, T.Y.
1991: Spontaneous breakdown of symmetry: Its rediscovery and integration into quantum field theory *Historical Studies in the Physical and Biological Sciences* **21** (2) pp 21–235
Brown, L.M., Dresden, M. and Hoddeson, L.
1989: *Pions to Quarks* (New York)
Brown, L.M. and Hoddeson, L. (eds)
1983: *The Birth of Particle Physics* (Cambridge)
Brown, L.M., Kawabe, R., Konuma, M. and Maki, Z.
1991: Elementary Particle Theory in Japan, 1930–1960 (Proceedings of the Japan-USA Collaborative Workshops) *Progress of Theoretical Physics* Supp. No. **105**
Brown, L.M. and Moyer, D.F.
1984: Lady or tiger? – the Meitner-Hupfeld effect and Heisenberg's neutron theory *Am. J. Phys.* **52** pp 130–6
Brown, L.M., Pais, A. and Pippard, B.
1995: *Twentieth Century Physics* Three volumes (Bristol and New York)

Brown, L.M. and Rechenberg, H.
 1987: Paul Dirac and Werner Heisenberg—a partnership in science, in Kursunoglu and Wigner (1987, pp 117–62)
 1988: Nuclear structure and beta decay (1932–1933) *Am. J. Phys.* **55** pp 982–8
 1990: Yukawa's heavy quantum and the mesotron (1935–1937) *Centaurus* **33** pp 214–52
 1991a: Quantum field theories, nuclear forces, and the cosmic rays (1934–1938) *Am. J. Phys.* **59** pp 595–605
 1991b: The development of the vector meson theory in Japan and Britain (1937–38) *Brit. J. Hist. Sci.* **24** pp 405–33
 1994: Field theories of nuclear forces in the 1930s: The Fermi-field theory *Historical Studies in the Physical and Biological Sciences* **25** (1) pp 1–24
Brown, L.M. and Rigden, J.S.
 1993: *Most of the Good Stuff—Memories of Richard Feynman* (New York)
Brown, L.M. and Telegdi, V.L.
 1958: On the spin of the muon *Nuovo Cimento* **7** pp 698–705
Brown, R., Camerini, U., Fowler, P.H., Muirhead, H. and Powell, C.F.
 1949: Observations with electron-sensitive plates exposed to cosmic radiation *Nature* **163** pp 47–51 and 82–7
Brueckner, K.A.
 1952: Meson-nucleon scattering and nucleon isobars *Phys. Rev.* **86** pp 106–9
Brueckner, K.A. and Case, K.M.
 1951: Neutral photomeson production and nucleon isobars *Phys. Rev.* **83** pp 1141–7
Bunge, M. and Shea, W.R.
 1979: *Rutherford and Physics at the Turn of the Century* (New York)
Burfening, J., Gardner, E. and Lattes, C.M.G.
 1949: Positive mesons produced by the 184-inch Berkeley cyclotron *Phys. Rev.* **75** pp 382–7
Butler, C.C.
 1985: Early cloud chamber experiments at the Pic-du-Midi, in Sekido and Elliot (1985, pp 177–89)

Carlson, A.G., Hooper, J.E. and King, D.T.
 1950: Nuclear transformations produced by cosmic ray particles of great energy V. The neutral mesons *Phil. Mag.* **41** pp 701–24
Carlson, J.F. and Oppenheimer, J.R.
 1937: On multiplicative showers *Phys. Rev.* **51** pp 220–31
Carlson, J.F. and Schein, M.
 1941: The production of mesotrons *Phys. Rev.* **59** p 840
Carmichael, H. and Chou, C.N.
 1939: Cosmic ray ionization bursts *Nature* **144** p 325–6
Cassen, B. and Condon, E.U.
 1936: On nuclear forces *Phys. Rev.* **50** pp 846–9
Cassidy, D.C.
 1981: Cosmic ray showers, high energy physics, and quantum field theories *Historical Studies in the Physical Sciences* **12** (1) pp 1–39
 1994: *Uncertainty: The Life and Science of Werner Heisenberg* (New York)
Chadwick, J.
 1914: Intensitätsverteilung in magnetischen Spektren der β-Strahlen von Radium B + C *Verh. Deutsch. Phys. Ges.* **16** pp 383–91
 1932: The existence of a neutron *Proc. Roy. Soc. (London)* A **136** pp 692–708
Chadwick, J. and Bieler, E.S.
 1921: The collisions of α particles with hydrogen nuclei *Phil. Mag.* **42** pp 923–40

Chakrabarty, S.K. and Majumdar, R.C.
    1944: On the spin of the meson *Phys. Rev.* **65** p 206
Chew, G.F.
    1954: Renormalization of meson theory with a fixed extended source *Phys. Rev.* **94** pp 1748–54
    1961: A unified dynamical approach to high and low energy strong interactions *Rev. Mod. Phys.* **33** pp 467–70
Chew, G.F. and Frautschi, S.C.
    1960: Unified approach to high- and low-energy strong interactions on the basis of the Mandelstam representation *Phys. Rev. Lett.* **5** pp 580–3
Chew, G.F., Goldberger, M.L., Low, F. and Nambu, Y.
    1957: Application of dispersion relations to low-energy meson scattering *Phys. Rev.* **106** pp 1337–44
Chew, G.F., Karplus, R., Gasiorowitz, S. and Zachariasen, F.
    1958: Electromagnetic structure of the nucleon in a local-field theory *Phys. Rev.* **110** pp 265–76
Chinowsky, W. and Steinberger, J.
    1954a: The mass difference of neutral and negative β mesons *Phys. Rev.* **93** pp 586–9
    1954b: Absorption of negative pions in deuterium: Parity of the pion *Phys. Rev.* **95** pp 1561–4
Christy, R.F. and Kusaka, S.
    1941a: The interaction of γ-rays with mesotrons *Phys. Rev.* **59** pp 405–14
    1941b: Burst production by mesons *Phys. Rev.* **59** pp 414–21
Clay, J. and von Gemert, A.
    1939: Decrease of the intensity of cosmic rays in the earth down to 1380 m water equivalent *Physica* **6** pp 497–510
Cline, D. and Riedasch, G.
    1986: *Fifty Years of Weak Interactions* (Madison, WI)
Close, F., Marten, M. and Sutton, C.
    1987: *The Particle Explosion* (New York)
Cocconi, G., Loverdo, A. and Tongiorgi, V.
    1943: Über das Vohandensein von Mesotron-Schauern in ausgedehnten Luftschauern *Naturwiss.* **31** pp 135–6
Cockcroft, J.
    1967: Homi Jehangir Bhabha (1909–1966) *Proc. Roy. Inst. of Great Britain* **41** 411–22
Cockroft, J.D. and Walton, E.T.S.
    1932: Experiments with high velocity ions II. The disintegration of elements by high velocity protons *Proc. Roy. Soc. (London)* A **137** pp 23–38
Collins, P.D.P and Squires, E.J.
    1968: *Regge Poles in Particle Theory* (Berlin)
Colloque International
    1982: *Colloque International sur l'Histoire de la Physique des Particules* (*J. Phys. Paris*, Supp. No. **12**)
Committee for YHAL [Yukawa Hall Archival Library]
    1982: YHAL Resources (I), Soryushiron Kenkyu **65** pp 239–69
    1985: YHAL Resources (II), Soryushiron Kenkyu **79** pp 289–306
Compton, A.H.
    1939: Foreword, in Symposium on Cosmic Rays held at the University of Chicago, June, 1939 *Rev. Mod. Phys.* **11** p 122
Condon, E.U.
    1938: A simple derivation of the Maxwell-Boltzmann Law *Phys. Rev.* **54** pp 937–40
Conversi, M.

1983: The period that led to the discovery of the leptonic nature of the 'mesotron', in Brown and Hoddeson (1983, pp 251–7)

1988: From the discovery of the mesotron to that of its leptonic nature, in Foster and Fowler (1988, pp 1–20)

Conversi, M., Pancini, E. and Piccioni, O.

1945: On the decay process of positive and negative mesons *Phys. Rev.* **68** p 232

1947a: On the disintegration of negative mesons *Phys. Rev.* **71** pp 209–210

1947b: Sull'assorbimento e sulla disintegrazione dei mesoni alla fine del loro percurso *Nuovo Cimento* **3** pp 1–19

Conversi, M. and Piccioni, O.

1943: Sulle registrazione di coincidenza a piccoli tempi di separazione *Nuovo Cimento* **1** pp 1–12

1944a: Misura diretta della vita media dei mesoni frenati *Nuovo Cimento* **2** pp 40–70

1944b: Sulla disintegrazione dei mesoni lenti *Nuovo Cimento* **2** pp 71–87

1946: On the mean life of slow mesons *Phys. Rev.* **70** pp 859–73

Corben, H.C. and Schwinger, J.

1940: The electromagnetic properties of mesotrons *Phys. Rev.* **58** pp 953–68

Corson, D.R. and Brode, R.B.

1938a: Evidence for a cosmic ray particle of intermediate mass *Phys. Rev.* **53** p 215

1938b: The specific ionization and mass of cosmic ray particles *Phys. Rev.* **53** pp 773–7

Critchfield, C.L.

1939: Spin dependence in the electron-positron theory of nuclear forces *Phys. Rev.* **56** 540–7

Critchfield, C.L. and Lamb, W.E. Jr

1940: Note on a field theory of nuclear forces *Phys. Rev.* **58** pp 46–9

Critchfield, C.L. and Teller, E.

1938: On the saturation of nuclear forces *Phys. Rev.* **53** 812–18

Critchfield, C.L., Teller, E. and Wigner, E.P.

1939: The electron-positron field of nuclear forces *Phys. Rev.* **56** pp 530–9

Cumming, J. and Osborn, H.

1971: *Hadronic Interactions of Electrons and Photons. Proc. of the Eleventh Session of the Scottish Universities Summer School in Physics* (London)

Curie, I. and Joliot, F.

1932: Émission de protons de grand vitesse par les substances hydrogénées sous l'influence des rayons γ très pénétrant *Comptes Rendus (Paris)* **194** pp 273–5

1934: Une nouveau type de radioactivité *Comptes Rendus (Paris)* **198** pp 254–6

Cushing, J.T.

1986: The importance of Heisenberg's S-matrix program for the theoretical high-energy physics of the 1950s *Centaurus* **29** pp 110–49

1990: *Theory Construction and Selection in Modern Physics* (Cambridge)

Dancoff, S.M.

1939a: On the radiative corrections for electron scattering *Phys. Rev.* **55** pp 959–63

1939b: Virtual state of He5 and the field theory of nuclear forces *Phys. Rev.* **56** pp 384–5

Dancoff, S.M. and Serber, R.

1942: Nuclear forces in the strong coupling theory *Phys. Rev.* **61** p 394

Darrigol, O.

1988: The quantum electrodynamical analogy in early nuclear theory or the roots of Yukawa's theory *Rev. Hist. Sci.* **XLI** pp 226–97

Davis, N.P.

1968: *Lawrence and Oppenheimer* (New York)

Dean, N.W.
    1976: *Introduction to the Strong Interactions* (New York)
De Benedetti, S.
    1934: Absorption measurements on the cosmic rays at 11°30′ geomagnetic latitude
        and 2370 meters elevation *Phys. Rev.* **45** pp 214–5
Debye, P. and Hardmeier, W.
    1926: Anomale Zerstreuung von α-Strahlen *Z. Phys.* **27** pp 196–9
Dirac, P.A.M.
    1927: The quantum theory of the emission and absorption of radiation *Proc. Roy.*
        Soc. (London) A **114** pp 243–65
    1936: Relativistic wave equations *Proc. Roy. Soc. (London)* A **155** pp 447–59
    1942: The physical interpretation of quantum mechanics [Bakerian Lecture 1941]
        Proc. Roy. Soc. (London) A **180** pp 1–40
Dirac, P.A.M., Fock, V. and Podolsky, B.
    1932: On quantum electrodynamics *Phys. Z. Sowj.* **2** pp 468–79
Dower, J.W.
    1978: Science, society and the Japanese atomic-bomb project during World War
        Two *Bull. Concerned Asian Scholars* **10** (2) pp 41–54
Dresden, M.
    1993: Renormalization in historical perspective — the first stage, in Brown (1993,
        pp 31–55)
Duffin, R.J.
    1938: On the characteristics of covariant systems *Phys. Rev.* **54** p 1114
Durandin, E. and Erschow, A.
    1937: Über einige Anwendungen der Supraquantelung in der Wellenmechanik des
        Elektrons *Phys. Zeit. Sowj.* **12** pp 466–71

Eddington, A.S.
    1923: The interior of a star *Nature* **111** — Supp. to issue of 12 May pp v–xii
Eden, R.J.
    1948: Analytic behavior of Heisenberg's S-matrix *Phys. Rev.* **74** p 982
    1949: The analytic behavior of Heisenberg's S matrix *Proc. Roy. Soc. (London)* A
        199 pp 256–71
    1971: Theorems of high-energy collisions of elementary particles *Rev. Mod. Phys.*
        43 pp 15–35
Ehmert, A.
    1937a: Die Absorptionkurve der harten Komponenten der kosmischen Ultra-
        strahlung *Z. Phys* **106** pp 751–73
    1937b: Über den Breiteneffekt der kosmischen Ultrastrahlung *Phys. Z.* **38** pp
        975–8
Ehrenfest, P., Jr
    1938: Sur deux clichés de rayons cosmiques pénétrants otténus dans le champ
        magnétique de Bellevue, et l'existence d'une particule lourd *Comptes rendus*
        (Paris) **206** pp 428–30
Ehrenfest, P., Jr and Fréon, A.
    1938: Désintegration spontanée des 'mésotrons,' particules composant les
        rayonnement cosmique pénétrant *J. Phys. Radium* **9** pp 529–38
Einstein, A.
    1905: Über einen die Erzeugung und Verwandlung des Lichtes betreffenden
        heuristischen Gesichtpunkt *Ann. Phys. (Leipzig)* **17** pp 132–48
Ellis, C.D. and Wooster, W.A.
    1927: The average energy of disintegration of Radium E *Proc. Roy. Soc. (London)*
        A **117** pp 109–23
Estermann, I. and Stern, O.

1933a: Über die magnetische Ablenkung von· Wasserstoffmolekülen und das magnetische Moment des Protons. II *Z. Phys.* **85** 17–24

1933b: Über die magnetische Ablenkung von isotopen Wasserstoffmolekülen und das magnetische Moment des 'Deuterons' *Z. Phys.* **86** 132–4

Estermann, I., Simpson, O.C. and Stern, O.

1937: The magnetic moment of the proton *Phys. Rev.* **57** 1004

Euler, H.

1937: Theoretische Gesichtspunkte zur Untersuchung der Ultrastrahlung *Z. Tech. Phys.* **18** pp 517–25

1938a: Die Erzeugung der Hoffmannschen Stöße durch Multiplikation *Z. Phys.* **110** 450–72

1938b: Über die durchdringende Komponente der kosmischen Strahlung und die von ihr erzeugten Hoffmannschen Stöße *Z. Phys.* **110** pp 692–16

Euler, H. and Heisenberg, W.

1938: Theoretische Gesichtspunkte zur Deutung der kosmischen Strahlung *Ergebnisse der Exakten Naturwissenschaften* **17** pp 1–69

Euler, H. and Wergeland, H.

1939: Über die ausgedehnten Schauer der kosmischen Strahlung in der Luft *Naturwiss.* **27** pp 484–5

Ezawa, H.

1992: Tomonaga in Leipzig, in *Werner Heisenberg als Physiker und Philosoph* ed. B. Geyer, H. Herwig and H. Rechenberg (Heidelberg, Berlin and Oxford) pp 78–86

Fajans, K.

1913: Die Stellung der Radioelemente im Periodischen System *Phys. Zeit.* **14** pp 136–42

Faraday, M.

1839: *Experimental Researches in Electricity* (London)

Feather, N.

1963: Rutherford at Manchester: an epoch in physics, in Rutherford (1963, pp 15–33)

Federbush, P., Goldberger, M.L. and Treiman, S.B.

1958: Electromagnetic structure of the nucleon *Phys. Rev.* **112** pp 642–65

Feinberg, E.

1989: Wissenschaft: Schicksale, Probleme, Hypothesen *Snamia Nr. 3* (German translation)

Fermi, E.

1929: Sopra l'eletrodinamica quantista *Rend. Accad. Lincei* **5** pp 881–7

1930a: Magnetic moments of atomic nuclei *Nature* **125** p 16

1930b: Über die magnetischen Momente der Atomkerne *Zeit. Phys.* **60** pp 320–33

1932a: Quantum theory of radiation *Rev. Mod. Phys.* **4** pp 87–132

1932b: Lo stato attuale della fisica del nucleo atomico *Ricerca Scientifica* **3** pp 101–13

1933: Tentativo di una teoria dell'emissione dei raggi β *Ricerca Scientifica* **4** pp 491–5

1934a: Tentativo di una teoria dei raggi β *Nuovo Cimento* **11** pp 1–19

1934b: Versuch einer Theorie der β-Strahlen. I *Zeit. Phys.* **88** pp 161–71

1939: The absorption of mesotrons in air and in condensed materials *Phys. Rev.* **56** p 1242

1940: The ionization loss of energy in gases and in condensed materials *Phys. Rev.* **57** pp 485–93

1962: *Collected Papers* ed. E. Amaldi, E. Persico, F. Raselti and E. Segrè, Vol. 1 (Chicago)

Fermi, E. and Teller, E.
    1947: Capture of negative mesons in matter *Phys. Rev.* **72** pp 399–407
Fermi, E., Teller, E. and Weisskopf, V.F.
    1947: Decay of negative mesotrons in matter *Phys. Rev.* **71** pp 314–5
Fermi, E. and Yang, C.N.
    1949: Are mesons elementary particles *Phys. Rev.* **76** pp 1739–43
Fermi, L.
    1954: *Atoms in the Family* (Chicago)
Ferretti, B.
    1943: Considerationi sulle forze nucleari, ed alcuni risultati sperimentali sulla diffusione per urto dei neutroni contro i protoni *Nuovo Cimento* **1** pp 25–32
Feynman, R.P. and Gell-Mann, M.
    1958: Theory of Fermi interaction *Phys. Rev.* **109** pp 192–8
Fierz, M.
    1939: Über die relativistische Theorie kräftefreier Teilchen mit beliebigen Spin *Helv. Phys. Acta* **12** pp 3–37
Fierz, M. and Pauli, W.
    1939a: On relativistic wave equations for particles of arbitrary spin in an electromagnetic field *Proc. Roy. Soc. (London)* A **173** pp 211–32
    1939b: Über relativistische Feldgleichungen von Teilchen mit beliebigen Spin *Helv. Phys. Acta* **12** pp 297–300
Fock, V.
    1936: Inconsistency of the neutrino theory of light *Nature* **138** pp 1011–2
    1937: The neutrino theory of light *Nature* **140** p 113
Follet, D.H. and Cranshaw, J.D.
    1936: Cosmic ray measurements under thirty meters of clay *Proc. Roy. Soc. (London)* A **155** pp 546–58
Foster, B. and Fowler, P.H. (eds)
    1988: *40 Years of Particle Physics (International Conference, University of Bristol, July 1987)* (Bristol)
Fowler, P.H.
    1988: The π discovery, in Foster and Fowler (1988, pp 35–54)
Franck, C.
    1993: *Operation Epsilon: The Farm Hall Transcripts* (Bristol)
Frazer, W.R. and Fulco, J.R.
    1959: Effect of a pion-pion scattering resonance on nuclear structure *Phys. Rev. Lett.* **2** pp 365–8
Freier, P., Lofgren, E.J., Ney, E.P., Oppenheimer, F., Bradt, H.L. and Peters, B.
    1948: Evidence for heavy nuclei in the primary cosmic radiation *Phys. Rev.* **74** pp 213–7
Frenkel, J.
    1926: Die Elektrodynamik des rotierender Elektrons *Zeit. Phys.* **37** pp 243–62
Frisch, O. and Stern, O.
    1933: Über die magnetische Ablenkung von Wasserstoffmolekülen und das magnetische Moment des Protons *Zeit. Phys.* **85** 4–16
Fritzsch, H., Gell-Mann, M. and Leutwyler, H.
    1973: Advantages of the color octet gluon picture *Phys. Lett* B **47** pp 365–8
Fröhlich, H.
    1985: The development of the Yukawa theory of nuclear forces *Prog. Theor. Phys. Supp.* No. **85** pp 9–10
Fröhlich H. and Heitler, W.
    1936: Time effects in the magnetic cooling method. II–The conductivity of heat *Proc. Roy. Soc. (London)* A **155** pp 640–52
    1938a: Über die Einstellzeit von Kernspins in Magnetfeld *Phys. Z. Sowj.* **10** pp 847–8

1938b: Magnetic moments of the proton and the neutron *Nature* **141** pp 37–8
Fröhlich, H., Heitler, W. and Kahn, B.
    1939a: Deviation from the Coulomb law for a proton *Proc. Roy. Soc. (London)* A **171** pp 269–80
    1939b: Deviation from the Coulomb law for a proton *Phys. Rev.* **56** pp 961–2
Fröhlich, H., Heitler, W. and Kemmer, N.
    1938: On the nuclear forces and the magnetic moments of the neutron and the proton *Proc. Roy. Soc. (London)* A **166** 154–71
Fubini, S.
    1965: Renormalization effects for partially conserved currents *Physics* **1** pp 228–47
Fussel, L., Jr.
    1937: Production and absorption of cosmic-ray showers *Phys. Rev.* **51** pp 1005–6

Galison, P.
    1983a: The discovery of the muon and the failed revolution against quantum electrodynamics *Centaurus* **26** pp 262–316
    1983b: How the first neutral current experiment ended *Rev. Mod. Phys.* **55** pp 477–509
Galison, P. and Assmus, A.
    1989: Artificial clouds, real particles, in Gooding *et al.* (1989, pp 225–74)
Gamow, G.
    1928a: The quantum theory of nuclear disintegration *Nature* **122** pp 805–6
    1928b: Zur Quantentheorie des Atomkernes *Z. Phys.* **51** pp 204–12
    1928c: Zur Quantentheorie der Atomzertrümmerung *Z. Phys.* **52** pp 510–5
    1931: *Constitution of Atomic Nuclei and Radioactivity* (London)
    1937: *Structure of Atomic Nuclei and Nuclear Transformations* (Oxford)
Gamow, G. and Teller, E.
    1936: Selection rules for the β-disintegration *Phys. Rev.* **49** pp 895–9
    1937: Some generalizations of the β transformation theory *Phys. Rev.* **51** p 289
Gardner, E., Barkas, W.H., Smith, F.M. and Bradner, H.
    1950: Mesons produced by the cyclotron *Science* **111** pp 191–7
Gardner, E. and Lattes, C.M.G.
    1948: Production of mesons by the 184-inch Berkeley cyclotron *Science* **107** pp 270–2
Gasiorowicz, S. and Geffen, D.A.
    1969: Effective Lagrangians and field algebra with chiral symmetry *Rev. Mod Phys.* **41** pp 531–73
Geiger, H.
    1935: Die Sekundäreffekte der kosmischen Ultrastrahlung *Ergebnisse der Exakten Naturwissenschaften* **14** pp 42–78
Geiger, M. and Heyden, M.
    1938: Experimentelles zur Strahlenmultiplikation in den Schauern *Z. Phys.* **110** pp 310–9
Geiger, M. and Marsden, E.
    1909: On a diffuse reflection of the α-particle *Proc. Roy. Soc. (London)* A **82** pp 495–500
Gell-Mann, M.
    1964a: A schematic model of baryons and mesons *Phys. Lett* **8** pp 214–5
    1964b: The symmetry group of vector and axial vector currents *Physics* **1** pp 63–75
Gell-Mann, M., Goldberger, M.L. and Thirring, W.E.
    1954: Use of causality conditions in quantum theory *Phys. Rev.* **95** 1612–27
Gell-Mann, M. and Lévy, M.
    1961: The axial vector current in beta decay *Nuovo Cimento* **16** pp 705–26

Gell-Mann, M. and Ne'eman, Y.
     1964: *The Eightfold Way* (New York)
Gentner, W., Maier-Leibniz, H. and Bothe, W.
     1954: *An Atlas of Typical Expansion Chamber Photographs* (New York)
Gooding, D., Pinch, T. and Schaffer, S.
     1989: *The Uses of Experiment* (Cambridge)
Goldhaber, M.
     1979: The nuclear photoelectric effect and remarks on higher multipole transitions:
         a personal history, in Stuewer (1979, pp 81–110)
Gordon, W.
     1928: Über den Stoß zweier Punktladungen nach der Wellenmechanik *Z. Phys.* **48**
         pp 180–91
Goudsmit, S.A.
     1947: *ALSOS* (New York)
Gross, D.J. and Wilczek, F.
     1973: Ultra-violet behavior of non-abelian gauge theories *Phys. Rev. Lett.* **30** pp
         1343–6
Groves, L.
     1962: *Now It Can Be Told* (New York)
Grythe, I.
     1982: Some remarks on the early S-matrix *Centaurus* **26** pp 198–203
Gurney, R.W. and Condon, E.U.
     1928: Wave mechanics and radioactive disintegration *Nature* **122** p 439
Gürsey, F.
     1968: Effective Lagrangians in particle physics, in Urban (1968, pp 185–225)
Gürsey, F. and Chang, P.
     1968: Nonlinear Lagrangian models under the generalized chiral groups SL(4,C)
         and SL(12C) *Phys. Lett* **26B** pp 520–3

Haken, H.
     1975: Herbert Fröhlich 70 Jahre alt *Phys. Blätter* **31** 664–5
Haken, H. and Wagner, M.
     1973: *Cooperative Phenomena* (Heidelberg)
Halpern, O. and Hall, H.
     1940 Energy losses of fast mesotrons and electrons in condensed materials *Phys.
         Rev.* **57** pp 459–60
Han, M.Y. and Nambu, Y.
     1965: Three-triplet model with double SU(3) symmetry *Phys. Rev.* **139B** pp 1006–
         10
Hara, O.
     1991: Theory of the C-meson, in Brown *et al.* (1991, pp 193–6)
Harkins, W.D.
     1920: The nuclei of atoms and the new periodic system *Phys. Rev.* **15** pp 73–94
Hayakawa, S.
     1969: *Cosmic Ray Physics* (New York)
     1983: The development of meson physics in Japan, in Brown and Hoddeson (1983,
         pp 82–107)
     1991: Sin-itiro Tomonaga and his contributions to quantum electrodynamics and
         high energy physics, in Brown *et al.* (1991, pp 157–67)
Heilbron, J.L.
     1968: The scattering of α and β particles and Rutherford's atom *Arch. Hist. Exact
         Sci.* **4** pp 247–307
     1974: *H.G. Moseley. The Life and Letters of an English Physicist, 1887–1915* (New
         York)

1977a: Lectures on the history of atomic physics, in *History of Twentieth Century Physics* (New York) pp 40–108

1977b: J.J. Thomson and the Bohr Atom *Physics Today* **30** (4) pp 23–30

Heisenberg, W.

1930: Die Selbstenergie des Elektrons *Z. Phys.* **65** pp 4–13

1931: Zum Paulischen Ausschließungsprinzip *Ann. Phys. (Leipzig)* **10** pp 888–904

1932: Über den Bau der Atomkerne. I, II *Z. Phys.* **77** pp 1–11; **78** pp 156–64

1933: Über den Bau der Atom Kerne. III *Z. Phys.* **80** pp 587–96

1934: Considerations générales sur la structure du noyau in *Institut Internationale de Physique Solvay* (1934, pp 289–335)

1935: Bemerkungen zur Theorie des Atomkerns, in *Zeeman Verhandelingen* (The Hague) pp 108–16

1936: Zur Theorie der 'Schauer' in der Höhenstrahlung *Z. Phys.* **101** pp 533–40

1937: Der Durchgang sehr energiereicher Korpuskeln durch den Atomkern *Sächs. Akad. Wiss. (Leipzig). Berichte d. math.-phys. Kl.* **89** pp 369–384

1938a: Über die in der Theorie der Elementarteilchen auftretende universelle Länge *Ann. Phys. (Leipzig)* **32** pp 20–33

1938b: Die Grenzen der Anwendbarkeit der bisherigen Quantentheorie *Z. Phys.* **110** pp 251–66

1939a: Das schwere Elektron (Mesotron) und seine Rolle in der Höhenstrahlung (report of Physikalisches Colloquium Hamburg) *Angewandte Chemie* **52** p 41

1939b: Zur Theorie der explosionartigen Schauer in der kosmischen Strahlung *Z. Phys.* **113** pp 61–86

1939c: On the theory of explosive showers in cosmic rays *Rev. Mod. Phys.* **11** p 241

1943a: *Kosmische Strahlung. Vorträge gehalten im Max-Planck Institut Berlin-Dahlem* (Berlin)

1943b: Übersicht über den jetzigen Stand unserer Kentnisse von der kosmischen Strahlung, in Heisenberg (1943a, pp 1–10)

1943c: Die 'beobachteren Größen' in der Theorie der Elementarteilchen *Zeit. Phys.* **120** pp 513–38

1943d: Die beobachtbaren Größen in der Theorie der Elementartilchen. II *Z. Phys.* **120** pp 673–702

1944: Die beobachtbaren Größen in der Theorie der Elementartilchen. III *Z. Phys.* **123** pp 93–112

1946 Der mathematischen Rahmen der Quantentheorie der Wellenfelder *Z. Naturforschung* **1** pp 608–22

1953: *Kosmische Strahlung: Vorträge gehalten in Max-Planck-Institut für Physik, Göttingen* (Berlin)

1984–93: *Gesammelte Werke/Collected Works* eds W. Blum, H.-P. Dürr, and H. Rechenberg, Series A,B, 1984–1993, Berlin; Series C 1984–1989, Munich

1992: *Deutsche und 'jüdische' Physik* ed. H. Rechenberg (Munich)

Heisenberg, W. and Pauli, W.

1929 Zur Quantendynamik der Wellenfelder *Z. Phys.* **56** pp 1–61

1930: Zur Quanten theorie der Wellen felder. II *Z. Phys.* **59** pp 168–90

Heitler, W.

1936: *The Quantum Theory of Radiation* (Oxford) (2nd ed 1944)

1937: On the analysis of cosmic rays *Proc. Roy. Soc. (London)* A **161** pp 261–83

1938a: Showers produced by the penetrating cosmic radiation *Proc. Roy. Soc. (London)* A **166** 529–43

1938b: Remarks on nuclear disintegrations by cosmic rays *Phys. Rev.* **54** pp 873–6

1938c: Cosmic rays *Rep. Prog. Phys.* **5** pp 361–89

1940: Scattering of mesons and the magnetic moments of proton and neutron *Nature* **145** pp 29–30

1941: The influence of radiation damping on the scattering of light and mesons by free particles. I *Proc. Camb. Phil. Soc.* **37** pp 291–300

1946: A theorem in the charge-symmetrical meson theory *Proc. Roy. Irish Acad.* **51A** pp 33–9

1947: The quantum theory of damping as a proposal for Heisenberg's S-matrix, in International Conference (1947, pp 189–94)

1973: Erinnerungen an die gemeinsame Arbeit mit Herbert Fröhlich, in Haken and Wagner (1973, pp 421–4)

1985: Personal recollections of early theoretical cosmic ray work, in Sekido and Elliot (1985, pp 209–11)

Heitler, W. and Herzberg, G.

1929: Gehorchen die Stickstoffkerne der Boseschen Statistik? *Naturwiss.* **17** pp 673–4

Heitler, W. and Peng, H.W.

1942a: Anomalous scattering of mesons *Phys. Rev.* **81** p 81

1942b: The influence of radiation damping on the scattering of mesons. II. Multiple processes *Proc. Camb. Phil. Soc.* **38** pp 296–312

1943: On the production of mesons by proton–proton collisions *Proc. Roy Irish Acad.* **49** pp 101–33

1945: On the production of mesons by proton–proton collisions. II *Proc. Roy. Irish Acad.* **50** pp 155–65

Heitler, W., Powell, C.F. and Fertel, G.E.F.

1939: Heavy cosmic ray particles at Jungfraujoch and sea level *Nature* **144** pp 283–4

Heitler, W. and Teller, E.

1936: Time effects in the magnetic cooling method.I *Proc. Roy. Soc. (London) A* **155** 629–39

Helmholtz, H.

1881: On the modern development of Faraday's conception of electricity (excerpt) *Nature* **23** pp 536–40

Hendry, J.

1984: *Cambridge Physics in the Thirties* (Bristol)

Hess, V.F.

1912: Beobachtungen der durchdringenden Strahlung bei sieben Freiballonfahrten *Sitz. ber. Akad. Wiss. (Wien)* **121** 2001–32

Hevesy, G. von and Paneth, F.

1914: Zur Frage der isotope Elemente *Phys. Z.* **15** pp 797–805

Hiebert, E.

1988: The role of experiment and theory in the development of nuclear physics in the early 1930's, in *Theory and Experiment* eds D. Batens and J.P. van Bendergern (Dordrecht) pp 55–76

Hirosige, T.

1974: Social conditions for prewar Japanese research in nuclear physics, in Nakayama *et al.* (1974, pp 202–20)

Hoch, P.

1990: Flight into self absorption and xenophobia. The plight of refugee theorists amongst British and American experimentalists in the 1930s highlights cultural and natural differences in science *Physics World* No 1, pp 23–6

Hoddeson, L., Brown, L.M., Dresden, M. and Riordan, M.

1996: *The Rise of the Standard Model* (New York)

Hoffmann, G.

1931: Über exakte Intensitätsmessung der Hess'schen Ultrastrahlung *Z. Phys.* **69** pp 703–18

1935: The connection between cosmic radiation and atomic disintegration, in *Conference International* (1935, pp 226–32)

Hoffmann, G. and Lindholm, F.
1928: Registrierbeobachtungen der Hessschen Ultra-Strahlung auf Muottas Muraigl (2456 m) *Gerlachs Beiträge zur Geophysik* **20** pp 12–54

Hofstadter, R.
1989: A personal view of nuclear structure as revealed by electron scattering, in L.own *et al.* (1989, pp 126–43)

Hofstadter, R. and McAllister, R.W.
1955: Electron scattering from the proton *Phys. Rev.* **98** pp 217–8

Houtermans, F.G.
1930: Neuere Arbeiten in der Quantentheorie des Atomkerns *Ergebnisse der Exakten Naturwissenschaften* **9** pp 123–221

Hu, N.
1945: The relativistic correction in the meson theory of nuclear forces *Phys. Rev.* **67** pp 339–46

Hulthén, L.
1943a: On the scattering of neutrons by protons *Phys. Rev.* **63** p 383
1943b: On the meson field theory of nuclear forces and the scattering of fast neutrons by protons. I, II *Arkiv för Matematik, Astronomi och Fysik.* A **29** No 33; A **30** No 9
1944a: On the meson field theory of nuclear forces and the scattering of fast neutrons by protons. III *Arkiv för Matematik, Astronomi och Fysik* A **31** No 15
1944b: Nuclear forces in a non-symmetrical scalar-pseudoscalar meson field theory *K. Fysiogr. Sällsk. Lund Förh.* **14** No 2

Institut International de Physique (E. Solvay)
1934: *Structure et Propriétés de Noyaux Atomiques: Rapports et Discussions du Septiéme Conseil de Physique* (Paris)
1950: *Les Particules Élémentaire. Rapports et Discussions du Huitième Conseil de Physique* (Brussels)

International Conference
1935: *International Conference on Physics, London, 1934* (Cambridge)
1947: *International Conference on Fundamental Particles and Low Temperatures at the Cavendish Laboratory, Cambridge, on 22–27 July 1946. Volume I. Fundamental Particles* (London)

International Institute of Intellectual Cooperation
1939: *New Theories in Physics Conference (Warsaw, 30 May–3 June, 1938)* (Paris)

Irving, D.
1967: *The German Atomic Bomb* (New York)

Itakura, K. and Yagi, E.
1974: The Japanese research system and the establishment of the Institute of Physical and Chemical Research, in Nakayama *et al.* (1974, pp 158–201)

Ito, D., Koba, Z. and Tomonaga, S.
1947: Correction due to the reaction of 'cohesive force field' for the elastic scattering of an electron *Prog. Theor. Phys.* **2** pp 216–7. *Errata: ibid.* pp 217–8

Iwanenko, D.
1932a: Sur la constitution des noyaux atomiques *Comptes rendus (Paris)* **195** pp 439–41
1932b: The neutron hypothesis *Nature* **129** p 798
1934: Interaction of neutrons and protons *Nature* **133** pp 981–2
1939: Classical dynamics of the meson *Nature* **144** pp 77–8
1940: Remarks on the meson theory *J. Phys. (USSR)* **3** pp 417–9

Iwanenko, D. and Sokolov, A.
   1940: Zur klassischen Mesodynamik *J. Phys. (USSR)* **3** pp 57–64
   1942: The dipole character of the meson and the polarization of the vacuum *J. Phys. (USSR)* **6** pp 175–9

Jackmann, J.C. and Borden, C.M. eds.
   1983: *The Muses Flee Hitler: Cultural Transfer and Adaptation, 1930–1945* (Washington, D.C.)
Jammer, M.
   1966: *The Conceptual Development of Quantum Mechanics* (New York)
Janóssy, L.
   1942: Penetrating cosmic ray showers *Proc. Roy. Soc. (London)* A **179** pp 361–376
Janóssy, L. and Ingelby, P.
   1940: Penetrating cosmic ray showers *Nature* **145** p 511
Johnson, T.H. and Pomerantz, M.A.
   1939: The difference in the absorption of cosmic rays in air and water and the instability of the barytron *Phys. Rev.* **55** pp 104–5
Johnson, T.H. and Shutt, R.P.
   1942: Cloud-chamber track of a mesotron stopped by gas *Phys. Rev.* **61** pp 380–1
Jordan, P.
   1928: Die Lichtquanthypothese *Ergebnisse der Exakten Naturwissenschaften* **7** pp 155–208
   1935: Zur Neutrinotheorie des Lichtes *Z. Phys.* **93** pp 464–72
   1936a: Lichtquant und Neutrino *Z. Phys.* **98** pp 759–67
   1936b: Beiträge zur Neutrinotheorie des Lichtes. I *Z. Phys.* **102** 43–252
   1937: Beiträge zur Neutrinotheorie des Lichtes. II *Z. Phys.* **105** pp 114–21
Jordan, P. and Klein, O.
   1927: Zum Mehrkörperproblem der Quantentheorie *Z. Phys.* **47** pp 751–63
Jordan, P. and Kronig, R. de Laer
   1936: Lichtquant und Neutrino *Z. Phys.* **98** pp 759–67
Jordan, P. and Wigner, E.
   1928: Über das Paulische Äquivalenzverbot *Z. Phys.* **47** pp 631–51
Jost, R.
   1983: Walter Heitler. Nekrolog *Vierteljahrschrift des Naturf. Ges. in Zürich* **128** 139–41
Juilfs, J.
   1943: Über die Lebensdauer von Mesonen *Naturwiss.* **31** pp 109–10

Kaneseki, Y.
   1974 The elementary particle theory group, in Nakayama *et al.* (1974, pp 221–52)
Kaufmann, W.
   1901: Die magnetische und elektrische Ablenkbarkeit der Becquerelenstrahlen und die scheinbare Masse der Elektron *Nachr. Ges. Wiss. Göttingen* **1901** pp 143–55
Kawabe, R.
   1991a: From meson theory to nonlocal field theory, in Brown *et al.* (1991, pp 289–94)
   1991b: Two unpublished manuscripts of Yukawa on the meson theory... Hideki Yukawa in 1937, in Brown *et al.* (1991, pp 262–9)
   1991c: A grand illusion, in Brown *et al.* (1991, pp 84–85)
Kellogg, J.B., Rabi, I.I., Ramsey, N.F. and Zacharias, J.R.
   1939: An electrical quadrupole moment of the deuteron *Phys. Rev.* **55** pp 318–9
Kellogg, J.B., Rabi, I.I. and Zacharias, J.R.
   1936: The gyromagnetic properties of hydrogen *Phys. Rev.* **56** 472–6

Kemmer, N.
  1937: Field theory of nuclear interaction *Phys. Rev.* **52** 906–10
  1938a: Nature of the nuclear field *Nature* **141** 116–7
  1938b: Quantum theory of Einstein-Bose particles and nuclear interaction *Proc. Roy. Soc. (London)* A **166** 127–53
  1938c: The charge-dependence of nuclear forces *Proc. Camb. Phil. Soc.* **34** pp 354–64
  1939: The particle aspect of meson theory *Proc. Roy. Soc. (London)* A **173** pp 91–116
  1965: The impact of Yukawa's meson theory on workers in Europe — a reminiscence *Supp. of the Prog. of Theor. Phys. (Commemoration Issue for the 30th Anniversary of the Meson Theory by Dr. H. Yukawa)* pp 602–8
  1971: Some recollections from the early days of particle physics, in Cumming and Osborn (1971, pp 1–16)
  1983a: Die Anfänge der Mesontheorie und das verallgemeinerten Isospins *Phys. Blätter* **39** pp 170–5
  1983b: Hideki Yukawa *Biographical Memoirs of Fellows of the Royal Society* **29** pp 661–6
Klein, O.
  1926: The atomicity of electricity as a quantum law *Nature* **118** p 516
  1927: Zur fünfdimensionalen Darstellung der Relativitätstheorie *Z. Phys.* **46** pp 188–208
  1929: Die Reflexion von Elektronen an einem Potentialsprung nach der relativistischen Dynamik von Dirac *Z. Phys.* **53** pp 157–65
  1939: On the theory of charged fields, in International Institute of Intellectual Cooperation (1939, pp 77–93)
Klein, O. and Nishina, Y.
  1929: Über die Streuung von Strahlen durch freie Elektronen nach der neuen relativistischen Quantendynamik von Dirac *Z. Phys.* **52** 853–68
Koba, Z. and Tomonaga, S.
  1947: Application of the 'self-consistent' subtraction method to the scattering of an electron *Prog. Theor. Phys.* **2** pp 218–9
Kobayasi, M.
  1941: On the meson theory of the penetrating component of cosmic radiation *Proc. Phys.-Math. Soc. Japan* **23** pp 891–914
Kobayasi, M. and Okayama, T.
  1939: On the creation and annihilation of heavy quanta in matter *Proc. PMSJ* **21** pp 1–13
Koizumi, K.
  1975: The emergence of Japan's first physicists: 1868–1900 *Historical Studies in the Physical Sciences* **6** pp 3–108
Konopinski, E.J. and Uhlenbeck, G.E.
  1935: On the Fermi theory of β-radioactivity *Phys. Rev.* **48** pp 7–12
Konuma, M.
  1989: Social aspects of Japanese particle physics in the 1950s, in Brown *et al.* (1989, pp 536–50)
Konuma, M. and Rechenberg, H.
  1985: Hideki Yukawa in Deutschland *Phys. Blätter* **41** pp 342–4
Kramers, H.A.
  1939: Limits of applicability of the present system of theoretical physics, in International Institute of Intellectual Cooperation (1939, pp 95–118)
Kramers, H.A. and Wannier, G.H.
  1941: Statistics of the two dimensional ferromagnet. I. and II. *Phys. Rev.* **60** pp 252–76
Kronig, R. de L.

1926: Spinning electrons and the structure of spectra *Nature* **117** p 550

1928: Der Drehimpuls des Stickstoffkerns *Naturwiss.* **16** p 335

1935: Zur Neutrinotheorie des Lichtes *Physica* **2** pp 491–8, 854–60 and 968–80

1936a: On a relativistically invariant formulation of the neutrino theory of light *Physica* **3** pp 1120–32

1936b: The neutrino theory of radiation and the emission of γ-rays *Nature* **137** p 149

Kronig, R. de L. and Fisch, S.

1931: Kernmomente *Phys. Z.* **32** pp 457–72

Kuhn, T.S., Heilbron, J.L., Forman, P. and Allen, L.

1967: *Sources for History of Quantum Physics* (Philadelphia)

Kulenkampff, H.

1938: Bemerkungen über die durchdringende Komponente der Ultrastrahlung (zum Teil nach Messungen von H. Kappler und H. Martin) *Ver. Deutsch. Phys. Ges.* **19** p 92

Kunze, P.

1932a: Magnetisches Spektrum der Höhenstrahlung *Zeit. Phys.* **79** pp 203–5

1932b: Magnetische Ablenkung der Ultrastrahlen in der Wilsonkammer *Z. Phys.* **80** pp 559–72

1933: Untersuchung der Ultrastrahlung in der Wilsonkammer *Z. Phys.* **83** pp 1–18

Kursunoglu, B. and E.P. Wigner (eds)

1987: *Reminiscences About a Great Scientist* (Cambridge)

Kusaka, S.

1941: β-decay with neutrino of spin 3/2 *Phys. Rev.* **60** pp 61–2

1943: The effect of radiation damping on burst production *Phys. Rev.* **64** pp 256–7

Kusch, P. and Foley, H.M.

1948: The magnetic moment of the electron *Phys. Rev.* **74** pp 250–3

Lamb, W.E.

1936: The unobservable decay of Na^{24} *Phys. Rev.* **50** pp 388–9

1939: Deviation from the Coulomb law for a proton *Phys. Rev.* **56** p 384

1940: Deviation from the Coulomb law for a proton *Phys. Rev.* **57** p 458

1983: The fine structure of hydrogen, in Brown and Hoddeson (1983, pp 311–328)

Lamb, W.E. and Retherford, R.

1947: Fine structure of the hydrogen atom by a microwave method *Phys. Rev.* **72** pp 136–8

Lamb, W.E. and Schiff, L.I.

1938: On the electromagnetic properties of nuclear systems *Phys. Rev.* **53** pp 651–61

Landau, L.D.

1941: The theory of superfluidity of helium II *J. Phys. (Moscow)* **8** pp 1–3 (also published in 1941 *Phys. Rev.* **60** pp 356–8)

1944: On the hydrodynamics of helium II *J. Phys. (Moscow)* **5** 71–90, 1941

Landau, L. and Rumer, G.

1937: Production of showers by heavy particles *Nature* **140** p 682

Laporte, O.

1938a: Scattering of Yukawa particles by protons *Nature* **142** p 432

1938b: Elastic scattering of Yukawa particles. I *Phys. Rev.* **54** pp 905–12

Lasarew, B.G. and Schubnikow, L.W.

1936: Über den magnetischen Moment des Protons *Phys. Z. Sowj.* **10** 117–8

1937: Über den magnetischen Moment des Protons *Phys. Z. Sowj.* **11** 445–57

Lattes, C.M.G.

1983: My work in meson physics with nuclear emulsions, in Brown and Hoddeson (1983, pp 307–10)

Lattes, C.M.G., Muirhead, H., Occhialini, G.P.S. and Powell, C.F.
1947a: Processes involving charged mesons *Nature* **159** pp 694–7

Lattes, C.M.G., Occhialini, G.P.S. and Powell, C.F.
1947b: Observations on the tracks of slow mesons in photographic emulsion *Nature* **160** pp 453–6 and 486–492
1948: A determination of the ratio of the masses of π and μ mesons by the method of grain-counting *Proc. Phys. Soc. (London)* **61** pp 173–83

Lawrence, E.O. and Livingston, M.S.
1932: The production of high speed light ions without the use of high voltages *Phys. Rev.* **40** pp 19–35

Lawrence, E.O., Livingston, M.S. and White, M.G.
1932: The disintegration of lithium by swiftly-moving protons *Phys. Rev.* **42** pp 150–51

Lee, T.D.
1954: Some special examples of a renormalizable field theory *Phys. Rev.* **95** pp 1329–34

Lee, T.D., Oehme, R. and Yang, C.N.
1957: Remarks on possible noninvariance under time reversal and charge conjugation *Phys. Rev.* **106** pp 340–5

Lee, T.D., and Yang, C.N.
1956: Question of parity conservation in weak interactions *Phys. Rev.* **104** pp 254–8

Lenz, W.
1918: Über ein invertierte Bohrsches Modell *Sitzungsber. Akad. Wiss. (Munich)* **1918** pp 355–65

Leprince-Ringuet, L.
1983: The scientific activities of Leprince-Ringuet and his group on cosmic rays: 1933–1953, in Brown and Hoddeson (1983, pp 177–82)

Leprince-Ringuet, L. and Crussard, J.
1937: Étude des particules de grande énergie dans le champ magnétique de l'electro-aimant de Bellevue *Comptes rendus (Paris)* **201** pp 1184–7

Leprince-Ringuet, L. and Lhéritier, M.
1944: Existence probable d'une particule de masse $990m_e$ dans la rayonnement cosmique *Comptes Rendus (Paris)* **219** pp 618–20
1946: Existence probable d'une particule de masse $(990 \pm 12 \text{ pour } 100)m_e$ dans la rayonnement cosmique *J. Phys. Rad.* **7** pp 65–9

Lewis, H.W., Oppenheimer, J.R. and Wouthuysen, S.A.
1948: The multiple production of mesons *Phys. Rev.* **73** pp 127–40

Lifschitz, E.
1941: Report on the nuclear physics conference *J. Phys. Acad. Sci. USSR* **4** pp 277–86

Livingston, M.S. and Bethe, H.A.
1937: Nuclear physics. C. Nuclear dynamics, experimental *Rev. Mod. Phys.* **9** pp 245–390

Lock, W.O.
1988: Early work with electron-sensitive emulsions, in Foster and Fowler (1988, pp 61–76)

Lorentz, H.A.
1916: *The Theory of Electrons* 2nd edn (Leipzig)

Lüders, G.
1954: On the equivalence of invariance under time reversal and under particle–antiparticle conjugation for relativistic field theories *Kgl. Danske Videskab. Selskab., Mat.-Fys. Meddelelser* **28** No 5

Ma, S.T.
    1942: Calculations of the scattering of mesons by the matrix method *Phys. Rev.* **62** pp 403–11
Ma, S.T. and Yu, F.C.
    1942: Electromagnetic properties of nuclei in the meson theory *Phys. Rev.* **62** pp 118–26
Maass, H.
    1936: Über eine harte Sekundärstrahlung *Ann. Phys.* **27** pp 507–31
Maglic, B.
    1973: *Adventures in Experimental Physics. Gamma Volume* (Princeton)
Maglic, B., Alvarez, L.W., Rosenfeld, A.M. and Stevenson, M.L.
    1961: Evidence for a $T = 0$ three-pion resonance *Phys. Rev. Lett.* **7** pp 178–82
Maier-Leibnitz, H.
    1938: Wilson-Aufnahmen schwerer Elektronen *Naturwiss.* **26** pp 677–8
Majorana, E.
    1933a: Über die Kerntheorie *Z. Phys.* **82** pp 137–45
    1933b: Sulla teoria dei nuclei *Ricerca Scientifica* **4** pp 559–65
Majumdar, R.C. and Kothari, D.S.
    1939: The meson and its transformation into heavy particles *Nature* **143** pp 796–7
Marsden, E.
    1962: Rutherford at Manchester, in *Rutherford at Manchester* ed. J.B. Birks (London, pp 1–16)
Marshak, R.E.
    1940: Heavy electron pair theory of nuclear forces *Phys. Rev.* **57** pp 1101–6
    1951: Meson reactions in hydrogen and deuterium *Rev. Mod. Phys.* **23** pp 137–46
    1952: *Meson Physics* (New York)
Marshak, R.E. and Bethe, H.A.
    1947: On the two-meson hypothesis *Phys. Rev.* **72** pp 506–9
Marshak, R.E. and Weisskopf, V.F.
    1941: On the scattering of mesons of spin 1/2 by atomic nuclei *Phys. Rev.* **59** pp 130–5
Massey, H.S.W. and Corben, H.C.
    1939: Elastic collisions of mesons with electrons and protons *Proc. Camb. Phil. Soc.* **35** pp 463–73
Matsui, M. and Ezawa, H. (eds)
    1995: *Sin-itiro Tomonaga — Life of a Japanese Physicist* (Tokyo)
McMillan, E.M.
    1979: Early history of particle accelerators, in Stuewer (1979, pp 11–155)
Mehra, J.
    1975: *The Solvay Conferences on Physics* (Dordrecht)
Mehra, J. and Rechenberg, H.
    1982: *The Historical Development of Quantum Theory, Vol. 1: The Quantum Theory of Planck, Einstein, Bohr, and Sommerfeld* (New York, Heidelberg and Berlin)
    1987: *The Historical Development of Quantum Theory, Vol. 5: Erwin Schrödinger and the Rise of Wave Mechanics* (New York, Heidelberg and Berlin)
Meitner, L.
    1921: Über die verschiedenen Arten des radioaktiven Zerfalls und die Möglichkeit ihrer Deutung aus der Kernstruktur *Phys. Z.* **4** pp 146–56
Meitner, L. and Orthmann, W.
    1930: Über eine absolute Bestimmung der Energie der primären β-Strahlen von Radium E *Z. Phys.* **60** pp 143–55
Messerschmidt, W.
    1936: Untersuchungen über Ultrastrahlungsstöße *Z. Phys.* **103** pp 27–56
Meyenn, K. von

1982: Pauli, das Neutrino und die Entdeckung des Neutrons vor 50 Jahren *Naturwiss.* **69** pp 564–73

1993: Die Princetoner Jahre und die Rückkehr nach Zürich, in [WPSC3] pp VII-LXIV

Meyer, P.

1953: Stöße in Ionizationkammern, in Heisenberg (1953, pp 105–10)

Miller, A.I.

1984: *Imagery in Scientific Thought* (Boston)

Millikan, R.A.

1939: Mesotron as the name of the new particle *Phys. Rev.* **55** p 105

Miyazima, T.

1941: On the mesotron of spin 1/2 *Sci. Papers IPCR* **39** pp 161–73

Miyazima, T. and Tomonaga, S.

1942: Zur Theorie des Mesons II *Sci. Papers IPCR* **40** pp 21–67

1943: On the mesotron theory of nuclear forces *Sci. Papers IPCR* **40** pp 274–310

Møller, C.

1937a: On the capture of orbital electrons by nuclei *Phys. Rev.* **51** pp 84–5

1937b; On the capture of orbital electrons by nuclei *Phys. Z. Sowj.* **11** pp 9–17

1938: The theory of nuclear forces *Nature* **142** pp 290–291

1945–46: General properties of the characteristic matrix in the theory of elementary particles I and II, Kgl. Danske Videskab. Selskab., Mat.-Fys. Meddelelser **23** Nr. 1 pp 1–48 (1945) and *ibid.* **24** Nr. 19 pp 1–46 (1946)

1946: New developments in relativistic quantum field theory *Nature* **158** pp 403–6

1947: The possible existence of mass spectra.–On the theory of the characteristic matrix, in International Conference (1947, Vol. I, pp 194–8)

Møller, C. and Rosenfeld, L.

1939a: Theory of mesons and nuclear forces *Nature* **143** pp 241–2

1939b: The electric quadrupole moment of the deuteron and the field theory of nuclear forces *Nature* **144** pp 476–7

1940: On the field theory of nuclear forces *K. Danske Vidensk. Selskab Mat.-fys. Meddelser* **17** No 8 pp 1–72

Møller, C., L. Rosenfeld, and S. Rozental

1939: Connexion between the life-time of the meson and the beta-decay of light elements *Nature* **144** p 629

Montgomery, C.G. and Montgomery, D.D.

1935a: The production of cosmic ray showers by lead at different elevations *Phys. Rev.* **47** p 339

1935b: The variation with altitude of the production of bursts of cosmic ray ionization *Phys. Rev.* **47** pp 429–34

1939: The behavior of high energy electrons in the cosmic radiation *Rev. Mod. Phys.* **11** pp 255–64

1941: The transition effect for large showers of cosmic rays *Phys. Rev.* **59** p 471

1947: The transition effect for large bursts of cosmic-ray ionization *Phys. Rev.* **72** pp 131–4

Moore, Ruth

1970: *Niels Bohr* (Munich)

Moseley, H.G.J.

1913: High-frequency spectra of the elements *Phil. Mag.* **26** pp 1025–34

1914: High-frequency spectra of the elements II *Phil. Mag.* **27** pp 703–13

Mott, N.

1928: The solution of the wave equation for the scattering of particles by a Coulombian centre of force *Proc. Roy. Soc. (London)* A **118** pp 542–9

1929: The exclusion principle and aperiodic systems *Proc. Roy. Soc. (London)* A **125** pp 222–30

1930: The collision between two electrons *Proc. Roy. Soc. (London)* A **126** pp 259–67

1984: Theory and experiment in the Cavendish circa 1932, in Hendry (1984, pp 125–32)

Mukherji, V.

1974: A history of the meson theory of nuclear forces from 1935 to 1952 *Archive for History of the Exact Sciences* **13** pp 27–102

Nagaoka, H.

1903: Motion of particles in an ideal atom illustrating the line and band spectra and the phenomena of radioactivity *Proc. Tokyo Math-Phys. Soc.* **2** pp 92–107 and 129–31

Nagaoka, H., Sugiura, Y. and Mishima, T.

1924: Isotopes of mercury and bismuth revealed in the satellites of their spectral lines *Nature* **113** pp 459–60

Nagendra Nath, N.S.

1937: Neutrino theory of light *Nature* **139** p 331

Nakayama, S.

1977: *Characteristics of Scientific Development in Japan* (New Delhi)

1984: *Academic and Scientific Traditions in China, Japan, and the West* translated by J. Dusenbury (Tokyo)

Nakayama, S., Swain, D.L. and Yagi, E.

1974: *Science and Society in Modern Japan* (Cambridge, MA)

Nambu, Y.

1957: Possible existence of a heavy neutral meson *Phys. Rev.* **106** pp 1366–7

1966: A systematics of hadrons in subnuclear physics in *Preludes in Theoretical Physics* eds A. De-Shalit, H. Feshbach, and L. van Hove (Amsterdam) pp 143–53

Neddermeyer, S.H.

1938: The penetrating cosmic ray particle *Phys. Rev.* **53** pp 102–3

Neddermeyer, S.H. and Anderson, C.D.

1937: Note on the nature of cosmic ray particles *Phys. Rev.* **51** pp 884–6

1938: Cosmic-ray particles of intermediate mass *Phys. Rev.* **54** pp 88–89

1939: Nature of cosmic-ray particles *Rev. Mod. Phys.* **11** pp 191–207

Neher, H.V. and Stever, H.G.

1940: The mean lifetime of the mesotron from electroscope data *Phys. Rev.* **58** pp 766–70

Nelson, E.C.

1939: Mass and mean life-time of the meson *Nature* **143** pp 761–2

1941: On the pseudoscalar mesotron theory of β-decay *Phys. Rev.* **60** pp 830–3

Nereson, N. and Rossi, B.

1943: Further measurements on the disintegration curve of mesotrons *Phys. Rev.* **64** pp 199–201

Nishina, Y., Sekido, Y., Miyazaki, Y. and Masudo, T.

1941: Cosmic rays at a depth equivalent to 1400 meters of water *Phys. Rev.* **59** p 401

Nishina, Y., Takeuchi, M. and Ichimiya, T.

1937: On the nature of cosmic ray particles *Phys. Rev.* **52** p 1198–9

Nishina, Y. and Tomonaga, S.

1936: A note on the interaction of the neutron and the proton *Sci. Papers IPCR* **30** pp 61–9

Nisio, S.

1965: α-rays and the atomic nucleus *Jap. Stud. Hist. Soc.* **4** pp 91–116

Nobel Lectures

1964: *Physics 1942–1962* (Amsterdam)

1965: *Physics 1922–1941* (Amsterdam)

1972: *Physics 1963–1970* (Amsterdam)

Nordheim, G., Nordheim, L.W., Oppenheimer, J.R. and Serber, R.

    1937: The disintegration of high energy protons *Phys. Rev.* **51** pp 1037–45

Nordheim, L.W.

    1938: A new analysis of cosmic radiation including the hard component *Phys. Rev.* **53** pp 694–706

    1939: Lifetime of the Yukawa particle *Phys. Rev.* **55** p 506

Nordheim, L.W. and Nordheim, G.

    1938: On the production of heavy electrons *Phys. Rev.* **54** pp 254–65

Nordsieck, A.

    1934: Neutron collisions and the beta-ray theory of Fermi *Phys. Rev.* **46** pp 234–5

Occhialini, G.P.S. and Powell, C.F.

    1947: Nuclear disintegrations produced by slow charged particles of small mass *Nature* **159** pp 168–90

Oehme, R.

    1989: Theory of the scattering matrix (1942–1946), Annotation, in Heisenberg, Vol. AII (1989, pp 605–10)

Onsager, L.

    1944: Crystal statistics. I. A two-dimensional model with an order-disorder transition *Phys. Rev.* **65** pp 117–49

Oppenheimer, J.R.

    1939: Discussion *Rev. Mod. Phys.* **11** pp 264–6

    1941: On the spin of the mesotron *Phys. Rev.* **59** p 462

Oppenheimer, J.R. and Schwinger, J.

    1941: On the interaction of mesotrons and nuclei *Phys. Rev.* **60** pp 150–2

Oppenheimer, J.R. and Serber, R.

    1937: Note on the nature of cosmic ray particles *Phys. Rev.* **51** p 1113

Oppenheimer, J.R., Snyder, H.S. and Serber, R.

    1940: The production of soft secondaries by mesotrons *Phys. Rev.* **57** pp 75–81

Pais, A.

    1977: Radioactivity's two early puzzles *Rev. Mod. Phys.* **49** pp 925–38

    1986: *Inward Bound* (Oxford and New York)

    1995: Introducing atoms and their nuclei, in Brown *et al.* (1995, pp 43–141)

Panofsky, W.K.H., Aamodt, R.L. and Hadley, J.

    1951: The gamma-ray spectrum resulting from capture of negative (-mesons in hydrogen and deuterium *Phys. Rev.* **81** pp 565–74

Pasternack, S.

    1938: Note on the fine structure of H_α and D_α *Phys. Rev.* **54** pp 1113

Pauli, W.

    1924: Zur Frage der theoretischen Dentung der Satelliten einiger Spektrallinien und ihrer Beeinflussung durch magnetische Felder *Naturwiss.* **12** pp 741–3

    1939: Bericht über allgemeinen Eigenschaften der Elementarteilchen, chapters I and II (chapters III and IV by Heisenberg), Report for 1939 Solvay Conference, in [WPSC3] pp 825–901. With commentary by M.G. Doncel.

    1940: The connection between spin and statistics *Phys. Rev.* **58** pp 716–22

    1941: Relativistic field theories of elementary particles *Rev. Mod. Phys.* **13** pp 203–12

    1943a: On Dirac's new method of field quantization *Rev. Mod. Phys.* **15** pp 85–108

    1943b: On applications of the λ-limiting process to the theory of the meson field *Phys. Rev.* **64** pp 332–44

1946: *Meson Theory of Nuclear Forces* (New York) 2nd edn 1948
1961: Zur älteren und neueren Geschichte des Neutrinos, in *Aufsätze und Vorträge über Physik und Erkenntnistheorie* ed. V. Weisskopf (Braunschweig) pp 156–80
Pauli, W. and Belinfante, F.J.
    1940: On the statistical behavior of known and unknown elementary particles *Physica* **7** pp 177–92
Pauli, W. and Dancoff, S.M.
    1942: The pseudoscalar meson theory with strong coupling *Phys. Rev.* **62** pp 85–108
Pauli, W. and Kusaka, S.
    1943: On the theory of a mixed pseudoscalar and a vector meson field *Phys. Rev.* **63** pp 400–16
Pauli, W. and Weisskopf, V.
    1934: Über die Quantisierung der skalaren relativistischen Wellengleichung *Helv. Phys. Acta* **7** pp 709–31
Peierls, R.
    1939: The meson *Rep. Prog. Phys.* **6** pp 78–94
Perkins, D.H.
    1947: Nuclear disintegration by meson capture *Nature* **159** pp 126–7
    1988: Concluding remarks, in Foster and Fowler (1988, pp 189–93)
    1989: Cosmic-ray work with emulsions in the 1940s and 1950s, in Brown *et al.* (1989, pp 89–110)
Pfotzer, G.
    1936: Dreifachkoinzidenzen der Ultrastrahlung aus vertikaler Richtung in der Stratosphäre *Z. Phys.* **102** pp 23–40 and 41–58
Piccioni, O.
    1982: Ideas and non ideas and the discovery of the leptonic property in Rome, in Colloque Internationale (1982, pp 207–14)
    1983: The observation and leptonic nature of the 'mesotron' by Conversi, Pancini, and Piccioni, in Brown and Hoddeson (1983, pp 222–41)
Pickering, A.
    1984: *Constructing Quarks* (Chicago)
    1989: From field theory to phenomenology: the history of dispersion relations, in Brown *et al.* (1989, pp 579–99)
Planck, M.
    1900: Zur Theorie des Gesetzes der Energieverteilung in Normalspektrum *Verh. Deutsch. Phys. Ges.* **2** pp 237–45
Politzer, H.D.
    1973: Reliable perturbative results for strong interactions? *Phys. Rev. Lett.* **30** pp 1346–9
Pomerantz, M.A.
    1940: The instability of the meson *Phys. Rev.* **57** pp 3–12
Pontecorvo, B.
    1947: Nuclear capture of mesons and the meson decay *Phys. Rev.* **72** pp 246–7
Powell, C.F.
    1950: The cosmic radiation, in *Nobel Lectures* (1964, pp 144–57)
    1972a: *Selected Papers* ed. E.H.S. Burhop, W.O. Lock, and M.G.K. Menon (Amsterdam)
    1972b: Fragments of autobiography, in Powell (1972a, pp 7–34)
Powell, C.F. and Fertel, G.E.F.
    1939: Energy of high-velocity neutrons by the photographic method *Nature* **144** pp 115–6
Powell, C.F., Fowler, P.H. and Perkins, D.H.
    1959: *The Study of Elementary Particles by the Photographic Method* (New York)

Powell, C.F., Occhialini, G.P.S., Livesy, D.L. and Chilton, L.V.
    1946: A new photographic emulsion for the detection of fast charged particles *J. Sci. Instrum.* **23** pp 102–6
Powers, T.
    1993: *Heisenberg's War* (New York)
Present, R.D.
    1936: Must neutron-neutron forces exist in the H3 nucleus? *Phys. Rev.* **50** pp 635–42
Proca, A.
    1936: Sur la théorie ondulatoire des électrons positifs et négatifs *J. Phys. Radium* **7** pp 347–53
Proca, A. and Goudsmit, S.
    1939: Sur le masse de particules élémentaires *J. Phys. Radium* **10** pp 209–14
Pryce, M.H.L.
    1937: Zur Neutrinotheorie des Lichtes *Z. Phys.* **105** pp 127–32
    1938a: Connection between electromagnetic and neutrino fields *Nature* **141** p 976
    1938b: On the neutrino theory of light *Proc. Roy. Soc. (London)* A **165** pp 247–71
Purcell, E.M.
    1964: Nuclear physics without the neutron; clues and contradictions, in *Proceedings of the Tenth International Congress of the History of Science (Ithaca, New York, 1962)* (Paris) pp 121–32

Rarita, W. and Schwinger, J.
    1941a: On the neutron-proton interaction *Phys. Rev.* **59** pp 436–52
    1941b: On a theory of particles with half-integral spin *Phys. Rev.* **60** p 61
Rasche, E.
    1980: Laudatio auf Prof. Walter Heitler *Arch. Int. Histoire des Sciences* **30** 159–66
Rasetti, F.
    1930: Über die Rotation-Ramanspektren von Stickstoff und Sauerstoff *Z. Phys.* **61** pp 598–601
    1941a: Evidence for the radioactivity of slow mesotrons *Phys. Rev.* **59** pp 706–8
    1941b: Disintegration of slow mesotrons *Phys. Rev.* **60** pp 198–204
    1962: Introduction to Fermi's papers on β-decay, in Fermi (1962, pp 538–40)
Rathgeber, H.D.
    1938: Der Barometereffekt der kosmische Ultrastrahlung und das Mesotron *Naturwiss.* **52** pp 842–3
Rechenberg, H.
    1988a: Vor Fünfzig Jahren *Phys. Blätter* **44** *November*
    1988b: 50 Jahre Kernspaltung *Phys. Blätter* **44** pp 453–9
    1989: The early S-matrix and its propagation, in Brown *et al.* (1989, pp 551–78)
    1993a: Heisenberg and Pauli. Their program of a unified quantum field theory of elementary particles (1927–1958), in Heisenberg (1984, Vol. AIII, pp 1–19)
    1993b: Die Theorie der Atomkerne in Leipzig, in *Werner Heisenberg in Leipzig, 1927–1942* ed C. Klein and G. Weiners (Berlin) pp 30–52
    1994: *Farm-Hall-Berichte. Die abgehörten Gespräche der 1945/46 in England internierten Atomwissenschaftler. Ein Kommentar* (Stuttgart)
    1995: Quanta and quantum mechanics, in Brown *et al.* (1995, pp 143–248)
Rechenberg, H. and Brown, L.M.
    1990: Yukawa's heavy quantum and the mesotron (1935–1937) *Centaurus* **33** pp 214–52
Regener, E.
    1937: Die kosmische Ultrastrahlung *Naturwiss.* **25** pp 1–11
Regge, T.
    1959: Introduction to complex orbital momenta *Nuovo Cimento* **14** pp 951–76

1960: Bound states, shadow states and Mandelstam representation *Nuovo Cimento* **18** pp 947–56

Reischauer, E.O.

1964: *Japan Past and Present* 3rd edn (Tokyo)

Rhodes, R.

1986: *The Making of the Atomic Bomb* (New York)

Rider, R.E.

1984: Alarm and opportunity: Emigration of mathematicians and physicists to Britain and the United States, 1933–1945 *Historical Studies in the Physical Sciences* **15** 107–76

Rochester, G.D.

1982: Observations on the discovery of the strange particles, in Colloque Internationale (1982, pp 169–76)

1985: The early history of the strange particles, in Sekido and Elliot (1985, pp 299–321)

1988: The early history of the strange particles, in Foster and Fowler (1988, pp 121–31)

1989: Cosmic-ray cloud-chamber contributions to the discovery of the strange particles in the decade 1947–1957, in Brown *et al.* (1989, pp 57–88)

Rochester, G.D. and Butler, C.C.

1947: Evidence for the existence of new unstable elementary particles *Nature* **160** pp 855–7

Rochester, G.D. and Wilson, J.G.

1952: *Cloud Chamber Photographs of the Cosmic Radiation* (New York)

Rosenfeld, L.

1948: *Nuclear Forces* (Amsterdam)

Rossi, B.

1931: Magnetic experiments on the cosmic rays *Nature* **128** pp 300–1

1934: Directional measurements on the cosmic rays near the geomagnetic equator *Phys. Rev.* **45** pp 212–4

1938a: Le attuali conoscenze sperimentali sulla radiazione cosmica *Nuovo Cimento* **15** pp 35–65

1938b: Further evidence of radioactive decay of mesotrons *Nature* **142** p 993

1939: The disintegration of mesotrons *Rev. Mod. Phys.* **11** pp 296–303

1949: Nuclear interactions of cosmic rays, in *Proc. Echo Lake Cosmic Ray Symposium, 23–28 June 1949* (Washington) pp 307–345

1964: *Cosmic Rays* (London)

1983: The decay of 'mesotrons' (1939–1943): experimental particle physics in the age of innocence, in Brown and Hoddeson (1983, pp 183–205)

1990: *Moments in the Life of a Scientist* (Cambridge)

Rossi, B., Greisen, K., Stearns, J.C., Froman, D.K. and Koontz, P.G.

1942: Further measurements of the mesotron lifetime *Phys. Rev.* **61** pp 675–9

Rossi, B. and Hall, D.B.

1941: Variation of the rate of decay of mesotrons with momentum *Phys. Rev.* **59** pp 223–8

Rossi, B., Hilberry, N. and Hoag, J.B.

1939: The disintegration of mesotrons *Phys. Rev.* **56** pp 837–8

1940: The variation of the hard component of cosmic rays with height and the disintegration of mesotrons *Phys. Rev.* **57** pp 461–9

Rossi, B. and Nereson, N.

1942: Experimental determination of the disintegration curve of mesotrons *Phys. Rev.* **62** pp 417–22

Rulhig, A.J. and Crane, H.R.

1938: Evidence for a particle of intermediate mass *Phys. Rev.* **53** p 266

Rutherford, E.

1911: The scattering of α and β particles by matter and the structure of the atom *Phil. Mag.* **21** pp 669–88

1914: The structure of the atom *Scientia* **16** pp 337–51

1919a: Collisions of α particles with light atoms. I. Hydrogen *Phil. Mag.* **37** pp 537–561.

1919b: Collisions of α particles with light atoms. IV. An anomalous effect in nitrogen *Phil. Mag.* **37** pp 581–7

1920: Bakerian Lecture: Nuclear constitution of atoms *Proc. Roy. Soc. (London)* A **97** pp 374–400

1929: Discussion on the structure of atomic nuclei *Proc. Roy. Soc. (London)* A **123** pp 373–90

1963: *The Collected Papers of Lord Rutherford, Vol. Two: Manchester* ed. J. Chadwick (London)

1965: *The Collected Papers of Lord Rutherford, Vol. Three: Cambridge* ed. J. Chadwick (London)

Rutherford, E., Chadwick, J. and Ellis, C.D.

1930: *Radiations from Radioactive Substances* (Cambridge)

Sachs, R.G.

1953: *Nuclear Theory* (Cambridge, MA)

Sakata, S.

1940: Connection between the meson decay and the beta-decay *Phys. Rev.* **58** 576

1941a: On the theory of the meson decay *Proc. Phys.-Math. Soc. Japan* **23** pp 283–91

1941b: On Yukawa's theory of the beta-disintegration and the lifetime of the meson *Proc. Phys.-Math. Soc. Japan* **23** pp 291–309

1956: On a composite model for the new particles *Prog. Theor. Phys.* **16** pp 686–8

1965: Reminiscences of research on meson theory (in Japanese), in Yukawa *et al.* (1965) (translated by Noriko Eguchi)

Sakata, S. and Inoue, T.

1942: On the relation between the meson and the Yukawa particle *Bull. Phys.-Math. Soc. Japan* **16** pp 232–4 (in Japanese)

1946: On the correlations between the meson and the Yukawa particle *Prog. Theor. Phys. (Kyoto)* **1** pp 143–50

Sakata, S. and Taketani, M.

1938: Versuch einer Theorie des β-Zerfalls *Proc. Phys.-Math. Soc. Japan* **20** pp 962–3

1940: Note on Casimir's method of the spin summation in the case of the meson *Sci. Papers IPCR* **38** pp 1–11

Sakata, S. and Tanikawa, T.

1940: The spontaneous disintegration of the neutral meson (neutretto) *Phys. Rev.* **57** p 548

Schein, M. and Gill, P.S.

1939: Burst frequency as a function of energy *Rev. Mod. Phys.* **11** pp 267–27

Schein, M. and Wilson, V.C.

1938: Evidence for the production of secondary cosmic ray particles in the atmosphere *Phys. Rev.* **54** pp 304–5

Scherzer, O.

1935: Zur Neutrinotheorie des Lichtes *Z. Phys.* **97** pp 724–39

Schmid, Christoph

1969: Duality and exchange degeneracy, in Zichichi (1969, pp 746–62)

Schönberg, E.M.

1939: β-ray selection rules and the meson theory *Phys. Rev.* **56** p 612

Schweber, S.S.

1961: *An Introduction to Relativistic Quantum Field Theory* (Evanston, IL)

1986: The empiricist temper regnant: Theoretical physics in the United States 1920–1950 *Historical Studies in the Physical and Biological Sciences* **17** (1) pp 55–98

1994: *QED and the Men Who Made It: Dyson, Feynman, Schwinger, and Tomonaga* (Princeton)

Schweber, S.S., Bethe, H.A. and de Hoffmann, F.

1955: *Mesons and Fields* Vol. 1 (New York)

Schwinger, J.

1942: On a field theory of nuclear forces *Phys. Rev.* **61** p 387

1958: (ed.) *Selected Papers on Quantum Electrodynamics* (New York)

1983a: Renormalization theory of quantum electrodynamics: an individual view, in Brown and Hoddeson (1983, pp 329–53)

1983b: Two shakers of physics: Memorial lecture for Sin-itiro Tomonaga, in Brown and Hoddeson (1983, pp 354–75)

1993: A path to quantum electrodynamics, in Brown and Rigden (1993, pp 59–76)

Segrè, E.

1970: *Enrico Fermi–Physicist* (Chicago)

1980: *From X-Rays to Quarks: Modern Physicists and their Discoveries* (San Francisco)

Sekido, Y. and Elliot, H.

1985: *Early History of Cosmic Ray Studies* (Dordrecht)

Serber, R.

1938: On the dynatron theory of nuclear forces *Phys. Rev.* **53** p 211

1939: Beta-decay and mesotron lifetime *Phys. Rev.* **56** p 1065

1950: Artificial mesons, in Institut Internationale de Physique (1950, pp 89–104)

1983: Particle physics in the 1930s: a view from Berkeley, in Brown and Hoddeson (1983, pp 206–21)

Serber, R. and Dancoff, S.M.

1943: Strong coupling mesotron theory of nuclear forces *Phys. Rev.* **63** pp 143–61

Shimizu, T.

1921a: A reciprocating expansion apparatus for detecting ionizing rays *Proc. Roy. Soc. (London)* A **99** pp 425–31

1921b: A preliminary note on branched α-ray tracks *Proc. Roy. Soc. (London)* A **99** pp 432–5

Shonka, F.R.

1939: New evidence for the existence of penetrating neutral particles *Phys. Rev.* **55** pp 24–6

Shutt, R.P., De Benedetti, S. and Johnson, T.H.

1942: Cloud-chamber track of a decaying mesotron *Phys. Rev.* **62** 552–553

Siegel, D.M.

1978: Classical-electromagnetic and relativistic approaches to the problem of nonintegral atomic masses *Hist. Studies in the Physical Sciences* **9** 323–360

Siegert, A.J.F.

1937: Note on the interaction between nuclei and electromagnetic radiation *Phys. Rev.* **52** pp 787–9

Silliman, R.H.

1963: William Thomson: smoke rings and nineteenth century atomism *Isis* **54** pp 461–74

Smith, A.K.

1965: *A Peril and a Hope* (Cambridge, MA)

Smyth, H. De Wolf

1945: *Atomic Energy for Military Purpose* (Princeton)

Soddy, F.

1913: Radio-elements and the periodic laws *Chemical News* **107** pp 97–9

Sokolov, A.

1937: Neutrino theory of light *Nature* **139** 1071
1938: Connection between electrodynamics and neutrino fields *Nature* **141** p 976
Sommerfeld, A.
1924: *Atombau und Spektrallinien* 4th edn (Braunschweig)
Souza Santos, M.D. de, Pompeia, P.A. and Wataghin, G.
1941: Showers of penetrating particles *Phys. Rev.* **59** pp 902–3
Stachel, J.
1995: History of relativity, in Brown *et al.* (1995, pp 249–356)
Steinberger, J.
1949: On the range of the electrons in meson decay *Phys. Rev.* **75** pp 1136–43
1989: A particular view of particle physics in the fifties, in Brown *et al.* (1989, pp 307–30)
Steinberger, J., Panofsky, W.K.H. and Stellar, J.
1950: Evidence for the production of neutral mesons by photons *Phys. Rev.* **78** pp 802–5
Steinke, E.G. and Schindler, H.
1932: Zertrümmerung von Blei durch Ultrastrahlung *Z. Phys.* **75** pp 115–8
Street, J.C. and Stevenson, E.C.
1937a: Penetrating corpuscular component of the cosmic radiation *Phys. Rev.* **51** p 1005 (Abstract of talk at the Washington Meeting of the American Physical Society, 29–30 April 1937)
1937b: New evidence for the existence of a particle of mass intermediate between the proton and the electron *Phys. Rev.* **52** pp 1003–4
Stueckelberg, E.C.G.
1936a: Austauschkräfte zwischen Elementarteilchen und Fermische Theorie des β-Zerfalls als Konsequenzen einer möglichen Feldtheorie der Materie *Helv. Phys. Acta* **9** pp 389–404
1936b: Invariante Störungstheorie des Elektron-Neutrino- Teilchens unter dem Einfluss von elektromagnetischem Feld und Kernfeld. (Feldtheorie der Materie II) *Helv. Phys. Acta.* **9** pp 533–54
1936c: Radiative β-decay and nuclear exchange force as a consequence of a unitary field theory *Nature* **137** p 1032
1937a: On the existence of heavy electrons *Phys. Rev.* **52** pp 41–2
1937b: Neutrino theory of light *Nature* **139** pp 198–9
1938a: Die Wechselwirkungkräfte in der Elektrodynamik und in der Feldtheorie der Kernkräfte. I, II und III *Helv. Phys. Acta* **11** 225–44, 299–328
1938b: Über die Energieverluste von Elementarteilchen mit ganzzahligem Spin *Helv. Phys. Acta* **11** pp 378–80
1938c: Rigorous theory of interaction between nuclear particles *Phys. Rev.* **54** pp 889–92
1939a: Theory of mesons and the nuclear forces *Nature* **143** pp 560–1
1939b: A new model of the point charge electron and of other elementary particles *Nature* **144** p 118
1942, 3: Une méthode nouvelle de la quantification des champs *(I, II, and III) Archive des Sciences physique et naturelles* **24** pp 1–70; 193–224 (1942); *ibid.* **25** pp 5–34 (1943)
1944: Un modèle de l'électron ponctuel *Helv. Phys. Acta* **17** pp 3–26
1945: La charge gravifique e le spin de l'électron classique *Helv. Phys. Acta* **18** pp 21–44
1946: Un propriété de l'operateur S en mechanique asymptotique *Helv. Phys. Acta* **19** pp 242–3
Stueckelberg, E.C.G. and Rivier, D.
1946: Operateurs non linéaire en théorie des quanta *Helv. Phys. Acta* **19** pp 240–2

Stuewer, R.H.
  1979: *Nuclear Physics in Retrospect* (Minneapolis)
  1983: The nuclear electron hypothesis, in *Otto Hahn and the Rise of Nuclear Physics* ed. W.R. Shea (Dordrecht) pp 19–67
  1984: Nuclear physicists in a new world. The emigrés of the 1930s in America *Berichte Wissenschaftsgesch.* **7** pp 23–40
Sudarshan, E.C.G. and Marshak, R.E.
  1958: Chiral invariance and the universal Fermi interaction *Phys. Rev.* **109** pp 1860–2
Swann, W.F.G.
  1936: A theoretical discussion of the deviation of high energy charged particles passing through magnetized iron *Phys. Rev.* **49** pp 574–82
Swann, W.F.G. and Ramsey, W.E.
  1938: Cosmic-ray electron showers in a mine 600 feet below sea level *Phys. Rev.* **54** pp 229–30

Takabayasi, T.
  1983: Some characteristic aspects of early elementary particle theory in Japan, in Brown and Hoddeson (1983, pp 294–303)
  1991: Nonlocal theories and related topics, in Brown *et al.* (1991, pp 270–86)
Takeda, G. and Yamaguchi, Y.
  1982: Role of institutions in research of high energy physics in Japan for the period 1930–1960, in Colloque Internationale (1982, pp 335–40)
Taketani, M.
  1971: Methodological approaches to the development of the meson theory of Yukawa Supp. *Prog. of Theor. Phys.* **50** pp 12–24
  1985: Theoretical studies of cosmic rays in Japan, in Sekido and Elliot (1985, pp 285–94)
Taketani, M. and Sakata, S.
  1940: On the wave equation of meson *Proc. Phys.-Math. Soc. Japan* **22** pp 757–70
Takeuchi, M.
  1975: Recollections on research in cosmic rays *Shizen* July pp 52–59 (translation by R. Yoshida)
  1985: Cosmic ray study in Nishina Laboratory, in Sekido and Elliot (1985, pp 137–43)
Tamm, Ig.
  1934: Exchange forces between neutrons and protons, and Fermi's theory *Nature* **133** p 981
  1940: Mesons in a Coulomb field *Phys. Rev.* **58** p 952
Tanikawa, Y.
  1943: On a theory of the β-decay mediated by the boson with larger mass than that of the heavy particle *Bull. Phys. Math. Soc. Japan* **17** 597–602 (in Japanese)
  1947: On the cosmic-ray meson and the nuclear meson *Prog. Theor. Phys. (Kyoto)* **2** pp 220–1
Tanikawa, Y. and Yukawa, H.
  1941: On the scattering of mesons by nuclear particles *Proc. Phys.-Math. Soc. Japan* **23** pp 445–454
Thomson, J.J.
  1897: Cathode-rays *Phil. Mag.* **44** pp 293–316
  1899: On the masses of ions in gases at low pressures *Phil. Mag.* **48** pp 547–67
  1904: On the structure of the atom *Phil. Mag.* **7** pp 237–65
  1906: On the number of corpuscles in an atom *Phil. Mag.* **11** pp 769–81
  1913: Rays of positive electricity *Proc. Roy. Soc. (London)* A **89** pp 1–20
Tomonaga, S.

1940: Über den Zusammenstoß des Mesotrons mit Elektronen *Sci. Pap Inst. Phys. Chem. Res.* **37** pp 319–413

1941: Zur Theorie des Mesons I *Sci. Papers IPCR* **39** pp 247–66

1942: Bemerkungen über die Streuung der Mesotronen am Kernteilchen *Sci. Pap IPCR* **40** pp 73–86

1946a: On a relativistically invariant formulation of the quantum theory of wave field I *Prog. Theor. Phys.* **1** pp 27–42

1946b: On the effect of the field reactions on the interaction of mesotrons and nuclear particles *Prog. Theor. Phys.* **1** pp 83–101, 109–24; *ibid.* **2** pp 6–24, 63–70

1972: Development of quantum electrodynamics – Personal recollections, in Nobel Lectures Physics, 1963–1970 (New York) pp 126–136

Tomonaga, S. and Araki, G.
1940: Effect of the nuclear Coulomb field on the capture of slow mesons *Phys. Rev.* **58** p 82

Trenn, J.T.
1977: *The Self-splitting Atom. The History of the Rutherford-Soddy Collaboration* (London)
1981: *Transmutation, Natural and Artificial* (London)

Tuve, M.
1933: The atomic nucleus and high voltages *J. Franklin Inst.* **216** pp 1–38

Tuve, M.A., Heydenburg, N.P. and Hafstad, L.R.
1936: The scattering of protons by protons *Phys. Rev.* **49** pp 806–25

Urban, P. (ed.)
1968: *Particles, Currents, Symmetries* (Vienna)

Urey, H., F.G. Brickwedde, and G.M. Murphy
1932: A hydrogen isotope of mass 2 *Phys. Rev.* **39** pp 164–5

Van de Graaff, R.J.
1931: A 1,500,000 volt electrostatic generator *Phys. Rev.* **38** pp 1919–20

Walker, M.
1989: *German National Socialism and the Quest for Nuclear Power, 1939–1949* (Cambridge)

Walker, R.L.
1989: Learning about nucleon resonances with pion photoproduction in Brown *et al.* (1989, pp 111–43)

Waller, C.
1988: British patent 580,504 and Ilford nuclear emulsions, in Foster and Fowler (1988, pp 55–60)

Wataghin, G., de Souza Santos, M.D. and Pompeia, P.A.
1940a,b: Simultaneous penetrating particles in the cosmic radiation *I. Phys. Rev.* **57** p 61; II. *Phys. Rev.* **57** 339

Watanabe, M.
1990: *The Japanese and Western Science* (translated by O.T. Bentley) (Philadelphia)

Watson, K.M.
1951: The hypothesis of charge independence for nuclear phenomena *Phys. Rev.* **85** pp 852–7

Watson, K.M. and Brueckner, K.A.
1951: The analysis of π-meson production in nucleon-nucleon collisions *Phys. Rev.* **83** pp 1–9

Weinberg, J.W.
1941: Scattering in the pair theory of nuclear forces *Phys. Rev.* **59** pp 776–80

Weinberg, S.
  1967: Dynamical approach to current algebra *Phys. Rev. Lett.* **18** pp 188–91
Weiner, C.
  1972: 1932 – Moving into the new physics *Physics Today* **25** pp 40–9
  1975: Cyclotrons and internationalism: Japan, Denmark and the United States, in *Proceedings of the XIVth International Congress of the History of Science 1974* (Tokyo) pp 353–65
  1978: Retroactive saber rattling. A note on nuclear physics in Japan *Bull. Atomic Scientists* **34** pp 10–12
Weisskopf, V.F.
  1983: Growing up with field theory: the development of QED, in Brown and Hoddeson (1983, pp 56–81)
  1985: Search for simplicity: The molecular bond *Am J. Phys.* **53** pp 399–400
Weizsäcker, C.F. von
  1934: Ausstrahlung bei Stößen sehr schneller Elektronen *Z. Phys.* **88** pp 612–25
  1937: *Die Atomkerne* (Leipzig)
Wentzel, G.
  1926: Zwei Bemerkungen über die Zerstreuung korpskularer Strahlen als Beugungsercheinung *Z. Phys.* **40** pp 590–3
  1934: Zur Frage der Äquivalenz von Lichtquanten und Korpuskelpaaren *Z. Phys.* **92** pp 337–58
  1937a,b: Zur Theorie der β-Umwandlung und der Kernkräfte. I.,II. *Z. Phys* **104** 34–47; *Z. Phys.* **105** 738–746
  1937c: Zur Frage der β-Wechselwirkung *Helv. Phys. Acta* **10** 107–111
  1938a: Schwere Elektronen und Theorien der Kernvorgänge *Naturwiss.* **26** pp 273–279.
  1938b: The angular spread of hard cosmic-ray showers *Phys. Rev.* **54** pp 869–72
  1940: Zum Problem des statischen Mesonfeldes *Helv. Phys. Acta* **13** pp 269–308
  1941a: Zur Hypothese der höheren Proton-Isobaren *Helv. Phys. Acta* **14** pp 3–20 and 633
  1941b: Beiträge zur Paartheorie der Kernkräfte *Z. Phys.* **118** pp 277–94
  1942: Zur Paartheorie der Kernkräfte *Helv. Phys. Acta* **15** pp 11–126
  1943a: Zur Vektormesontheorie *Helv. Phys. Acta* **16** pp 551–96
  1943b: *Einführung in die Quantentheorie der Wellenfelder* (Vienna) English translation by C.M. Houtermans and J.M. Jauch: *Quantum Theory of Fields* (New York 1949)
  1947: Recent research in meson theory *Rev. Mod. Phys.* **19** pp 1–18
  1950: μ-pair theories and the π-meson *Phys. Rev.* **79** pp 710–6
Wheeler, J.A.
  1937: On the mathematical description of light nuclei by the method of resonating group structure *Phys. Rev.* **52** pp 1107–22
  1979: Some men and moments in the history of nuclear physics, in Stuewer (1979, pp 217–322)
White, M.G.
  1936: Scattering of high-energy protons in hydrogen *Phys. Rev.* **49** pp 309–316
Whittaker, E.
  1951: *A History of the Theories of Aether and Electricity: The Classical Theories* (London)
Wick, G.C.
  1934: Sugli elementi radioattivi di F. Joliot e I. Curie *Rend. Accademia dei Lincei* **19** pp 319–24
  1935: Teoria dei raggi β e momento magnetico del protone *Rend. Acad. Lincei* **21** 170–3
  1938: Range of nuclear forces in Yukawa's theory *Nature* **142** pp 993–4

Wightman, A.
  1989: The general theory of quantized fields in the 1950s, in Brown *et al.* (1989, pp 608–29)
Wigner, E.
  1933: On the mass defect of helium *Phys. Rev.* **43** pp 252–7
  1937: On the consequences of the symmetry of the nuclear Hamiltonian on the spectroscopy of nuclei *Phys. Rev.* **51** pp 106–19
Wigner, E.P., Critchfield, C.L. and Teller, E.
  1939: The electron-positron field theory of nuclear forces *Phys. Rev.* **56** pp 530–9
Williams, E.J.
  1933: Applications of the method of impact parameter in collisions *Proc. Roy. Soc. (London)* A **139** pp 163–86
  1935: Nature of high energy particles of penetrating radiation and status of ionization and radiation formulae *Phys. Rev.* **48** pp 49–54
  1938: Cosmic rays *Nature* **141** pp 1085–7
  1939: Some observations on cosmic rays using a large randomly operated cloud chamber *Proc. Roy. Soc. (London)* A **172** pp 194–211
  1940: The average number of electrons accompanying a cosmic-ray meson due to collisions of the meson with atomic electrons *Proc. Camb. Phil. Soc.* **36** pp 183–92
Williams, E.J. and Pickup, E.
  1938: Heavy electrons in cosmic rays *Nature* **141** pp 684–5
Williams, E.J. and Roberts, G.E.
  1940: Evidence for transformation of mesotrons into electrons *Nature* **145** p 102–3
Williams, R.C.
  1938: The fine structure of H_α and D_α under varying discharge conditions *Phys. Rev.* **54** pp 558–67
Wilson, J.G.
  1938: The energy loss of penetrating cosmic-ray particles in copper *Proc. Roy. Soc. (London)* A **166** pp 482–501
  1939: Absorption of penetrating cosmic ray particles in gold *Proc. Roy. Soc. (London)* A **172** pp 517–29
  1940: The scattering of mesotrons in metal plates *Proc. Roy. Soc. (London)* A **174** pp 73–85
  1985: The 'Magnet House' and the muon, in Sekido and Elliot (1985, pp 145–159)
Wilson, V.C.
  1938: On the nature of the penetrating rays *Phys. Rev.* **53** pp 908–9
  1939a: Nature of cosmic rays below ground *Rev. Mod. Phys.* **11** pp 230–1
  1939b: The nature of the penetrating cosmic rays *Phys. Rev.* **55** pp 6–10

Xu, Q. and Brown, L.M.
  1987: The early history of cosmic ray research *Am. J. Phys.* **55** pp 23–33

Yagi, E.
  1964: On Nagaoka's Saturnian atomic model (1903) *Jap. Stud. Hist. Sci.* **3** pp 29–47
  1990: Nishina, Yoshio, in *Dictionary of Scientific Biography* Supp **II**
Yang, C.N. and R. Mills
  1954a: Isotopic spin conservation and a generalized gauge invariance *Phys. Rev.* **95** p 631
  1954b: Conservation of isotopic spin and a generalized gauge invariance *Phys. Rev.* **96** pp 191–5
Yearian, M.R. and Hofstadter, R.
  1958: Magnetic form factor of the neutron at 600 MeV *Phys. Rev.* **111** pp 934–9

Yukawa, H.
  1933a: Introduction to W. Heisenberg, Über den Bau der Atomkerne *J. Phys.-Math. Soc. Japan* **7** pp 195–205
  1933b: A comment on the problem of electrons in the nucleus *Bull. Phys.-Math. Soc. Japan* **7** p 195
  1935: On the interaction of elementary particles. I. *Proc. Phys.-Math. Soc. Japan* **17** pp 48–57
  1937: On a possible interpretation of the penetrating component of the cosmic ray *Proc. Phys.-Math. Soc. Japan* **19** pp 712–3
  1941: Outline of the meson theory *Scientia* **70** pp 97–102
  1942: Bemerkungen über die Natur des Mesotrons *Z. Phys.* **119** pp 201–5
  1971: *Self-Selected Essays* Vol. 5 (Tokyo) (in Japanese)
  1979: *Hideki Yukawa: Scientific Works* ed. Y. Tanikawa (Tokyo)
  1982: *Tabibito (The Traveler)* trans. L.M. Brown and R. Yoshida (Singapore)
Yukawa, H. and Okayama, T.
  1939: Note on the absorption of slow mesotrons in matter *Sci. Papers IPCR* **36** pp 385–9
Yukawa, H. and Sakata, S.
  1935a: On the theory of internal pair production *Proc. Phys.-Math. Soc. Japan* **17** pp 397–407
  1935b: On the theory of β-disintegration and the allied phenomena *Proc. Phys.-Math. Soc. Japan* **17** pp 467–79
  1936: Supplement *Proc. Phys.-Math. Soc. Japan* **18** pp128–30
  1937a: On the theory of β-disintegration and the allied phenomena *Proc. Phys.-Math. Soc. Japan* **17** pp 467–469
  1937b: Note on Dirac's generalized wave equations *Proc. Phys.-Math. Soc. Japan* **19** pp 91–5
  1937c: On the interaction of elementary particles, II. *Proc. Phys.-Math. Soc. Japan* **19** pp 1084–93
  1939a: The mass and the lifetime of the mesotron *Proc. Phys.-Math. Soc. Japan* **21** pp 138–9
  1939b: Mass and life-time of the meson *Nature* **143** pp 761–2
Yukawa, H., Sakata, S. and Taketani, M.
  1938a: On the interaction of elementary particles, III *Proc. Phys.-Math. Soc. Japan* **20** pp 319–40
  1965: *Quest for Elementary Particles. At the Battlefield of Truth* (Tokyo) (in Japanese)
Yukawa, H., Sakata, S., Kobayashi, M. and Taketani, M.
  1938b: On the interaction of elementary particles, IV. *Proc. Phys.-Math. Soc. Japan* **20** pp 720–45

Zichichi, A. (ed)
  1969: *Subnuclear Phenomena* (New York)
Zweig, G.
  1964: An SU(3) model for strong interaction symmetry and its breaking *Preprint* CERN 8182/TH 401 (17 Jan. 1964)

Index

Aamodt, R.L., 359
Abraham, Max (1904), 3, 335
Accelerators *see* particle acceleration
Ageno, M., 271, 335
Alichanian, A.I., 271, 335
Alichanow, A.I., 187, 199, 268, 271, 335
Allen, R., 354
Alpha- (α-) decay, 17, 39
Alpha- (α-) particles (or rays), 4, 5, 10, 11–4, 17, 19, 20
  anomalous scattering, 15, 37
  binding energy, 60
  electrons in, 66
  emission, 34
  radioactivity, 16
  scattering, 8, 12, 15–7, 24, 27, 38
  scintillation method to study scattering, 13, 15
  structure, 45
Alvarez, Luis W., 118, 284, 291, 318, 333, 335, 356
Amaldi, Eduardo, 45, 46, **55**, 268, 271, 282–4, 290, 297, 299, 313, 314, 335, 345
Analytic S-matrix, 276, 317
Anderson, Carl David, 25, 29, 30, 49, 56, 63, 69, 70, 73, 74, 80, 82, 85, 90, 92, 112, 115, 121–3, **124**, 125, 131, 132, 136, 138–40, 142, 170, 178, 179, 187, 188, 200, 204, 205, 222, 224–7, 238–40, 296, 301, **319**, 332, 335, 358
  Nobel Prize, 122, 123, 204
  and uranium, 281
  and V-particles, 296, 297
Anderson, H.L., 138–40, 281, 320, 322, 333, 335, 336
annus mirabilis, 1932 as, 30, 31
"Anomalous scattering", 12, 14, 15, 38, 195
Anti-neutrino, 51
Arakatsu, Bunsaku, 222, 281, 284
Araki, G, 196, 197, 201, 298, 299, 367
Ariyama, Kanetake, 281
Arley, N, 213, 226, 236
Assmus, A., 23, 347
Aston, Francis William, 11, 12, 15, 336
"Atom of electricity", 2

Atomic bombs, 253, 254, 282, 284, 318
Atomic models,
  Bohr–Rutherford nuclear, 6, 11
  Bohr's 6, 10
  Nagaoka's "Saturnian", 7, 23
  nuclear, 5
  Rutherford's, 6–9
  Thomson's ("raisin pudding"), 6, 8, 23
  vortex, 6
Atomic nucleus, 9
  discovery, 1, 8
  source of radioactivity, 9, 10
Atomic number, 9, 10
Atomic radiation, 5
  line spectra, 6
Atomic weight, 11
Auger, Pierre, 69, 73, 88, 90, 92, 183–5, 187, 188, 204, 336
Aydelotte, F., 270

Bacher, Robert F., 59, 60, 67, 90, 123, 155, 172, 337
Bagge, E.R., 198, 199, 273, 336
Bainbridge, Kenneth T., 118, **126**, 224, 336
Balmer, J.J., 8
Barnóthy, Jenö, 69, 84, 92, 205, 213, 225, 226, 336
Bartholomew, J., 112, 336
Barton, H., 113
Barut, A.O., 66, 336
Baryons, 323, 325, 331
  number, 211
  see also nucleons
Barytrons, 183, 215
Beck, Guido 38, 47, 52, 65, 137, 215, 226, 261, 336
  in Japan, 117
  letter, **214**
Beck–Sitte theory of β-decay, 38, 47, 52, 53, 56, 58
Becker, Herbert, 28, 339
Becker, Richard, 287
Becquerel, Henri, 1, 4, 5, 336
Belinfante, Frederik Josef, 232, 233, 241, 248, 249, 336, 360
Berestetski, V., 272

Berkeley Conference (December 1960), 327

Bernardini, Gilberto, 200, 239, 282, 290, 298, 313, 337

Beta- (β-) decay, 33, 38, 44, 51, 57
  "allowed", 52
  angular momentum conservation, 27, 32
  "artificial radioactivity" involving positrons, 66
  Bohr on, 38
  and charge-exchange interaction, 49, 50
  and conservation laws, 31
  experiments on, 15, 33
  "forbidden", 52
  Heisenberg on, 35, 39, 41, 42, 49, 50, 57
  mean lifetime, 52
  and meson decay, 189–94, 257
  radioactivity, 4, 44
  Yukawa on, 106, 177, 183, 209
  see also Beck–Sitte theory of β-decay; Fermi, Enrico, β-decay theory; Konopinski–Uhlenbeck derivative forces in (or theory of) β-decay

Beta- (β-) rays (or radiation), 4, 5, 9, 10, 12
  emission, 35, 54, 59
  scattering, 23
  spectrum of electron energies, 18, 23, 52

Bethe, Hans A., 19, 25, 56–60, 64, 67, 68, 74, 84, 90, 122, 123, 155, 172, 187, **191**, 192, 194, 200, 204, 215, 291–21, 226, 227, 240, 253, 255, 258, 259, 261, 263, 270, 279, 280, 289, 293, 309, 312, 321, 322, 333, 337, 355, 356, 363
  and "Fermi-field theory", 56–8
  and meson theory, 192, 255, 258
  on neutral mesons, 220
  and track curvature, 293

Bethe–Heitler theory, 76, 84, 85, 295

"Bethe's Bible", 59

Bhabha, Homi Jehangir, 68, 80–4, 87, 89, 91, 92, 110, 112, 137, 141, 151, 154, 155, 157, 158, 160, 161, 164–9, 171, 172, 174, 179, 180, 189, 198, 205, 207–9, 214, 222, 225–7, 230, 233–7, 240, 243, 244, 249, 250, 254, 255, 261, 263, 266, 270, 271, 333, 337, 338, 342
  and application of meson theory to cosmic ray phenomena, 166–9, 179
  background, 157, 158
  classical meson theories, 237
  et al. (1939), 244
  first vector meson theories, 157–62
  Heisenberg on, 235, 236
  Kemmer on, 209
  on lifetime of mesons, 179
  meson theory application to cosmic ray phenomena, 166–9

Bieler, E.S., 14, 24, 341

Binding energies (BE), 37, 38, 56, 59

Bjorklund, R., 68, 310, 315, 338

Blackett, Patrick Maynard Stuart, 15, 24, 25, 38, 70, **72**, 73, 82, 84, 85, 89, 91, 121, 135, 136, 139, 140, 142, 160, 161, 167, 172, 183, 184, 186–8, 199, 205, 209, 233, 238, 240, 266, 295, 296, 303, 313, 314, 338
  on γ-rays, 38
  and cloud chamber experiments, 127
  and cosmic rays, 121
  and Dirac, 135, 136
  on heavy electrons, 85, 135, 136, 205
  and Heisenberg, 183
  and mesotron decay, 183, 184, 188
  and penetrating cosmic rays, 135, 136, 160

Blanpied, W.A., 172, 338

Blatt, J.M., 25, 338

Blau, M., 93, 301, 302, 313, 338

Bloch, Felix, 53, 66, 90, 236, 240, 338

Blum, W., 349

Bocciarelli, D., 335

Bøggild, Jørgen, 88, 93, 338

Bohr, H., 127

Bohr, Niels, 8–10, 12, 18, 19, 21, 23, 25, 31, 38, 39, 40, 41, 44, 48, 50, 52, 53, 55, 65, 66, 84, 90–2, 97, 100, 101, 127, **128**, 135, 140, 143, 155, 158, 186–8, 205, **206**, 207, 211, 230, 239, 255, 287–9, 291, 332, 339, 356
  on β-decay, 38, **41**
  atomic innovation, 8, 9
  correspondence principle, 74
  on cosmic rays, 155
  and electrons, 39, 40
  on energy conservation, 18, 25, 50, 112
  Heisenberg on, 40
  on isotopes, 9
  in Japan, 127
  on mesons, 127, 187
  and neutrinos, 53
  and neutrons, 21
  on radioactivity, 9
  world tour, 127

Bohr–Sommerfeld quantum theory of atomic structure, 15
Booth, F., 246, 250, 339
Bopp Fritz, 273, 274, 286, 288, 333
Borden, C.M., 270, 352
Born, Max, 78, 91, 92, 100, 153, 236, 288, 290, 291, 333, 339, 368, 386
Born–Infeld nonlinear electrodynamics, 92
"Bose electron", 105
Bose–Einstein statistics, 18, 44, 49, 144, 162, 233
Bosons, 49, 110, 279
Bothe, Walther, 28, 141, 188, 222, 240, 339, 348
Bowen, I.S., 186, 226, 339
Bradt, H.L., 346
Bragg, William Henry, 5, 23, 339
Bragg, William Lawrence, 5
Bramley, Arthur, 136, 339
Breit, Gregory, 21, 60, 63, 67, 290, 339
Bremsstrahlung process, 73, 74, 80, 205, 236, 237, 243, 244, 246, 247
Brickwedde, F.G., 367
Bridge, H.S., 89, 312, 324, 339
Brillouin, Léon, 206, 207, 225, 227, 340
Brink, D.M., 44, 340
Britain
    attitude to physics, 152
    meson physics (1941–45), 266, 267
    theorists, 160
    vector field theory papers, 162–9
    wartime, 355
Brode, Robert, 92, 121, 187, 200, 204, 225, 338, 340, 343
Broek, Jan Abram van den, 24
Broglie, Louis de, 7, **7**, 77, 78, 91, 101, 240, 241, 340
    on electrons, 101
    see also light, neutrino theory of
Bromberg, J., 44, 65, 340
Brown, L.M., 19, 20, 22, 23, 25, 42, 44–6, 56, 65, 66, 89, 90, 100, 113, 137–9, 169, 171, 172, 199, 225, 270, 289, 307, 313, 332–4, 335, 340, 343, 350, 352, 360, 361, 363, 365, 368, 370
Brown, R., 307, 341
Brueckner, Keith, 320, 332, 341, 367
Buchta, J.W., 134, 135
Bunge, M., 22, 23, 341
Burfening, J.L., 309, 341
Bursts (cosmic ray bursts, "Hoffmann bursts"), 76, 87–9, 116, 194, 244
    Bhabha *et al.* on, 244

causes, 89
cosmic ray, 87–9
    Oppenheimer on, 239
    production of, 88, 89
Butenandt, Adolf, 288
Butler, Clifford C., 295, 297, 313, 323, 341, 362

C-meson (cohesive meson), 321
Cacciapuoti, B.N., 335, 337
Camerini, U., 304, 341
Cameron, G.H., 199
Campbell, N.R., 23
Cao, T.Y., 334, 340
Carlson, A.G., 68, 315
Carlson, J. Franklin, 80–3, **81**, 92, 123, 138, 254, 341
Carmichael, H., 243, 244, 314, 338, 341
Cascade showers, 20, 73, 76, 83, 84, 155, 158
    electromagnetic, 70, 80, 81
    Heisenberg on, 84–6, 274, 275
    radiative processes contributing to, 74
Case, K.M., 320, 341
Casimir, Hendrik, 185, 255, 268, 270–2
Cassen, B., 62, 67, 68, 341
Cassidy, David C., 71, 74, 85, 90, 239, 291, 341
Cathode rays *see* electrons
Chadwick, James, 5, 14, 18, 20, 21, 24, 25, 28, 31, 32, **32**, 33, 38, 39, 42, 49, 50, 65, 102, 239, 253, 341, 363
Chakrabarty, S.K., 250, 342
Champion, Frank Clive, 24, 303, 338
Chang, P., 329, 348
Charap, J., 327
Charge
    coordinate ρ, 49
    conjugation, 232, 248
    invariance, 232, 233
    see also nuclear forces; meson theories
Chemical
    bond, 31, 34
    valence, 7
Chew, Geoffrey I., 326, 327, **327**, 333, 342
Chicago Symposium (June 1939), 188, 237–9
Chilton, L.V., 361
Chinowsky, W., 311, 332, 342
Chiral symmetry, 328
Chou, C.N., 244, 338, 341
Christy, Robert, F., 194, **247**, 247, 248, 250, 254, 278, 342
Clay, J., 141, 226, 238, 342

Cline, D., 333, 342

Close, F., 333, 342

Cloud ("Wilson") chambers, 4, 15, 19, 27, 46, 69, 70, 83, 93, 179, 196, 293, 294
    with magnets, 123

Cocconi, G., 283, 290, 342

Cockcroft, John, 29, 37, 172, 253, 342
    on deuterons, 37

Cockcroft–Walton machine, 29

Collins, P.D.P., 334, 342

Compton, Arthur H., 73, 141, 146, 168, 188, 213, 222, 224, 237, 238, 240, 342

Compton scattering, 146

Compton wavelength, 53, 96

Condon, Edward, U., 16, 62, 67, 68, **191**, 217, 227, 339, 341, 342, 348

Conferences
    Cambridge 1947, 287, 288
    Chicago 1939, 199, 237–9
    London 1934, 21, 52, 56, 73, 74
    Marienbad 1939 (planned), 203, 222–4, 237, **238**
    Moscow 1940, 272
    Rochester 1952, **319**
    Shelter Island 1947, 322
    Solvay 1933, 32, 35–41, 48, 53, 221
    Solvay 1939 (planned), 176, 222, 239–41
    Solvay 1948, 288
    Warsaw 1938, 205–7
    Zurich 1939 (planned), 203, 222–4, 239

Conservation (or invariance) laws
    angular momentum, 32, 42, 45
    breakdown, 52, 58
    electric charge, 45, 50
    energy, 18, 19, 23, 32, 42, 45
    fermion, 51
    momentum, 19, 45
    parity, 64, 119, 161, 171, 324
    see also Beck–Sitte theory of β-decay; β-decay; Bohr on energy conservation

Constants
    Einstein's gravitational, 207
    Fermi's coupling, 42, 43, 54, 190, 200
    Heisenberg's fundamental (universal) length, 71, 86, 87, 173, 207, 212
    Planck's, 8

Conversi, Marcello, 197, 251, 282, 297, 298, **299**, 304, 313, 337, 342, 343

Cooper, Mr., 254

Corben, Herbert Charles, 243, 245, 246, 250, 343, 356

Corson, Dale, 204, 205, 225, 343

Cosmic ray mesons, 64, 294, 301
    discovery, 292
    problems posed by, 294
    at sea level, 298, 299

Cosmic ray phenomena, meson theory applied to, 166–9, 179

Cosmic rays, 1, 2, 3, 6, 22, 69
    anomolous penetrating tracks, 125, 126
    barometer effect, 184, 200, 238
    burst problem, 83, 87–9
    cloud chamber studies with high magnetic fields, 27
    components, 70, 73, 142, 155, 160, 180
    cosmic ray and nuclear physics, 42
    "cosmic ray physicists" versus "field theorists", 71, 85, 90
    discovery, 5, 6, 122
    explosive events, 73
    Geiger–Müller counter studies, 27
    "heroic research", 60–70
    high-energy γs, 315
    history, 23
    inclination anomaly, 184
    mass absorption anomaly, 184
    momentum spectrum, 127
    name, 6
    and nuclear interactions, 72–6
    origin, 199
    penetrating (hard) component, 73, 74, 126, 160
    primary, 23, 73, 174
    showers, 73, 225
    soft component, 73, 74
    temperature effect, 184
    theory, 123
    see also bursts; cascade showers; Heisenberg's "explosive showers"; mesotrons; vector meson theories

Coulomb force, 14, 15
    discrepancy at small distances due to meson presence, 217, 218
    repulsion, 33, 34, 40, 53

Coulomb scattering, 17

Counter methods (coincidence, autocoincidence, telescope), 70, 188, 198, 293, 294
    see also Geiger–Müller counter

Coupling
    constants, 177
    spin–orbit of electrons in atoms, 18
    strength of meson, 56
    strong, 229, 230, 268
    strong and intermediate, 260–6
    weak, 45, 229, 230, 260, 266

see also constants
Crandall, W.E., 338
Crane, Horace Richard, 204, 205, 362
Cranshaw, J.D., 183, 348
Critchfield, Charles Louis, 63, 64, 68, 259, 260, 343, 353
Cross-sections, 216, 235, 244, 246, 247, 248, 257, 281, 309
Crussard, J., 90, 121, 138, 355
Curie, Irène, 21, 28, 66, 222, 240, 253, 343
Currents,
  algebra, 328
  charge changing, 42, 51
  conserved vector, 328
  exchange, 62
  light-particle (e-ν), 51
  neutral, 62, 68, 139
  nucleon, 62
  partially conserved axial-vector, 328
Cushing, J.T., 333, 334, 343
Cut-off procedure (in quantum field theories), 83, 169, 194, 208
Cyclotrons, 28, 29, 112, 308, 316, 318
  Japanese destroyed, 285, 289

Dancoff, S.M., 254, 261, **262**, 265, 271, 321, 343, 360, 363
Darrigol, O., 89, 113, 343
Davis, N.P., 25, 343
Davis, Watson, 122, 123
De Benedetti, Sergio, 184, 344, 364
De Hoffmann, Frederic, 261, 263, 293, 312, 333, 337, 363
Debye, Peter, 14, 15, 180, 274, 344
Delayed coincidence circuits, 297
Dempster, Arthur Jeffrey, 11
Deuterium (heavy hydrogen), 217, 218
  isotope, 28
Deuterons, 14, 31, 37, 45, 220, 258
  accelerated, 25
  Bethe on, 192
  binding energy, 60, 206
  quadrupole moment, 64, 171, 220, 258, 265
  states, 222
  see also meson theory, nuclear forces
Dirac, Paul A.M., 18, 27, 47–50, 54, 71, 100, 101, 111, 113, 114, 127, 135, 136, 140, 144, **147**, 163, 169, 170, 211, 234, 248, 266, 268, 288, 330, 340, 344, 359, 370
  on electrons, 136
  general spinor theory, 119, 136, 163, 231, 255

hole theory, 20, 27, 117, 231
  in Japan, 100
  on penetrating cosmic rays, 136
  on protons, 111, 112
  and quantum electrodynamics, 330
  relativistic electron theory, 18, 100, 101, 103, 104, 144, 155, 217
  and spontaneous decay rates, 101
  see also Fermi–Dirac statistics
Dirac field, 51
Dirac spinors, 232
Dirac's density matrix, 120
Dirac's general spinor calculus, 119, 146, 163, 171
Dirac's hole theory, 20, 27, 33, 111, 117, 231
Dirac's matrices, 103, 132
Dirac's relativistic electron equation, 18, 51, 68, 100, 103, 104, 155, 170, 233
Dispersion relations
  in high-energy pion (kaon)–nucleon interactions, 326
  Kramers–Kronig, 326
Dobrotin, N., 272
Doncel, M., 335, 359
Döpel, Robert, 274
Dower, J.W., 290, 344
Dresden, M., 23, 340, 344, 350
Duffin, R.J., 233, 256, 344
Dürr, H.P., 349
Dusenbury, J., 358
Dyson, Freeman, 321, 322, 363

e-ν joint production, Fermi on, 42, 43
e–p model, 17, 33
  problems with, 17–22
Eckart, Carl, 227
Eddington, Arthur Stanley, 12, 344
Eden, Richard John, 288, 291, 328, 344
Eguchi, N., 362
Ehmert, Alfred, 181, 183, 344
Ehrenfest, Paul, Jr., 69, 185, 186, 199, 204, 336, 344
Einstein, Albert, 3, 5, 22, 101, 113, 206, 224, 269, 344, 356
  quantum hypothesis, 101
Electrical counting tubes, 15
Electric charges in atoms, early theory, 5
Electromagnetic field, relativistic quantum theory of, 101
Electromagnetic form factors of nucleons, 326
Electromagnetic phenomena and nuclear phenomena, 77
Electromagnetic scattering, 195

Electromagnetism, Yukawa and, 107
Electron–muon universality, 301
Electron–neutrino exchange *see*
    "Fermi-field" theory
Electron–neutrino (e–ν) pairs, 42, 47, 52
    creation, 42
    exchange, 43
Electron-free nucleus, first steps towards,
    19–21
Electron–positron (e$^+$–e$^-$) pairs, 42, 52,
    53, 117, 314
    discovery, 20, 25
    production, 18, 20, 25, 38, 73, 117
    Yukawa and, 117
Electron–proton (e–p) model of atomic
    nuclei, 13, 24, 33, 49
    problems, 17–22
Electron–proton (e–p) model of the
    neutron, 20, 31 ·
    confinement problem, 18, 19
    difficulties, 40
Electrons, 2, 3, 33
    accelerated, 21
    deformation of, 3
    discovery, 6, 22
    early models, 6, 7
    electromagnetic mass, 3, 313
    field components, 51
    gravity and, 3, 4
    interaction between, 37
    knock-on, 244, 245, 248
    Lorentz theory, 1–4, 85, 112
    magnetic moment, 322
    meson scattering by, 243–5
    name, 22
    near nucleus, 100, 101
    negative energy states, 18
    in nuclei, Heisenberg on, 40
    showers, 294
    spin, 16, 18
    and their stability, 1–4
    within nucleus, 44, 46, 47, 50, 52, 102,
        113
    see also "heavy electrons"
Electroweak theory, 33, 330, 331, 332
Elementary (fundamental) particles, 31,
    33, 34, 49, 229, 323–5
    creation and annihilation, 33
    discovered 1944, 293
    free particles, 242, 243
    general properties (Heisenberg–Pauli
        report 1939), 240, 241
    resonances, 317, 320, 324, 327
    theory, 22

see also electrons; hadrons; heavy
    particles; interactions of elementary
    particles; kaons; leptons; mesons;
    mesotrons; neutrons; nuclear
    emulsion technique; pions; positrons;
    protons
Elliot, C.D., 89, 169, 350, 363, 366
Ellis, C.D., 15, 19, 56, 344, 363
Energy conservation problems of e–p
    model, 18, 19
Erschow, A., 171, 344
Estermann, Immanuel, 66, 172, 344, 345
"Ether", 3
Euler, Hans, 76, 88, 93, 180–3, **182**, 185,
    186, 193, 199, 200, 203, 205, 225,
    238, 239, 249, 288, 345
    on Hoffmann bursts, 88, 89
    on mesotrons, 180, 186
Euler–Heisenberg analysis, 298
Ewald, Peter Paul, 227
Exchange forces, 31, 32, 36, 95
    in atomic theory, 45
    see also "Fermi field" theory,
        Heisenberg's exchange force in
        nuclear physics; meson theory;
        nuclear forces; Yukawa meson theory
        of nuclear forces
Ezawa, H., 113, 200, 201, 345, 356

Fajans, Kasimir, 11, 24, 345
Faraday, Michael, 2
Feather, N., 23, 24, 345
Federbush, P., 326, 345
Feenberg, Eugene, 63, 67, 339
Feinberg, Evgeni, 272, 273, 345
Fermi, Enrico, 21, 25, 32, 33, 41–3, 47,
    48, **48**, 50–4, 56, 58, 62, 65, 66, 70,
    75, 90, 101, 102, 107, 110, 113, 117,
    134, 137, 139, 158, 177, 200, 201,
    222, 224, 240, 253, 255, 271, 274,
    281, 282, 290, 299–301, 313, 318,
    319, 325, 336, 345, 346, 363
    on e–ν joint production, 42, 43
    and Heisenberg, 54
    as improvable following Yukawa
        theory?, 182
    and nuclear chain reaction, 274
"Fermi field" theory, 30, 44, 46, 47–68,
    74, 75, 76, 83, 95, 96, 123, 167, 217
    Bethe and Bacher on, 90
    and charge independence of nuclear
        forces, 57–63
    combined with Yukawa theory, 64, 65
    criticism of, 58
    currents, 51

demise of, 96
Fermi's interaction (four-fermion interaction), 56–8
origins, 53–7
and related theories (1938–41), 63–5
Sakata and, 117, 118
as "standard model", 30
Tamm and Iwanenko on, 105
theories competing with, 52
Wentzel and, 209
Yukawa and, 107, 117, 118, 177, 217
Fermi, L., 290
Fermi motion, 307, 308
Fermi–Dirac statistics, 18, 36, 233
Fermi–Thomas method (self-consistent field), 34, 36, 41
Fermions, 49, 279
fractionally charged *see* quarks
scattering of identical, 24
Fermi's theory of β-decay, 30, 32, 33, 41–6, 51, 52, 56, 70, 75, 118, 134, 139, 182, 192
Bethe–Peierls derivatives, 57, 58
ft-value, 43
Gamow–Teller modification (selection rule), 129, 192, 21
low-energy spectrum, 56, 57, 58, 199
and Yukawa's theory, 182
Ferretti, B., 259, 313, 335, 337, 346
Fertel, G.E.F., 301, 302, 313, 350, 361
Feynman, Richard P., 321, **322**, 323, 333, 334, 341, 346, 363
Fierz, Markus, 137, 176, 212, 230–2, **232**, 240, 248, 255, 258, 321, 346
Flerov, G.N., 272, 273
Flügge, Siegfried, 273, 274, 281, 288
Fock, Vladimir, 78, 79, 264, 344, 346
Foley, H.M., 321, 354
Follet, D.H., 183, 346
Forces
electromagnetic, 1, 3
gravitation, 1
magnetic, 15
non-Coulombic attractive, 15
non-electromagnetic, 13
see also Coulomb force; nuclear forces
Forman, P., 354
Forró (Barnóthy), Eva, 84, 92, 213, 226, 239, 336
Foster, B., 314, 343, 346
Four-vector potential, 54
electromagnetic, 51
Fournier, A., 336
Fowler, Peter H., 314, 341, 343, 346, 360
Fowler, Ralph Howard, 171

France
cloud chamber and counter experiments on cosmic rays, 90
Frank, C., 291, 346
Frautschi, Steve C., 327, 342
Frazer, W.R., 326, 346
Freier, P., 313, 346
Frenkel, Jacov, 16, 154, 346
Fréon, A., 185, 186, 336, 344
Frisch, Otto, 66, 172, 346
Frisch, S., 25, 354
Fritzsch, H., 331, 346
Fröhlich, Herbert, 148, **153**, 154–7, 159, 161–6, 169, 170–4, 207, 208, 217, 218, 227, 266, 267, 346, 347, 358
Heitler on, 164
on magnetic moment deviation, 156, 157
on vector meson theory, 155, **156**, 157, 161, 162
wartime career, 266, 267
and Yukawa, Hideki, 148
Froissart, M., 327
Froman, D.K., 362
Fubini, S., 328, 347
Fulco, J.R., 326, 346
Fundamental theory of nuclear forces *see* nuclear forces
Fussel, L., Jr., 92, 294, 347

Galison, Peter, 23, 71, 74, 82, 137–9, 347
Gamma- (γ-) rays (nuclear), 4, 5, 6, 28
absorption, 23
anomalous scattering, 38
Blackett on, 38
crystal spectrometer, 46
early work on, 5
emission, 34
line spectra, 15
origin, 37
scattering, 34
Gamow, George, **16**, 16, 17, 37, 38, 56, 59, 63, 68, 90, 129, 192, **206**, 207, 222, 224, 240, 279, 347
liquid drop model, 17, 40
on nuclear constituents, 37, 38
Gardner, Eugene, **308**, 309, 315, 341, 347
Gavriola, E., 61
Geffen, D.A., 334, 347
Geiger, Hans, 7, 8, 19, 41, 92, 226, 275, 347
Geiger–Müller counter, 27
Gell-Mann, Murray, 313, **322**, 324–6, 328, 330, 334, 346, 347, 348
Gell-Mann–Nishijima relation, 325

Gemert, A. von, 226, 342
Gentner, W., 312, 348
Germany
    cosmic ray physics (1941–44), 274, 275
    Heisenberg's S-matrix, 245–7
    and Japan: wartime contacts, 280, 281
    meson physics (1940–43), 273
    uranium project, 241, 273, 287
    vector meson theory, 207–13
Gill, P.S., 238, 243, 248
Gluons, 329, 330
    see also quantum chromodynamics
Goldberger, Marvin L., 326, 342, 345,
    347, 363
Goldhaber, Maurice, 229, 248
Gooding, D., 348
Gordon, Walter, 17, 348
Goudsmit, Samuel, 42, 207, 221, 291,
    348, 361
Greisen, K., 361
Gross, D.J., 330, 348
Grotrian, W., 223, 228
Groves, L., 270, 348
Grythe, I., 289, 348
Gurney, R.W., 16, 348
Gürsey, Feza, 328, 329, 334, 348

Haas, Arthur Erich, 8
Hadrons, 229, 317, 325, 326, 328
    interaction, 331
Hafstad, L.R., 367
Hahn, Otto, 223, 239, 253, 288
Haken, H., 172, 348, 350
Hall, D.B., 189, 362
Hall, H., 200, 348
Halpern, O., 200, 348
Han, M.Y., 334, 348
Hansen, W.W., 48
Hara, Osamu, 321, 333, 348
Hardmeier, W., 14, 15, 344
Harkins, W.D., 15, 348
Hayakawa, Satio, 68, 129, 137, 138, 169,
    276, 277, 285, 289, 291, 312, 348
Hazen, W.E., 339
"Heavy boson proton" and "boson
    neutrons" of Wentzel, 216
"Heavy charge", 211
"Heavy electrons", 132, 301
    described, 64
    discrepancy with mesotron, 169
    experiments, 115
    and mesons, 246
    oppositely charged, 259
    pair theory, 64
    see also meson theory; mesotrons

Heavy hydrogen *see* deuterium
Heavy particles, 51, 108, 205
    interaction of *U*-field with, 107, 150
    see also baryons
"Heavy quanta" *see* U-quanta
Heilbron, John, 8, 9, 23, 24, 348, 349,
    354
Heisenberg, Werner, 19, 20, 25, 31–6, **35**,
    39–42, 44–6, 47–60, 62, 65–8, 70,
    71, 74–8, 83–9, **85**, 90–3, 95, 96,
    100–5, 107, 108, 110, 112, 114, 135,
    140, 155, 167, 173, 176, 178–83,
    185–8, 193, 194, 198–200, 203, 207,
    210, 212, 217, 222–8, 230, 233–41,
    244, 249, 251, 253, 258, 261, 263,
    264, 268, 273–6, 281, 283, 284,
    287–91, 295, 312, 313, 340, 344,
    345, 349, 350, 359, 361, 370
    and atomic bomb, 253
    and Berlin group, 274
    and β-decay, 35, 39, 41, 42, 49, 50, 57
    on Bethe, 220
    and Bhabha, 235, 236
    and Blackett, 183
    on cascade showers, 84–6, 274
    and chain reactor, 274
    and charge-exchange interaction, 49, 40
    classical meson theories, 237
    Eighth Solvay Conference report, 176,
        240–3
    on electrons, 40, 52, 95
    and exchange forces, 31, 32, 40–42, 40,
        130, 150
    "exchange operator", 145
    and "Fermi field" theory, 47, 53–8, 70,
        74–6, 83
    on general particle theory, 275, 276
    on Göttingen, 287
    in Japan, 100
    and meson decay, 179–83
    on mesotrons, 135, 183, 186
    and Möller, 287
    n–p (neutron–proton) model, 31–3, 36,
        49
    and neutrino theory of light, 77, 78
    on neutrinos, 39, 52, 53, 84
    on neutron's properties, 31, 34, 35, 50
    and nuclear chain reaction, 274
    nucleus: model (1932–33), 33–6, 49,
        95, 102
    and observable quantities in elementary
        particle theory, 275, 276
    opposition to ideas of, 239
    and Pauli, 39, 57, 74, 75
    post-war research, 287

and quantum electrodynamics, 76, 78, 86
response to cascade shower papers, 84–6
and S-matrix theory, 252, 275, 276
simplified model theories, 86, 87
Solvay Conference report (1934), 39–41
and Sommerfeld, 217, 273, 274
and Tomonaga, 194
and Wick, 54, 55, 283, 284
and wrong statistics, 49
and Yukawa, 102, 105, 107, 155, 156, 181–3
Heisenberg's (charge) exchange force in nuclear physics, 51, 56, 60, 96, 103, 105, 107, 130, 150, 162, 167
Heisenberg's "explosive showers", 71, 74–6, 207, 212, 239
Bhabha on, 167
Euler on, 88, 89
Heitler on, 168
Kemmer on, 173, 208
Kobasi and Okayama on, 215
Pauli on, 87
Wentzel on, 209
Heisenberg's "Platzwechsel", 44, 104, 108
see also Heisenberg's exchange force in nuclear physics
Heisenberg's S-matrix theory, 274–6, 281, 287, 288
Japan and, 281
Heisenberg's theory of nuclear forces (Heisenberg forces), 49, 51
neutron–proton symmetry, 49
Heitler, Walter, 18, 71, 74, 80–5, 90, 112, 122, 126, 127, 139, 148, 153–62, 164–6, 168, 169, 171–4, 205, 207, 208, 213–8, 225–7, 234, 236, 240, 254, 255, 263, 266–8, **267**, 279, 286, 288, 295, **296**, 301, 302, 313, 319, 336, 337, 338, 346, 347, 350
background, 153
and Bhabha, 166
on energy loss of electrons, 74
on Fröhlich and Kemmer, 164
letter, **296**
on mesons, 216
on muons, 82
on quantum theory, 74
and vector meson theory, 155, 156, 160, 161
wartime career, 266–8
Heitler's "isobaric states", 263, 271
Heitler's "radiation damping", 267, 268

Helmholtz, Hermann von, 2, 350
Hendry, J., 25, 350
Hermann, A., 335
Herzberg, G., 18, 350
Herzfeld, Karl, 153, 227
Hess, Victor Franz, 6, 69, 122, 141, 188, 199, 350
Hevesy, G. von, 24, 350
Heyden, M., 226, 347
Heydenburg, N.R., 367
Hiebert, Erwin, 30, 350
High-energy nuclear physics, 22
Hilbury, N., 362
Hirosige, T., 112, 350
Hoag, J.B., 362
Hoch, Paul, 152, 154, 171, 172, 350
Höcker, Karl-Heinz, 283
Hoddeson, L., 90, 138, 334, 340, 343, 350, 363, 368
Hoffmann, Gerhard, 70, 73, 76, 87, 180, 223
see also bursts
Hofstadter, Robert, 326, 333, 334, 351, 369
Holborn, Mr., 183
Hondo, Kotaro, 97
Hooper, E.J., 341
Houtermans, Fritz, 17, 24, 351
Hoyer, U., 339
Hu, N., 250, 271, 351
Hulthén, L., 268, 271, 351
Hupfeld, H.H., 34, 45
Husimi, K., 170
Hydrogen, 154, 217
isotope of mass 2: discovery, 28
nucleus, 11–14
Rutherford on, 12, 13
spectrum anomaly, 321, 322
Hyperfine structure, 18, 42, 100, 102, 105, 113

Ichimiya, Tarao, 127, 134, 358
Impact parameters method, 245
"Infra-red catastrophe", 234
Infled, Leopold, 92, 339
Ingelby, P., 295, 312
Inoue, Takesi, 68, 197, 251, 278, 279, 285, 289, 294, 363
"Interaction picture", 211
Interactions of elementary particles, 106, 107, 112, 113, 116, 141, 229, 230
constants, large and small, 100
electromagnetic, 1, 3, 317
electroweak (gauge), 33
four-fermion contact, 51

gravitation, 1
magnetic, 15
non-Coulombic, attractive, 15
strong, 24, 317, 325–9
weak, 317
see also Coulomb force; nuclear forces
Intermediate particles
Anderson and discovery of, 122, 123
boson, 110
designations, 239
discovery confirmed, 123
earliest evidence, 178
fields, 62
as of numerous types, 204
see also meson decay; meson theory;
mesons; mesotrons;
Ireland, 255
meson physics (1941–45), 267, 268
Irving, D., 270, 351
Ishiwara, Jun, 97
Isospin, 36
amplitudes, 318–20
formalism, 34, 49
invariance *see* nuclear forces,
charge-independence
matrices, 108
name, 68
of nucleus, 261
operators, 51, 62, 65
Isotopy, 9, 10
Italy
(1945–47), 284
experiments on capture of mesons by
nuclei (1943–47), 297–301
nuclear and cosmic ray physics
(1942–44), 282–4
physics before the war, 282
Itakura, K., 351
Ito, D., 321, 351, 352
Itoh, J., 138
Iwanenko, Dmitri, 31, 41, 46, 56, 57, 58,
66, 105–7, 123, 234, 249, 250, 268,
271, 272, 351

Jackmann, J.C., 270, 352
Jammer, M., 113, 352
Janóssy, Lajos, 88, 188, 252, 295, 312,
352
Japan
β-mesotron decay, 193, 194
early physics, 97, 98
and Germany: wartime contacts, 280,
281
Meson Club, 277, 278
meson physics (1941–45), 277

natural philosophy 1920s, 111
nuclear energy programme, 281
physicists, 99
post-war, 284–6
Progress of Theoretical Physics, 285
quantum mechanics, introduction, 97,
98, 100
universities, 202
vector meson theory, 146–51
Jauch, Joseph Maria, 260, 271
Jensen, H., **232**
Jews, persecution of, 217, 255, 282
Johnson, T.H., 195, 200, 201, 352, 364
Joliot, Frédéric, 21, 28, 66, 91, 222, 239,
240, 253, 282, 343
Jordan, Edward B., 118, 336
Jordan, Pascual, 46, 65, 78, 79, 91, 230,
231, 352
Jost, Res, 172, 288, 352
Juilfs, J., 283, 290, 352

K-electron capture, 118, 137
K-mesons (kaons), 295–7, 324
Kahn, B., 217, 218, 347
Kaneseki, Y., 138, 286, 352
Kaons *see* K-mesons
Kapitza, P., **16**
Karplus, R., 342
Kaufmann, Walther, 3, 22, 352
Kawabe, Rokuo, **113**, 137–40, 169, 225,
289, 290, 340, 352
Kay, William, 13
Kellogg, J.B., 172, 220, 227, 352
Kelvin (Lord) *see* Thomson, William
Kemmer, Nicholas, 62, 63, 68, 113, 137,
148, 150, 153, 154, 157–67, **159**,
169, 171–3, 180, 192, 194, 198,
207–12, 215, 218, 219, 221, 225–7,
233, 241, 248, 256, 257, 266, 309,
316, 320, 347, 353
background, 153, 154
and charge independence, 316
et al., 210
and Fermi field, 62
first vector meson theories, 157–64
on Fröhlich, Heitler and Kahn, 218
and Heitler, 161, 164
symmetric theory, 221
on vector field theory, 162–4
Kemmer–Duffin formalism, 233, 242, 256
Kikuchi, M., **119**
Kikuchi, Seishi, 98, 118, 119, **119**, 123,
128, 138, 222, 241
Kimura, Masamichi, 100, 222
King, D.T., 341

Kinoshita, Toichiro, 286
Kleeman, R., 5, 339
Klein, Oscar, 18, 46, 65, 170, **206**, 206, 207, 213, 214, 225, 231, 250, 252, 253
"Klein paradox", 18
Kleint, C., 361
Koba, Z., 321, 351, 353
Kobayashi, Minoru, 99, 116, 142, 202, 213, 214, 226, 250, **279, 280**, 281, 353, 370
Koizumi, K., 112, 353
Kolhörster, W., 141
Konopinski, E.J., 58, **59**, 67, 90, 92, 117, 134, 182, 190, 199, 353
Konopinski–Uhlenbeck (K–U) derivative forces in β-decay, 59, 60, 86, 118, 130, 134, 183, 190
Konuma, M., 113, 137, 170, 198, 228, 289, 340, 353
Koontz, P.G., 362
Korsching, Horst, 274
Kothari, D.S., 221, 356
Kramers, Hendrick A., **133**, 206, 207, 225, 232, 269, 270, 276, 289, 322, 326, 353
and S-matrix, 276
Kronig, Ralph de Laer, 18, 25, 78, 79, 91, 207, 276, 326, 352–4
Kuhn, Thomas, 197, 198, 354
Kulenkampff, Helmuth, 181, 198, 354
Kunze, Paul, 115, 121, 138, 185, 354
Kurchatov, I., 272
Kursunoglu, B., 341, 354
Kurti, Nicholas, 154
Kurz, Marietta, 305
Kusaka, Shuichi, 194, **247**, 247, 248, 250, 254, 259, 261, 265, 270, 271, 278, 342, 354, 360
Kusch, Polycarp, 321, 333, 354

Lamb, Willis E., 68, 118, 137, 216, 218, 226, 227, 254, 321, 333, 343, 354
Lambda (Λ), 323
Landau, Lev D., 205, 268, 271, 272, **273**, 340, 354
Landé, Alfred, 227
Langer, R.A., 126
Laporte, Otto, 216, 217, 224, 226, 243, 354
Lasarew, B.G., 154, 354
Laslett, L. Jackson, 118
Lattes, Cesar M.G., 197, 304, 305, **306**, **308**, 309, 314, 315, 341, 347, 354, 355

"Lattice-world" model, 86, 87
Laue, Max von, 287
Lawrence, Ernest O., 21, 28, 29, **29**, 37, 112, 291, 318, 343, 355
Lee, T.D., 324, 326, 333, 355
Length, new fundamental, 86
Lenz, Wilhelm, 12, 355
Leprince-Ringuet, Louis, 70, 73, 90, 121, 138, 293, 312, 336, 355
Leptons ("light particles"), 66, 229, 325, 331
universality, 301
see also electrons; muons; neutrinos
Leutwyler, H., 346
Lévy, Maurice, 328
Lewis, H.W., 315, 355
Lhéritier, Michel, 293, 312, 355
Lifschitz, E., 272, 355
Light, neutrino theory of, 76–80, 91
Lindholm, F., 87, 351
"Liquid drop" model, 17, 40
see also α-particles
Livesy, D.L., 361
Livingston, M. Stanley, 28, 29, 59, 355
"Loading" technique, 305
Lock, W.O., 314, 355
Lofgren, E.J., 346
London, Fritz, 153, 227
Lorentz, Hendrik Antoon, 1–3, **2**, 22, 23, 112, 225, 340, 355
Loverdo, A., 342
Low, Francis, 342
Lüders, G., 333, 355

Ma, S.T., 250, 279, 288, 356
Maass, H., 198, 213, 226, 356
Maglic, Bogdan, 326, 333, 356
"Magnetic lens", 283
Magnetic moments
anomalous, 58, 146, 148, 162
Dirac, 58, 66
electrons (anomalous), 321
mesons (heavy quanta), 149
nuclei, 25, 41
nucleons, 19, 55, 57, 64, 131, 147, 154, 155, 162, 172, 258
see also vector meson theories; neutrons,
Maier-Leibnitz, Heinz, 226, 348, 356
Majorana, Ettore, 32, 33, 36, 40, 49, 60, 134, 167, 241, 335, 356
and exchange forces, 40, 41, 53, 57, 60, 130, 145, 150
Majumdar, R.C., 221, 250, 342, 356
Maki, Z., 340

Manhattan Project, 253
Marsden, Ernest, 7, 8, 13, 23, 347, 356
Marshak, Robert E., 64, 68, 195, 226,
    259, 260, 270, 271, **273**, **279**, 280,
    290, 323, 332–4, 337, 356, 366
Marston, W.V., 227
Marten, M., 342
Martin, R., 336
"Mass spectrograph", 11
"Mass-defect" problem, 12
Massey, H.S.W., 243, 356
Masudo, T., 358
Matsui, M., 113, 356
Maxwell, James Clerk, 2, 23, 57, 58, 148,
    149, 152, 165, 206
McAllister, R.W., 333
McMillan, Edward M., 28, 318, **319**, 332,
    356
McMillan, Edwin, 310
Mehra, J., 22–4, 36, 113, 356
Meitner, Lise, 12, 19, 41, 46, 49, 50, 222,
    240, 356
Meitner–Hupfeld effect, 34, 45
Mendeleev, Dimitri, 6, 9
Mercier, André, 276
Meson decay, 177–201
    and β-decay (1938–41), 189–94, 201
    Conversi and Piccioni on, 297, 298
    estimating lifetime, 178–83
    first decay, **306**
    Rome experiment, 297, 298
    and theory/experiment confrontation,
        197
    versus meson capture (1939–42), 194–7
    Yukawa and Sakata on, 203
    see also meson theories
Meson theories, 44, 109, 214, 321
    application to cosmic ray phenomena,
        166–9, 179
    Beck's, 215
    Brillouin on, 207
    charge independence, 192, 221, 316
    classical nonperturbative approach
        (Bhabha), 192, 234
    four types, 194
    Kemmer's four theories, 256
    mixture of fields, 254
    neutral, 192, 219, 220, 259, 279
    Pauli in 1940 on, 255
    production of electron pairs by
        mesotrons, 245
    pseudoscalar, 170, 173, 194, 227, 248
    renormalized, 222, 223
    and renormalized QED, 320–3
    scalar, 173, 248

secondary literature, 71
self energy, 261
two-meson, 68, 197, 278, 279, 280, 289
various experimental data, **279**
see also Møller–Rosenfeld "patent
    mixture"; vector meson theories;
    Yukawa's meson theory
Meson–proton scattering, 237
Mesons, 56, 175, 197
    absorption and scattering, 233
    artificial production (1948–50), 307–12
    Beck on meson hypothesis, 117
    as having β-decay interaction, 110
    Bethe and Marshak on, 64, 65
    bremsstrahlung, 244
    capture by nuclei, 196, 297–301
    charged, 130
    cohesive (C-mesons), 321
    Coulomb scattering, 246
    creation, 110, 111
    designation, 70, 187, 221
    early 1950s, 317
    enthusiasm for study of, 204
    first intimations of, 56
    instability, 177
    interaction: classical approach, 233–7
    interaction strength, 216
    as "intermediate boson", 110
    isospin amplitudes, 318–20
    lifetime, 110, 111, 175
    meaning, 111, 112
    mixed theory, 219–21
    multiple production of, 230
    negative, 309
    neutral (neutrettos), 64, 68, 166, 192,
        209, 213, 215, 219–21, 226, 258,
        278, 310, 311, 326
    photoproduction cross section, 216
    π see pions
    positive, 309
    Proca on, 191,
    production in collision processes, 189
    pseudoscalar, 201
    quantum of nuclear force, 56
    radiative interaction, 247
    Sakata and Inoue on, 278, 279
    as same as mesotrons?, 197, 198, 258
    scattering of, 243–5, 318, 319
    slow, 203, 304
    spin, from electromagnetic effects
        (1939–41), 243–8
    stability, 198
    in Standard Model, 329–32
    "stopping", 298, 305
    types, 64, 292

see also cosmic ray mesons; mesotrons; muons; pions; pseudoscalar mesons
Mesotrons ("heavy electrons", "cosmic ray particles"), 70, 71, 115, 132, 141, 177
  absorption anomaly, 200
  Blackett on, 135, 136
  capture of negative, 195
  cloud chamber studies, 195
  cosmic ray particles, 63, 74, 177, 178
  decay, and cosmic ray puzzle resolution, 92, 137, 178, 183–7, 189
  designation, 70, 187, 221
  discovery, 82, 83, 115, 116, 121–9, 131
  discrepancies with heavy electrons, 169
  disintegration, 248
  European conferences and, 204
  and "heavy electrons", 91, 96
  "inclination anomaly", 184
  Japanese work on, 202, 203
  lifetime determinations, 175, 180–9
  lifetime measurements of Rossi and Rasetti, 187–9
  mass absorption anomaly, 183–7
  mass determination, 127, 128, 204, 205
  mesotron–meson problem, 197, 198
  new experiments and calculations on production of (1938–39), 213–5
  in nuclear physics, 159
  orbit, 299, 300
  pairs, 259
  reacting with nuclei and electrons 257
  as same as mesons?, 197, 198, 258
  as same as Yukawa's meson?, 178
  slow negative, 197
  "stopping", 189, 195, 196, 197
  terminology, 135, 187, 199, 227, 238, 249
  and *U*-quanta, 106–11
  and Yukawa's meson theory: hopes and doubts in 1937, 129–35
  see also β-decay; cosmic rays
Messerschmidt, W., 88, 356
Meyenn, K. von, 19, 66, 67, 270, 356, 357
Meyer, Lothar, 6
Meyer, P., 93, 357
Michel, Louis, 335
Michels, Walter, 285
Mie, Gustav, 148, 333
Migdal, A., 272
Miller, A.J., 44, 357
Millikan, Robert A., 23, 30, 73, 123, **124**, 128, 140, 141, 186, 187, 199, 200, 213, 224, 225, 339, 357

on cosmic rays, 73
on mesotron absorption, 186, 187
Mills, R., 329, 369
Mishima, T., 358
Mixed meson theory, 219–21
Miyazaki, Y., 358
Miyazima, Tatsuoki, 259, 264, 271, 277, 357
Molière, Gerhard, 274
Møller, Christian, 118, 191, 192, 194, 219–22, 226, 227, 240, 258, 259, 270, 276, 278, 279, 287–9, 291, **302**, 357
  and S-matrix, 276, 287
Møller–Rosenfeld "patent mixture", 191, 192, 219–21, 259, 265, 320, 321
Montgomery, C.G., 88, 89, 197, 239, 249, 357
Montgomery, D.D., 88, 89, 239, 249, 357
Moore, R., 139, 357
Morrison, Philip, 25, 254, 284, 332
Morse, P., **133**
Moseley, Henry G.J., 10, 357
Mott, Nevill, 172, 357, 358
Moyer, B.J., 338
Moyer, D.F., 20, 44, 45, 66
Muirhead, H., 341, 354
Mukherji, Visvapriya, 67, 68, 71, 137, 169, 198, 271, 289, 290, 358
Multiple-particle production, 83
Muons (μ-mesons), 64, 70, 71, 92, 187, 260, 292, 305, 307, 331
  decay: photographic observation, 307
  discovery and properties, 82, 307
Murphy, G.M., 367

n–p (neutron–proton) model
  Bhabha on, 167
  Heisenberg's, 31–3, 36, 49
  nuclear force theories and, 60
  Yukawa on, 106
Nagaoka, Hantaro, 23–5, 97, 100, 358
  model, 9
  "Saturnian atom" (1903), 7
Nagendra Nath, N.S., 78, 79, 91, 339, 358
Nakamura, Seitaro, 289
Nakano, Tadao, 324
Nakayama, Shigeru, 97, 112, 350, 358
Nambu, Yoichiro, 112, 286, 326, 330, 331, 334, 342, 348, 358
  and QCD, 330, 331
Neddermeyer, Seth H., 56, 63, 73, 74, 80, 82, 90, 92, 112, 115, 121–3, **124**, 125, 127, 129, 131, 132, 138–40,

142, 170, 178, 187, 188, 200, 204, 205, 224–7, 238, 301, 335, 358
Ne'eman, Yuval, 325, 348
Neher, N.V., 199, 339, 358
Nelson, E.C., 194, 254, 270, 358
Nereson, N., 189, 200, 298, 358, 362
Neutral countries, wartime research, 253
Neutral pions *see* pions, neutral
Neutrettos *see* mesons, neutral
Neutrino–electron exchange *see* "Fermi field" theory
Neutrinos, 19, 24, 27, 31, 32, 41, 47–53, 65, 76–80
  Beck on, 38, 52
  Bohr on, 38, 52
  experiments to detect, 50, 92
  Fermi on, 50
  field components, 51
  first references to, 32, 33, 36
  Heisenberg on, 39, 52, 53, 84
  hole theory 137
  hypothesis of Pauli, 7, 19, 25, 27, 36
  name, 50
  Pauli on, 27, 31–3, 39, 41, 42, 47, 50, 52, 77, 111
  Sitte on, 52
  see also light, neutrino theory of
"Neutron" of Pauli, 19
"Neutrone" of Fermi, 46, 50
Neutron–proton transformation, 157
Neutrons, 13, 31, 42, 49, 65
  capture, 21
  Chadwick on, 20, 21, 28, 31, 38, 39, 42
  composite structure, 31, 33–5, 39, 49
  discovery, 18, 20, 21, 25, 28, 31, 49
  elementary nature, 31, 33, 44, 50, 102
  first observation, 310
  Heisenberg on, 31, 34, 35, 50
  Heitler and Fröhlich on, 156, 157
  magnetic moment, 156, 157, 166, 245
  Majorana and Wigner on, 32
  mass defect, 34, 44
  name, 28
  nature of, 229
  Pauli on, 49, 50
  Rutherford's prediction, 21
  scattered by protons, 303
  "singlet neutron", 61
  slow, 60
  see also electron–proton model of the neutron; electron–proton model of atomic nuclei; magnetic moments, n–p model
Nicholson, John William, 23
Nishijima, Kazuhiko, 313, 324

  see also Gell-Mann–Nishijima relation
Nishikawa, Masaharu, 98, 138
Nishina, Yoshio, 21, 56, 92, 97–100, 105, 119, 120, 127, 128, **128**, 134, 139, 142, 143, 170, 178, 204, 222, 251, 281, 290, 353, 358
  and cloud chamber, 127
  and uranium, 281
Nisio, S., 358
Nomaguchi, M., 281, 290
Nordheim, G., 92, 190, 214, 216, 226, 358
Nordheim, Lothar W., 121, 190–2, **191**, 200, 214, 216, 226, 255, 337, 358
Nordsieck, Arnold, 56, 58, 236, 338, 359
Nuclear atom discovered, 5
  binding energy, 37, 38
Nuclear bombs, 253, 254, 282, 284, 318
Nuclear chain reaction, 274
Nuclear collisions, particles and, 55, 56
Nuclear emulsion technique, 294, 301–4
Nuclear fission, 239, 253, 281
Nuclear fluorescence, 38
Nuclear forces (potentials), 21, 22
  Bartlett's, 132
  charge exchange, 36, 41, 53
  charge independence, 48, 61, 64, 68, 138, 158
  electron-pair theory of Bethe and Marshak, 226
  exchange, 33–6, 41, 54, 57
  field of, 19
  fundamental theories, 33, 36, 47, 103
  high-energy behaviour, 65
  Majorana's, 36, 40, 41, 45, 57, 145, 150, 167, 200
  and neutral mesons, 300, 301
  neutron–neutron, 34, 60
  neutron–proton, 33–6, 40, 60, 148, 166
  non-central, 17
  non-Coulombic, attractive, 17
  non-exchange, attractive, 17
  nucleon–nucleon, 208
  pair-exchange theories, 61, 62, 64, 259, 260, 270
  proton–proton, 33, 60, 135
  quantum field theory, 44
  repulsion (Coulomb), 40
  saturation, 64
  short-range (attractive, repulsive), 34, 36, 37, 46, 60, 104
  spin dependence, 60, 64
  in standard model, 329–32
  strong, 53
  strong after pion, 53

Stueckelberg's field theory in 1938, 211
theory with heavy-electron pairs, 259
Wigner's 132
see also electrons within nucleus,
    nuclear models, range of nuclear
    forces; saturation; Yukawa's meson
    theory
Nuclear isobars, 54, 320
Nuclear mass defect, 11, 24
Nuclear mass number, 8
Nuclear models, 11, 12, 17
    alpha-particle structure, 45
    Gamow's "liquid drop", 40
    Heisenberg's neutron–proton model,
    44, 49
Nuclear particles, accelerated, 27
Nuclear phenomenology, 1, 22, 25, 49
Nuclear reactions
    first artificially induced, 13, 14
    production and study of, 31
Nuclear structure
    Heisenberg on, 49
    lines of inquiry in 1920s, 15
    physics launched, 41
    systematics, 49
Nuclear transformations, 11, 21, 31, 109
    artificial, 24
Nucleon–nucleon force/collisions, 208,
    308
Nucleons, 33, 49, 61, 62, 168, 331
    coupling of meson fields to, 261
    electromagnetic structure and
        properties, 245–8, 326
    heavy, 61
    isobars, 320
    nucleon–nucleon scattering, 28
    number, 211
    resonances, 320
Nuclei, 60
    atomic, 9
    Coulomb field of, 196
    discovery, 1, 8
    as envisaged in 1932, 31
    Heisenberg and, 33, 34, 39
    like-particle forces, 60, 61
    mass defects, 60
    mesons and, 195, 217
    and quantum theory, 15, 17
    as source of radioactivity, 9, 10
    stability, 60
    see also nuclear structure

Occhialini, Giuseppi P.S., 25, 121, 142,
    197, 290, 303–5, 309, 338, 355, 359,
    361

Ogawa, Hideki, *see* Yukawa, Hideki
Ogawa, K., 99
Ogawa, T., 99, 100
Okayama, T., 196, 201–3, 213, 214, 226,
    353, 370
Okubo, Susumu, 286
Omega- (ω-) mesons, 326
Onsager, Lars, 269, 359
Operators of creation and annihilation
    (destruction), 42, 46
Oppenheimer, J. Robert, 21, 56, 63, 71,
    74, 80–5, **81**, 90, 92, 123, 130–3,
    135, 138, 140, 144, 170, 178, 179,
    186, 194, 199, 215, 239, 244, 245,
    247, 248, 250, 253, 254, 259, 260,
    264, 265, 268, 270, 271, 278, 279,
    288, 341, 343, 355, 359
    and atomic bomb, 253
    on bursts, 239
    on Christy and Kusaka, 248
    on intermediate-mass particles, 131
    on mesotron absorption, 186, 187
    and Pauli, 259
    and Schwinger, 264, 265
Orthmann, W., 19, 356
Osborn, H., 343, 353

p–p (proton–proton) force, 166
    Bhabha on, 167
p–p (proton–proton) scattering, 60
Pair production, 18
    discovery, 20, 205
    process, 73
    Yukawa and, 117
Pair theories, 259, 260
Pais, Abraham, 4, 22–5, 307, 314, 329,
    334, 335, 340, 359
Pancini, Ettore, 252, 298, 337, 343
Paneth, Fritz, 24, 350
Panofsky, W.K.H., 311, **311**, 318, 332,
    359, 365
Particle accelerators, 27, 292, 293, 301
    (early 1950s), 317
    first observation of new particle, 310
    giga-electronvolt, 324, 325
    meson experiments; isospin amplitudes,
    318–20
    research before days of, 293
    in USA, 318, 326
    see also Cockcroft–Walton machine;
        cyclotrons: Van de Graaff generator
Particles *see* elementary particles
Pasternack, Simon. 217, 359
Pauli, Wolfgang, 19, 20, **20**, 25, 27, 31,
    32, 36, 39, 41, 42, 45–50, 53, 54, 56,

57, 60, 62, 65–7, 71, 74, 75, 77, 78,
84, 86, 87, 90–2, 101, 102, 111, 112,
114, 115, 129, 137, 139, 144, 145,
151–4, 158, 161, 169–73, 176, 179,
180, 198, 203, 210, 212, 217, 218,
220, 222, 226, 227, 230, 231, **232**,
233, 236, 240–3, 248–50, 254, 255,
258, 259, 260, 261, 265–72, 288,
289, 291, 321, 333, 346, 357, 359,
361
  on developments of meson theory
    during the war, 268–70
Pauli's exclusion principle, 45
Pauli's spin matrices, 65
Pauli–Weisskopf (scalar) field theory, 92,
  170
  "anti-Dirac" theory, 92, 144, 145, 231
Pauling, Linus, 227
Peierls, Rudolf E., 25, 50, 56–8, 67, 226,
  227, 243, 250, 253, 266, 267, 339,
  360
  on "Fermi field theory", 56–8
Penetrating radiation, 1, 5, 6
  see also cosmic rays
Penetrating showers, 81–4, 294, 295
Peng, H.W., 267, 268, 350
Perkins, Donald H., 304, 305, 313, 314,
  360
Perrin, Francis, 50, 207
Perrin, Jean, 22
Perturbation expansion, Wentzel on, 261,
  263
Peters, B., 346
Petrzhak, K., 272
Pfotzer, G., 92, 360
Phenomenological theories of nuclear
  physics, 21, 22
Photography
  film, 303, 304, 307
  nuclear emulsion technique, 294, 301–4
Photoproduction, 168
Pi- (π-) mesons *see* pions
Piccard, Auguste, 69
Piccioni, Oreste, 201, 251, 297, 298, 313,
  343, 360
Pickering, A., 139, 333, 360
Pickup, E., 204, 369
Pinch, T., 348
Pion–pion scattering, 327
Pions (π-mesons), 64, 96, 178, 187, 260,
  288, 292, 311, 331
  decay, 305–7, 309
  discovery (1947), 304–7, 323
  Fermi and Yang on, 325
  lifetime of charged, 307

  lifetime of neutral, 309
  masses, 304, 314
  multiple production, 309
  neutral, 91, 309
  and nucleons, 320, 326
  production, 308–11
  production threshold, 310
  and protons, 319
  spin, 311, 312
  strong interaction with nucleons, 319
  why not discovered in USA, 307
Pions (π-mesons), neutral
  discovery, 307–12
  observation difficulties, 309, 310
Pippard, Brian, 340
Planck, Max, 5, 8, 87, 101, 113, 212, 356,
  360
Plutonium bombs, 284
Podolsky, Boris, 344
Politzer, H.D., 330, 360
Pomeranchuk, I., 272
Pomerantz, M.A., 200, 352, 360
Pompeia, P.A., 365, 367
Pontecorvo, Bruno, 282, 300, 301, 313,
  360
Positrons, 31
  discovered, 20, 25, 27, 29, 30, 40, 49
  first observation, 310
  predicted, 27
  Yukawa on, 119, 120
Potentials
  electromagnetic, 3
  Newtonian, 3
  spin dependent, 32
  see also Coulomb force; nuclear forces
Powell, Cecil Frank, 197, 201, 251, 301,
  302, **302**, 303, 304, 307, 309, 312–4,
  341, 350, 355, 359, 360, 361
Powers, T., 270, 361
Present, R.D., 60, 67, 339, 361
Proca, Alexandru, 115, 146, 148, 158,
  165, 169, 171, 173, 185, 186, 199,
  208, 221, 222, 240, 361
  and mesotron decay, 185, 186
Protons, 12–5, 23, 33, 49
  artificially accelerated, 3
  cloud chamber studies, 195
  dipole moment, 54
  first observation, 310
  Heitler and Fröhlich on, 156, 157
  magnetic moment, 154–7, 166
  nature, 229
  name, 11, 28
  negative states, 57
  range, 28

"singlet proton", 61
see also n–p model; p–p force; p–p
    scattering
"Protons"
    Pauli's, 111, 112
    Rutherford's, 11, 13
Prout, William, 11, 12
Prout's hypothesis, 11
Pryce, M.H.L., 91, 361
Pseudoscalar mesons, 254
    coupling, 261
    theory, 194
    see also vector and pseudoscalar fields
Purcell, Edward M., 15, 23, 24, 361

QCD *see* quantum chromodynamics
QED *see* quantum electrodynamics
QFT *see* quantum field theory
Quantum chromodynamics (QCD),
    329–31
    valence quarks, 331
    see also quarks ("aces")
Quantum condition, 8
Quantum electrodynamics (QED), 33, 37,
    42, 46, 85, 86, 104, 106, 141, 329
    Bhabha on, 234
    breakdown (validity) at high energies,
        71, 76, 138, 160
    and cascade showers, 74
    and cosmic rays, 72, 73
    and coupling, 260, 261
    divergence, 176
    discrepancy in fine structure, 217
    electron self energy, 86
    Fermi and, 42
    Heisenberg and, 76
    high-energy behaviour, 69, 153
    renormalized, and meson theories, 317,
        320–3
    systematic formulation, 86
    Yukawa on, 120
Quantum field theory (QFT), 25, 33, 36,
    45
    of β-decay, 47, 51
    constructing, 36
    difficulties, 95
    history, 25
    limits, 216
    local, 57
    nonlinear, 45
    non-local, 45
    perturbation theory, limits, 86, 176,
        214, 326
    relativistic, 86
    supersymmetric, 45

unified, 206
Yang–Mills type, 45
Yukawa on, 120
Quantum mechanics, 1, 15, 47
    Heisenberg on application to nuclei, 39,
        54
    and nucleus, 16
    role within atomic nucleus, 47
    tunnelling, 16
Quantum numbers
    isospin (g), 46
    strangeness, 297, 324
Quantum theory and nucleus, 15, 17
Quarks ("aces"), 325, 329–32

Rabi, Isidore I., 155, **191**, 220, 224, 266,
    321, 333, 352
"Radiation damping", 267, 268
Radiation theory, 56
Radiative capture, 168
Radiative energy losses, 74
Radioactivity, 4, 5
    decay, 10, 12, 23
    detection of, 1, 4
    nucleus as source of, 9, 10
    three early puzzles, 4
Radius
    atomic, 8
    neutron, 58
    nuclear, 33
    proton, 58
    see also electrons
Ramsey, N.F., 220, 352
Ramsey, W.E., 226, 366
Range of nuclear forces, 54, 55
Rarita, W., 194, 227, 248, 254, 256, 361
Rarita–Schwinger method, 256
Rasche, E., 172, 361
Rasetti, Franco D., 18, 42, 46, **55**, 65,
    139, 187, 189, 197, 200, 239, 282,
    345, 361
    on mesotron lifetimes, 189
Rathgeber, H.D., 200, 361
Rayleigh, Lord *see* Strutt, J.W.
Rechenberg, H., 22–5, 65, 100, 113, 170,
    198, 199, 227, 228, 248, 249, 270,
    289, 341, 349, 353, 356, 361
Regener, Erich, 69, 76, 82, 290, 361
Regge, Tullio, 327, 361, 362
Regge theory, 328
Reiche, Fritz, 113
Reischauer, E.O., 170, 362
Relativity special theory, 3, 22
Research apparatus, 1930s, 27
Resonances *see* elementary particles

Retherford, R., 321, 354
Rhodes, R., 270, 361
Richtmeyer, R.D., 249
Rider, R.E., 172, 362
Ridgen, J.S., 341
Riedasch, G., 333, 342
Riordan, M., 350
Rivier, D., 289, 365
Roberts, G.E., 196, 369
Rochester, George D., 251, 252, 295, 296,
    296, 313, 323, 362
Röntgen, Wilhelm Conrad, 4, 22
Rosenfeld, A.M., 356
Rosenfeld, Léon, 191, 192, 194, **206**, 207,
    219–22, 226, 227, 240, 258, 259,
    267, 268, 270, 276, 278, 279, 307,
    351, 361
Rossi, Bruno, 70, 73, 89, 142, 178, 184,
    185, 186–9, 199, 200, 205, 222, 224,
    225, 231, 253, 255, 282, 283, 290,
    295, 298, 313, 333, 339, 358, 362
    mesotron lifetime measurements, 187–9
    on mesotron lifetimes, 187–9
Rozental, Stefan, 191, 357
Rubinowicz, A., 227
Rulhig, A.J., 204, 205, 362
Rutherford, Ernest, 1, 5, **7**, 7–18, 21,
    23–5, 28, 29, 31, 37, 97, 172, 318,
    341, 345, 356, 362, 363
    and α-particle scattering, 7, 8, 12, 13,
    16
    and deuterons, 37
    moves to Cambridge, 14
    and the neutron, 28
    nuclear atom discovered by, 5, 8
    on uranium, 17

Sachs, R.G., 25, 254, 363
Sagane, Ryokichi, 128, 281, 284
Sakata, Shoichi, 68, 99, 116–9, **119**, 132,
    137, 139, 141–6, **143**, 159, 164–6,
    169–71, 179, 193, 194, 197, 201–3,
    222, 225, 251, 254, 256, 270, 271,
    278, 279, **280**, 285, 286, **286**, 289,
    309, 321, 325, 340, 363, 366, 370
    on β decay and mesotron decay, 194
    and cosmic ray mesons, 278
    and heavy quanta, 143, 146
    model, 325
    and pseudoscalar meson, 254
    and Yukawa, Hideki: joint fields of
    study, 118, 119
Saturation, 37, 38, 63, 64
"Saturnian atom" *see* atomic models
Scalar field theory, 144–6

quantum condition, 231
Scattering, anomalous *see* "anomalous
    scattering"
Schaffer, S., 348
Schein, Marcel,38., 238, 243, 248, 341,
    363
Scherzer, O., 78, 363
Schiff, Leonard I., 216, 226, 354
Schindler, H., 88, 365
Schmid, C., 334, 363
Schoenberg, Mario, 200, 221, 227, 363
Schrödinger equation, scalar relativistic,
    159
Schrödinger, Erwin, 100, 101, 108, 153,
    255, 264, 267, 356
Schubnikow, L.W., 154
Schweber, Silvan S., 25, 176, 332, 333,
    363, 364
Schwinger, Julian, **59**, 194, 227, 245, 246,
    248, 250, 254, 256, **257**, 258, 264,
    265, 270, 290, 321, 322, 329, 333,
    343, 359, 361, 363, 364
Scintillation study method, 13, 15
Scrocco, E., 337
Second quantization, 42, 51, 65
Segrè, Emilio, 22, 23, 46, 113, 282, 345,
    363
Sekido, Yataro, 69, 89, 169, 350, 358,
    363, 366
Self-consistent field *see* Fermi–Thomas
    method
Serber, Robert, 63, 74, 90, 130–3, 135,
    140, 144, 170, 178, 179, 186, 190–2,
    201, 215, 239, 254, 261, 265, 271,
    284, 315, 343, 359, 363
    on β-decay and mesotron decay, 190,
    191
Shankland, R.S., 136
Shea, W.R., 22, 23, 341, 366
Shimizu, Takeo, 15, 97, 363
Shonka, Francis R., 213, 363
Short-range forces, fundamental theory
    for, 36, 37
Short-range interactions, 34
Showers, 121, 122
    Carlson on theory, 254
    as explosions initiated by strong nuclear
    reaction, 74
    formed at great depths, 84
    interpretations, 73
    processes giving rise to, 168
    types, 83
    see also cascade showers; electrons;
    Heisenberg's "explosive showers";
    penetrating showers

Shutt, Ralph P., 195–7, 201, 352, 363
Siegel, D.M., 24, 364
Siegert, A.J.F., 66, 364
Sigma charged particles, 324
Sigma- (σ-) mesons, 305, 307
Silliman, R.H., 6, 364
Simon, Francis, 154
Simpson, O.C., 345
Singh, V., 327
Sitte, Kurt, 38, 47, 52, 65, 336
Smith, A.K., 270, 364
Smith, F.M., 347
Smolochowski, M. von, 22
Smorodinski, I., 272
Smyth, H. DeWolf, 268, 270, 364
Snyder, Hartland S., 239, 359
Soddy, Frederick, 9–11, 24, 364
Sokolov, A., 78, 79, 91, 123, 234, 250, 268, 271, 272, 352, 364, 365
Sommerfeld, Arnold, 8, 11, 12, 15, 78, 100, 153, 154, 216, 217, 273, 274, 356, 365
    Festschrift for, 217, 273, 274
    see also Bohr–Sommerfeld quantum theory of atomic structure
Souza Santos, M.D. de, 294, 295, 365, 367
Soviet Union, meson physics (1940–43), 272, 273
Spin
    field theories with higher, 255–8
    integral values requirements, 231
    nuclear, 15
    relationship with statistics, 230
    relativistic spin theories
    theory, 254
    see also electrons; mesons; spin-statistics relation
Spin-statistics relation, 18, 230
    non-relativistic, 231
    Pauli on, 242
    problems of e–p model, 18, 19
    theorem (1938–39), 230–3
Squires, E.J., 334, 342
Stability curve of nuclei, 36, 41
Stachel, J., 22, 365
Standard model of elementary particles, 225, 317, 325
    electroweak sector, 331, 332
    see also quantum chromodynamics
Standard model of nuclear forces, 30
    see also mesons; nuclear forces
Starr, M.A., 93, 346
"Stars", 87, 301, 302, 304

Steinberger, Jack, 68, 310, 311, **311**, 314, 315, 332, 333, 365
Steinke, E.G., 88, 365
Stellar, J., 365
Stern, Otto, 66, 155, 172, **185**, 224, 344–6
Stevenson, E.C., 56, 82, 92, 125, 127, 128, 131, 132, 139, 170, 178, 204, 225, 365
Stevenson, M.L., 356
Stever, H.G., 199, 358
Stoney, G.J., 22
Störmer, T.F., 142
Strange particles, 294–7, 324, 325
Strassmann, Fritz, 253
Street, Jabez C., 56, 71, 82, 92, 125, **126**, 127, 128, 131, 132, 138, 139, 170, 178, 204, 224, 225, 365
Strong interactions, 24, 329, 330
    without pion field, 325–9
    see also coupling; dispersion relations; quantum chromodynamics; Regge theory
Strutt, J.W., 152
Stueckelberg, Ernst Carl Gerlach, 63, 68, 78, 79, 91, 130, 132, **133**, 135, 137, 140, 144, 170, 178, 198, 210, 211, 216, 226, 249, 276, 286, 288, 289, 322, 365
    on intermediate particle, 178
    on matter/"radiation" interaction, 216, 217
Stueckelberg's "unitary (field) theory", 68, 132, 216
Stuewer, R.H., 15, 25, 65, 70, 172, 270, 348, 366
Subtraction physics, 57
Sudarshan, E.C.G., 334, 366
Sugiura, Y., 358
Swain, D.L., 358
Swann, W.F.G., 136, 187, 188, 226, 366
    and saturation, 37
Switzerland
    vector meson theory, 207–13
    and Yukawa theory, 211, 212
Symmetry groups, 325
    chiral, 328

Takabayasi, Takehito, 113, 285, 286, 291, 366
Takamine, Toshio, 97
Takeda, G., 112, 366
Taketani, Mituo, 45, 99, 116, 120, 127, 132, 136, 138, 139, 142–4, 146–9, 164, 165, 169–71, 179, 193, 198, 200, 202, 256, 363, 366, 370

scientific methodology ("three-stages theory"), 114
Takeuchi, Masa, 127, 128, 134, 139, 281, 358, 366
Tamaki, Kajuro, 100, 101, 202
Tamm, Igor, 56–8, 66, 105–7, 246, 250, 272, 366
Tanikawa, Yasutaka, 119, 137, 202, 278, **280**, 285, 289, 309, 363, 366, 367
Telegdi, V.L., 270, 340, 341
Teller, Edward, 21, 63, 64, 68, 129, 154, 192, 224, 227, 253, **257**, 259, 260, 279, 288, 299–301, 343, 346, 347, 350, 369
Theory of atomic structure and spectra, 6–10
Theory/experiment confrontation, meson decay and, 197
Thermodynamics, laws of, 18
Theta–tau (θ–τ) puzzle, 324
Thirring, W.E., 347
Thomson, Joseph John, 2, 6–9, 11, 14, 22, 23, 152, 366
see also atomic models
Thomson, William, 6, 152
Timoféef, N., 288
Tomonaga, Sin-itiro, 98–102, 113, 116, 119, 120, 138, 142, 143, 170, 180, 193, 194, 196, 197, 200, 223, 234, 245, 250, 255, 263, 264, 271, 277, 278, 281, 285, 286, **286**, 290, 291, 298, 299, 321, 333, 340, 351, 353, 357, 358, 363, 366
  on β-decay and mesotron decay, 194
  on coupling method, 263, 264
Tomonaga's intermediate coupling, 265, 266
Tomonaga's renormalized QED, 285
Tomonaga's super-many-time formalism, 285
Tomonaga–Araki theory, 298, 299
Tongiorgi, V., 342
Trabacchi, G.C., 335
Treimann, S.B., 345
Trenn, J.T., 22, 23, 367
Tuve, Merle A., 28, 60, 61, 222, 224, 240, 367
Two-meson theories
  see meson theories

U-quanta, 70, 96, 108–10, 115, 123, 139, 141, 148
  annihilation, 151
  Bhabha on scattering of, 168
  creation, 151

and cosmic ray mesotrons, 106–11
decay, 160, 161, 164, 170
designation, 70, 187, 221
magnetic moment, 149
mean lifetime, 96, 151
neutral, 148, 150, 161
see also mesons; mesotrons
Udgaonkar, B.M., 327
Uhlenbeck, G.E., 58, **59**, 67, 90, 92, 117, 134, 182, 270, 273, 353
"Undors", 232
Unified field theory of strong and weak nuclear forces, 43, 70, 106
see also Yukawa's meson theory of nuclear forces
United States of America (USA)
  in 1939, 254
  and atomic bomb, 284
  and nuclear physics, 21
  post-war, 318
  and wartime, 254
  and Yukawa's meson theory, 207, 215–21, 254
Uranium, 274
  bomb, 282
  nuclear fission, 239, 253, 281
Urban, P., 367
Urey, Harold C., 28, 224, 367

V-particles, 295–7, 323
  see also kaons, strange particles
Van de Graaff generator, 27, 28, 224
Van de Graaff, Robert J., 28, 367
Vector (Proca) fields, 130, 148
  Proca's equation, 158, 163, 173
Vector and pseudoscalar fields, theory of mixed, 258, 259
  see also Møller–Rosenfeld "patent mixture"
Vector meson theories, 235
  applications and modifications, 203
  of Bhabha, 166–8
  British papers, 162–6
  charge symmetry, 164, 209
  and cosmic ray problems, 168
  emergence, 204
  fourth-order pp-interactions, 208
  in Japan, 146–51
  of Kemmer 164–6
  magnetic moments of proton and neutron, 156, 162, 173
  neutral particles, 166, 167
  neutron–proton interactions, 167, 168
  origin, 146
  proton–neutron bound states, 163–6

in Switzerland and Germany, 207–13
symmetric, 173
see also Yukawa, Sakata and Taketani
Veksler, V., 272, 332

Wagner, M., 348, 350
Walker, M., 270, 367
Waller, C., 303, 314, 367
Waller, I., **48**
Walton, Ernest T.S., 29, 342
Wambacher, Hertha, 93, 301, 302, 313, 338
Wang, E.C.G., **133**
Wannier, G.H., 269, 353
Waraghin, G., 290, 294, 295, 314, 365, 367
Watanabe, M., 112, 367
Watase, T., 138
Watson, Kenneth M., 320, 367
Wave mechanics, 15
Wave–particle duality, 101
"Weak interactions" *see* interactions of elementary particles
Weinberg, Joseph, W., 260, 271, 367
Weinberg, Steven, 328, 329
effective Lagrangian, 328, 329
Weiner, C., 19, 27, 114, 290, 291, 313, 368
Weisskopf, Victor F., 21, 25, 44, 52, 68, 92, 115, 129, 139, 144, 145, **147**, 148, 151, 152, 169–72, 195, 210, 231, 240, 253, 260, 271, 299, 301, 338, 346, 356, 360, 368
Weizsäcker, Carl Friedrich von, 59, 67, 74, 76, 90, 123, 240, 245, 253, 274, 283, 287, 288, 368
Weizsäcker–Williams method, 76, 90, 245
Wentzel, Gregor, 53, 60–3, 66–8, 78, 91, 153, 154, 158, 179, 180, 198, 207, 209, 210, **210**, 211, 212, 216, 225, 226, 231, 254, 259–71, 276, 321, 332, 333, 368
and coupling constant, 263
on coupling theory, 261–4
and Fermi field theory, 61, 62, 209
on pair theories, 259, 260
simplified model of 1940, 260–3
strong-coupling theory, 260–3, 268
and Yukawa's theory, 179, 180
Wergeland, Harald, 88, 92, 345
Wheeler, John, 21, 172, 224, 289, 368
White, M.G., 60, 67, 368
Whittaker, E., 22, 368

Wick, Gian Carlo, 54, 55, **55**, 58, 66, 110, 117, 137, 155, 169, 275, 282–4, 289, 290, 313, 337, 368
and Heisenberg, 54, 55, 283, 284
Wiechert, E., 22
Wiemers, G., 361
Wien, Willy, 11
Wightman, A., 333, 369
Wigner, Eugene, 21, 32, 33, 36, 49, 60, 63–5, 67, 68, **206**, 207, 253, 270, 333, 341, 343, 352, 354, 369
nuclear forces ("Wigner forces") *see* nuclear forces
Wilczek, F., 330, 348
Williams, E.J., 74, 76, 90, 195, 196, 204, 225, 245, 246, 250, 369
Williams, F.E., **153**
Williams, Robley C., 217, 225, 369
Wilson, A.H., 246, 339
Wilson, Charles Thomson Rees, 4
cloud chamber, 4, 179
Wilson, John G., 82, 121, 127, 138, 167, 172, 226, 233, 234, 249, 266, 312, 338, 361, 369
Wilson, V.C., 226, 250, 363, 369
Winter, J., 78, 346
Wirtz, Karl, 273, 274, 288
Wooster, W.A., 19, 344
Wouthuysen, S.A., 355

Xi-minus ($\Xi-$) particle, 324
X-rays, 4, 5, 10
discovery, 1
scattering, 2, 23
Xu, Q., 23, 199, 369

Yagi, E., 23, 97, 351, 358, 369
Yamaguchi, Y., 112, 366
Yamanashi, Mr, 143
Yang, Chen Ning, 271, 324, 325, 329, 346, 355, 369
Yearian, M.R., 333, 369
York, H.F., 338
Yoshida, R., 366
Yoshida, Y., 114, 370
Yu, F.C., 250, 356
Yukawa, Hideki, 36, 45, 56, 59, 63, 67, 70, 95–111, **99**, 113, 115–21, **119**, 123, 127, 129–51, **143**, 155–62, 164–7, 169–71, 175–86, 190–4, 196–204, 207, 211, 213–5, 219–30, 236, 237, **238**, 239, 240, 248, 249, 251, 254, 259, 270, 277, 278, **280**, 281, 282, 285, 286, **286**, 288, 289,

292, 316, **319**, 340, 341, 343, 352, 353, 366
associates, 116
background and ambition, 99, 100
on β-decay, 106, 177, 183, 209
on cosmic rays, 141, 142
on couplings, 229, 230
"defects", 116
density matrix in positron theory, 144
meets Einstein, 224
and Einstein–Bose statistics, 162
in Europe, 222–4
and "Fermi field" theory, 177
and Fröhlich and Heitler, 148
and Heisenberg, 56, 102, 105, 107, 155, 156, 181–3
on himself, 202
"*Interaction, I*" (1935), 106, 107, 115, 175, 179
as journal editor, 285
Marienbad table, 237, **238**
mass–range formula, **109**
on meson theory (various versions), 146, 162, 277, 278, **279**
on meson/mesotron decay, 178–83, 186
new fundamental theory, 103–6, 116–21, 135, 136
Nobel Prize, 98
on nuclear electrons, 102
and nuclear fission, 281, 282
and nuclear forces to 1933, 99–103
Osaka talk (28-11-36), 129
and quantum field theory, 102
reputation grows, 63, 115, 130, 155, 156, 175, 176, 202–28
returns to meson theory, 141, 142
school of, formed, 142–6
U-quanta work resumed by, 129
and uranium, 281
in USA, 224, 225
and vector meson theory, 146, 162
Yukawa, Hideki and Sakata, Shoichi, "*Interaction, II*" (28-11-36), 116, 120, 121, 130, 141, 142, 144–6, 149, 150, 159, 165, 179
joint fields of study, 118, 119
Yukawa, Hideki, Sakata, Shoichi and Taketani, Mituo
"*Interaction, III*" (6-1-37), 121, 129, 130, 142, 145–51, 164, 165, 179, 203, 209, 211, 212
"*Interaction, IV*", 203, 213, 236
"on theory of new particle in cosmic rays" (22-10-37), 132–5

"Yukawa interactions", 279
Yukawa's meson (U-quantum) theory of nuclear forces, 30, 36, 44, 48, 59, 63, 64, 65, 70, 82, 87, 96, 98–114, 116, 117, 120, 123, 128, 141–4, 151, 167, 175–7, 180 189, 244, 279, 316
acceptance of, 65
beginnings, 56, 97–114
Bethe and Nordheim on, 192, 193
Bohr on, 155
combined with Fermi field theory, 64, 65
East–West competition and cooperation, 148
emergence, 103–11
emission of "light particles", 107
in Europe, 183, 204, 217–9
and experiment, 197
first article on, 106, 107, 115, 116, 175, 179
fundamental unified theory, 189
Heisenberg on, 241
intermediate-mass particles, 125, 138
intermediate particle in β-decay, 164, 167, 177
magnetic moments, 170, 246
mass of new particle, 105
and mesotrons, 129–35, 316
as neglected at first, 70, 96, 120
original features, 98
as paradigm, 70, 112
Pauli on, 151, 152
principal assertion, 175
reformulation, 115
and scalar particles, 107, 142–6, 166
scope, 257
significance, 98
spin 0 and spin 1 versions, 142
two forms, 142
U-field of force, 107, 129, 140
U-particle mass, 11
U-particle production
USA and, 215–21
in the West, 125
see also nuclear forces, U-quanta, vector meson theories
"Yukawa potential", 120
Yukons, 187, 208, 209, 212

Zacharias, Jerrold Reinach, 220, 352
Zachariason, F., 342
Zeeman, Pieter, *Festschrift*, 57, 59
Zichichi, A., 335, 370
Zweig, George, 325, 334, 370